Irene Katzela

MULTIACCESS, MOBILITY AND TELETRAFFIC FOR PERSONAL COMMUNICATIONS

THE KLUWER INTERNATIONAL SERIES IN ENGINEERING AND COMPUTER SCIENCE

COMMUNICATIONS AND INFORMATION THEORY

Consulting Editor

Robert Gallager

Other books in the series:

PERSONAL AND WIRELESS COMMUNICATIONS: ***Digital Technology and Standards,*** Kun Il Park
ISBN: 0-7923-9727-4

WIRELESS INFORMATION NETWORKS: ***Architecture, Resource Managment, and Mobile Data,*** Jack M. Holtzman
ISBN: 0-7923-9694-4

DIGITAL IMAGE COMPRESSION: ***Algorithms and Standards,*** Weidong Kou
ISBN: 0-7923-9626-X

CONTROL AND PERFORMANCE IN PACKET, CIRCUIT, AND ATM NETWORKS, XueDao Gu, Kazem Sohraby and Dhadesugoor R. Vaman
ISBN: 0-7923-9625-1

DISCRETE STOCHASTIC PROCESSES, Robert G. Gallager
ISBN: 0-7923-9583-2

WIRELESS PERSONAL COMMUNICATIONS: ***Research Developments***, Brian D. Woerner, Theodore S. Rappaport and Jeffrey H. Reed
ISBN: 0-7923-9555-7

PLANNING AND ARCHITECTURAL DESIGN OF INTEGRATED SERVICES DIGITAL NETWORKS, A. Nejat Ince, Dag Wilhelmsen and Bülent Sankur
ISBN: 0-7923-9554-9

WIRELESS INFRARED COMMUNICATIONS, John R. Barry
ISBN: 0-7923-9476-3

COMMUNICATIONS AND CRYPTOGRAPHY: ***Two sides of One Tapestry***, Richard E. Blahut, Daniel J. Costello, Jr., Ueli Maurer and Thomas Mittelholzer
ISBN: 0-7923-9469-0

WIRELESS AND MOBILE COMMUNICATIONS, Jack M. Holtzman and David J. Goodman
ISBN: 0-7923-9464-X

INTRODUCTION TO CONVOLUTIONAL CODES WITH APPLICATIONS, Ajay Dholakia
ISBN: 0-7923-9467-4

CODED-MODULATION TECHNIQUES FOR FADING CHANNELS, S. Hamidreza Jamali, and Tho Le-Ngoc
ISBN: 0-7923-9421-6

WIRELESS PERSONAL COMMUNICATIONS: ***Trends and Challenges***, Theodore S. Rappaport, Brian D. Woerner, Jeffrey H. Reed
ISBN: 0-7923-9430-5

ELLIPTIC CURVE PUBLIC KEY CRYPTOSYSTEMS, Alfred Menezes
ISBN: 0-7923-9368-6

SATELLITE COMMUNICATIONS: Mobile and Fixed Services, Michael Miller, Branka Vucetic and Les Berry
ISBN: 0-7923-9333-3

WIRELESS COMMUNICATIONS: Future Directions, Jack M. Holtzman and David J. Goodman
ISBN: 0-7923-9316-3

DISCRETE-TIME MODELS FOR COMMUNICATION SYSTEMS INCLUDING ATM, Herwig Bruneel and Byung G. Kim
ISBN: 0-7923-9292-2

APPLICATIONS OF FINITE FIELDS, Alfred J. Menezes, Ian F. Blake, XuHong Gao, Ronald C. Mullin, Scott A. Vanstone, Tomik Yaghoobian
ISBN: 0-7923-9282-5

WIRELESS PERSONAL COMMUNICATIONS, Martin J. Feuerstein, Theodore S. Rappaport
ISBN: 0-7923-9280-9

SEQUENCE DETECTION FOR HIGH-DENSITY STORAGE CHANNEL, Jaekyun Moon, L. Richard Carley
ISBN: 0-7923-9264-7

DIGITAL SATELLITE COMMUNICATIONS SYSTEMS AND TECHNOLOGIES: Military and Civil Applications, A. Nejat Ince
ISBN: 0-7923-9254-X

IMAGE AND TEXT COMPRESSION, James A. Storer
ISBN: 0-7923-9243-4

VECTOR QUANTIZATION AND SIGNAL COMPRESSION, Allen Gersho, Robert M. Gray
ISBN: 0-7923-9181-0

MULTIACCESS, MOBILITY AND TELETRAFFIC FOR PERSONAL COMMUNICATIONS

EDITED BY

Bijan Jabbari

George Mason University
Fairfax, VA, US

Philippe Godlewski

Ecole Nationale Supérieure des Télécommunications
Paris, France

Xavier Lagrange

Ecole Nationale Supérieure des Télécommunications
Paris, France

KLUWER ACADEMIC PUBLISHERS
BOSTON / DORDRECHT / LONDON

A C.I.P. Catalogue record for this book is available from the Library of Congress

ISBN 0-7923-9742-8

Published by Kluwer Academic Publishers,
P.O. Box 17, 3300 AA Dordrecht, The Netherlands.

Kluwer Academic Publishers incorporates
the publishing programmes of
D. Reidel, Martinus Nijhoff, Dr W. Junk and MTP Press.

Sold and distributed in the U.S.A. and Canada
by Kluwer Academic Publishers,
101 Philip Drive, Norwell, MA 02061, U.S.A.

In all other countries, sold and distributed
by Kluwer Academic Publishers Group,
P.O. Box 322, 3300 AH Dordrecht, The Netherlands.

Printed on acid-free paper

Printed in the Netherlands

Table of Contents

Preface

The success of first and second generation wireless systems has paved the way for further research opportunities towards the next generation systems. The two standards GSM and IS-95 based on TDMA and CDMA respectively, have deeply influenced our system-level understanding, bringing new perspectives on the problems associated with wireless networks and potential for innovations.

This volume presents the proceedings of the second workshop on multiaccess, mobility and teletraffic for personal communications held in May 1996 in Paris, France where some important subjects on the next generation systems have been treated. These include topics dealing with information theoretic aspects, channel modeling, diversity, interference control, resource allocation, power control, packet multi-access, stochastic modeling of mobility and traffic, and wireless network control.

The selected topics in this workshop and their presented set of solutions reflect the richness of the problems in wireless communications. Indeed, development of theoretical frameworks with considerable attention to the peculiar environment of wireless communications has been the prime objective of this workshop. To elaborate, consider the problem of multi-access methods which remains a challenge for researchers. A complete evaluation of an access scheme must consider different aspects such as propagation, interference, mobility and traffic modeling. Some common bases, paradigms and models are needed. For example, today, we do not have a common archetype like the AWGN channel as in classical statistical communication. Clearly, there is a need for justified assumptions and models. Some progress, however, can be noticed. Similarly, such a framework can be also useful in evaluating the effect of mobility and system capacity.

The next generation wireless systems as well as the evolved second generation systems will certainly benefit from these useful results. As in the past, we can gain experience from the existing implemented systems and develop novel features for the next generation systems. Indeed, it is not surprising to observe some common features in the second generation systems, not specifically present in the first generation systems. One example is channel encoding done in the framework of GSM and IS95 systems, though maybe viewed in different scales, to mitigate various interference and propagation effects through interleaving, equalization, spreading or hopping, synchronization and channel sounding. Other examples include the enhanced handover techniques and location management being deployed in these systems.

There is no doubt that further work is still needed in these areas, in particular in microcellular and picocellular environment. We have already witnessed wide-spread and successful use of microcells in the second generation mobile cellular systems and we are awaiting possible system implementation by public cordless system. The third generation systems may have to address the optimal way to achieve a goal of 10,000

Erlang/km2. This is to be considered in a heterogeneous traffic environment necessitating the need for a highly adaptive system with different access modes incorporating sophisticated resource allocation algorithms and various control procedures within the wireless channel and the network. Furthermore, they will have to provide user mobility in a larger scale. Certainly giving rise to interesting problems in mobility control and the signaling traffic.

We hope this workshop has made a small contribution to the field and further stimulated the ideas for research.

Bijan Jabbari

Philippe Godlewski

Xavier Lagrange

Acknowledgement

We wish to extend our gratitude to the following people who helped us in the review of the papers.

A. Baiocchi
K. Basu
T. Benkner
V. Brass
J. Chuang
G. Colombo
L. Decreusefond
K. Demetrios
C. Despins
U. Dropmann
J. Dunlop
D. Everitt
W. Fuhrmann
M. Gagnaire
B. Goode
P. Humblet
A. Jalali
H. Jiang
G. Kawas-Kaleh
V. Kumar
A. Levy
J-P. Linnartz
G. Maral
D. McMillan
G. Omidyar
K. Pahlavan
R. Pickholtz
R. Rom
C. Rose
M. Rumsewicz
H. Sari
S. Tabbane
G. Taricco
R. Vallet
B. Vojcic
P. Wirth
M. Yabusaki
P. Yegani
G. Zemor

DISTRIBUTED DISCRETE POWER CONTROL IN CELLULAR PCS

M. Andersin, Z. Rosberg* and J. Zander

Radio Communication Systems, KTH, Stockholm, Sweden

** Haifa Research Lab., Science and Technology, IBM Israel*

ABSTRACT

Transmitter power control has proven to be an efficient method to control cochannel interference in cellular PCS, and to increase bandwidth utilization. Power control can also improve channel quality, lower the power consumption, and facilitate network management decisions such as mobile removals, hand-off and admission control. Most of the previous studies have assumed that the transmitter power level can be controlled in a continuous domain. In systems where power is digitally controlled, power levels are discrete. In this paper we study transmitter power control algorithms in a cellular PCS, which use only a finite set of discrete power levels, and where the receiver is subject to background noise. We characterize an optimal discrete power vector, and present a Distributed Discrete Power Control (DDPC) algorithm which converges to it. We evaluate the algorithm in a microcellular environment, and investigate the impact of the discrete power level spacing.

1 INTRODUCTION

Future PCS cellular networks will require high quality channels, high bandwidth utilization, low power consumption and efficient network management. Power control (up-link and down-link) is a major technique which addresses these goals. In PCS, cell sizes are small and transmission power is limited, exposing the receiver to a more severe noise compared to a system with larger cells and higher transmission power. This has been recently incorporated into the model, in [8], where a constrained *continuous power level* control problem in a cellular network with cochannel interference, has been studied. The model there (and in this study) assumes a stationary link gain matrix, which is reasonable when the power control converges much faster than the link gain changes. The results in [8], extend previous results of centralized and distributed power control schemes obtained in [1, 5, 6, 7, 11, 14, 15].

In all the studies above, transmitter power level has been controlled in a *continuous power range*. However, in *digital systems* as ours, transmission power can be updated only at discrete levels ranging between a minimum and a maximum level (*dynamic range*). For example, in GSM, the uplink and downlink transmission power level may vary from 13 to 39 *dBm*, at values which are equally spaced by 2 *dB*. In Qualcomm's CDMA proposal for IS-95, [12], the power levels are equally spaced by 0.5 dB, within a dynamic range of 85 dB in the uplink, and 12 dB in the downlink. This practical limitation has been the main drive of this study, where we consider the discrete power control problem.

We define the optimality of a discrete power vector, and derive a distributed discrete power control algorithm which converges to it. Preliminary results in [13] show that rather coarse power level grid and limited dynamic range can provide useful control. This is further substantiated in this study.

In Section 2, we introduce the model and general definitions. In Section 3, we define several distributed discrete power control algorithms, and derive their convergence properties. In Section 4, we present the numerical example, and in Section 5 our conclusions. The proofs of all the assertions are given in [4].

2 SYSTEM MODEL

We restrict our definitions to the uplink case (from mobile to base). The downlink is modeled in the same way, with the appropriate notational changes. Consider a cellular radio system and focus on a generic channel (a specific frequency or time slot). Assume that channels are orthogonal, so adjacent channel interference is negligible. Let $\mathcal{N} = \{1, 2, \ldots, N\}$, be the set of active transmitters using this generic channel, and $\boldsymbol{p} = (p_1, p_2, \ldots, p_N)^T$, be the transmission powers used by the mobiles to communicate with their base stations. We will add a time index to the powers, whenever necessary, i.e., $\boldsymbol{p}(t) = (p_1(t), p_2(t), \ldots, p_N(t))^T$.

The power levels of each transmitter is restricted to an arbitrary set of discrete values (not necessarily on an equally spaced grid), $\mathcal{D} = \{x_1, x_2, \ldots, x_n\}$. The maximum power level x_n, is also denoted by $\bar{p}$.

Denote the link gain matrix (see e.g., [7] and [14]) by $\boldsymbol{G} = [g_{ij}]$, where g_{ij} is the gain of the radio link from transmitter j to base i, $1 \leq i, j \leq N$. All link gains assume values in the semi-open interval $(0, 1]$. Let $\boldsymbol{\nu} = (\nu_1, \nu_2, \ldots, \nu_N)^T$, be the receiver noise vector at the base stations. We require that at least one element in $\boldsymbol{\nu}$ is positive. The link quality is measured by the Carrier to Interference Ratio (CIR). For a given power vector $\boldsymbol{p}$, the CIR at the base station used by transmitter i, is given by

$$\gamma_i = \frac{p_i g_{ii}}{\nu_i + \sum_{j:j \neq i} g_{ij} p_j} \ , \quad 1 \leq i \leq N \ .$$

It is well known (see e.g. [9]) that there is a monotonically increasing relationship between the CIR and the channel symbol error rate. Thus, driving the CIR to a target value is the same as driving the channel to a target capacity. For notational convenience, we represent γ_i by

$$\gamma_i = \frac{p_i}{\eta_i + \sum_{j \in \mathcal{N}} a_{ij} p_j} \ , \quad 1 \leq i \leq N \ , \tag{1.1}$$

where $\eta_i = \nu_i / g_{ii}$, and

$$a_{ij} = \begin{cases} g_{ij}/g_{ii} \ , & \text{if } \ i \neq j \ , \\ 0 \ , & \text{if } \ i = j \ . \end{cases}$$

The matrix and the vector of the transformed gains and noises are denoted by $\boldsymbol{A} = [a_{ij}]$ and by $\boldsymbol{\eta} = (\eta_1, \eta_1, \ldots, \eta_N)^T$, respectively. Let $\boldsymbol{\gamma}^t = (\gamma_1^t, \gamma_2^t, \ldots, \gamma_N^t)^T$ and $\overline{\boldsymbol{p}} = (\bar{p}, \bar{p}, \ldots, \bar{p})^T$. Further, for every two vectors $\boldsymbol{v} = (v_1, v_2, \ldots, v_n)$, $\boldsymbol{w} = (w_1, w_2, \ldots, w_n)$, and square matrix $\boldsymbol{M} = [m_{ij}]$, define $\boldsymbol{v} \otimes \boldsymbol{w}$ as a componentwise product, and $\boldsymbol{v} \otimes \boldsymbol{M} \stackrel{\text{def}}{=} [v_i \cdot m_{ij}]$. To denote the ith element of a vector $\boldsymbol{v}$, we also use the notation $[\boldsymbol{v}]_i$.

In [8] all transmitters use the same CIR target, where here each transmitter may use a different CIR target γ_i^t. We only remark (without proving) that all the results which have been obtained in [8], hold true also in this more general setting.

We say that a power vector $\boldsymbol{p}$ supports transmitter i at its target CIR γ_i^t, if and only if

$$p_i \geq [\boldsymbol{\gamma}^t \otimes (\boldsymbol{A}\boldsymbol{p} + \boldsymbol{\eta})]_i \ . \tag{1.2}$$

That is, receiver i has a CIR γ_i satisfying $\gamma_i \geq \gamma_i^t$.

Next, we describe the *continuous* power updates made by the algorithms DCPC and ADCPC, when the target CIR vector is $\boldsymbol{\gamma}^t$. Given the power vector at time t, $\boldsymbol{p}(t)$, and the set of transmitters updating their powers at time $t + dt$, $U(t)$, then the power level of transmitter i at time $t + dt$ is given by

$$p_i(t+dt) = \begin{cases} \min\{\bar{p},\ \gamma_i^t \cdot \frac{p_i(t)}{\gamma_i(t)}\}\ , & \text{if } i \in U(t), \\ p_i(t)\ , & \text{otherwise}\ . \end{cases} \tag{1.3}$$

Note that $p_i(t)g_{ii}/\gamma_i(t) = \eta_i + \sum_{j\in\mathcal{N}} a_{ij}p_j(t)$ is the interference experienced by receiver i. From [8], it follows that for any given $\boldsymbol{\gamma}^t$, the power vectors under DCPC and under ADCPC, converge to a unique positive power vector determined by the solution to (i.e., a fixed-point solution of dynamic equations)

$$\boldsymbol{p} = \min\{\overline{\boldsymbol{p}},\ \boldsymbol{\gamma}^t \otimes (\boldsymbol{A}\boldsymbol{p} + \boldsymbol{\eta})\}\ . \tag{1.4}$$

A power vector $\boldsymbol{p}$ which satisfies the fixed point equations in (1.4), will be referred to as the *continuous stationary power vector.* For notational clarity, we will denote the continuous stationary power vector by $\tilde{\boldsymbol{p}}^*$. It has been shown in [8], that when the power level is continuous, then there is a unique stationary power vector $\tilde{\boldsymbol{p}}^*$.

When the power levels are discrete, the continuous DCPC algorithm in (1.3) cannot be applied, as the right-hand side may assume values outside the discrete set. Therefore, discrete algorithms are required, which are the subject of our next section.

3 DISTRIBUTED DISCRETE POWER CONTROL

A natural "discretization" of the continuous DCPC is to take the 'ceiling' or the 'floor' of the right-hand side of (1.3). For a power value x, let $\lceil x \rceil$ ($\lfloor x \rfloor$) be the smallest value in $\mathcal{D}$ which is larger than or equals x (the largest value in $\mathcal{D}$ which is smaller than or equals x).

We define two Distributed Discrete Power Control (DDPC) algorithms where the objective is twofold. First, they will be used to construct a convergence "envelope", or a "fixed-envelope" solution (paraphrasing on the fixed-point solution term). Second, they will be used as building blocks of a discrete algorithm which will drive the powers to the optimal convergence power vector (to be defined below). These are the following Ceiling and Floor DDPC algorithms. To differentiate between the powers under Ceiling DDPC and Floor DDPC, we will use $\boldsymbol{p}$ for the Ceiling, and $\boldsymbol{q}$ for the Floor.

Ceiling DDPC Algorithm

For any given power vector at time t, $\mathbf{p}(t)$, and a set of transmitters updating their powers at time $t + dt$, $U(t)$, the Ceiling DDPC updates the new powers according to

$$p_i(t+dt) = \begin{cases} \min\{\bar{p},\ \lceil \gamma_i^t \cdot \frac{p_i(t)}{\gamma_i(t)} \rceil\}\ , & \text{if } i \in U(t), \\ p_i(t)\ , & \text{otherwise}\ . \end{cases} \tag{1.5}$$

Floor DDPC Algorithm

For any given power vector at time t, $\mathbf{q}(t)$, and a set of transmitters updating their powers at time $t + dt$, $U(t)$, the Floor DDPC updates the new powers according to

$$q_i(t+dt) = \begin{cases} \min\{\bar{p},\ \lfloor \gamma_i^t \cdot \frac{q_i(t)}{\gamma_i(t)} \rfloor\}\ , & \text{if } i \in U(t), \\ q_i(t)\ , & \text{otherwise}\ . \end{cases} \tag{1.6}$$

Fixed-point solutions (i.e., stationary power vectors), even when not unique, can be defined as in the continuous case. A vector $\boldsymbol{p}^*$ is a *ceiling stationary power vector* if and only if, it is a fixed-point solution to

$$\boldsymbol{p} = \min\{\overline{\boldsymbol{p}},\ \lceil \boldsymbol{\gamma}^t \otimes (\boldsymbol{A}\boldsymbol{p} + \boldsymbol{\eta}) \rceil\}\ . \tag{1.7}$$

Similarly, a vector $\boldsymbol{q}^*$ is a *floor stationary power vector* if and only if, it is a fixed-point solution to

$$\boldsymbol{q} = \min\{\overline{\boldsymbol{p}},\ \lfloor \boldsymbol{\gamma}^t \otimes (\boldsymbol{A}\boldsymbol{q} + \boldsymbol{\eta}) \rfloor\}\ . \tag{1.8}$$

Although, the Ceiling and Floor DDPC algorithms are simple "discretizations" of the continuous DCPC, they may not converge, nor have a unique fixed point solution. This can be seen from the following counterexample. Consider two transmitters ($N = 2$) using power levels from $\{0.5, 1\}$. Let $a_{12} = a_{21} = 0.5$, $\eta_1 = \eta_2 = \epsilon/10$, and $\gamma_1^t = \gamma_2^t = 2 - \epsilon$ for Ceiling DDPC (and $\gamma_1^t = \gamma_2^t = 2 + \epsilon$ for Floor DDPC). Here, ϵ is some arbitrary small and positive number. Starting with $p(0) = (0.5, 1)$, the power vectors will oscillate between $(0.5, 1)$ and $(1, 0.5)$. Also, there are two fixed point solutions which are $(0.5, 0.5)$ and $(1, 1)$. This counterexample illustrates the convergence and uniqueness problems in a discrete power control.

Potentially, there are many discrete power update algorithms, each with its own set of update equations. An elementary requirement of any update algorithm is the following. Given that all transmitters can be supported at their CIR targets, then the power update equations will have at least one fixed-point solution which also supports all transmitters. The next Lemma shows that the set of power update equations defined by the Ceiling DDPC, and the Ceiling stationary power vectors satisfy this basic requirement.

Lemma 3.1 *If there is a discrete power vector under which all transmitters are supported, then there is a ceiling stationary power vector which also supports them.*

Thus, the set of Ceiling DDPC stationary power vectors contains the desirable power vector, in the case where all the mobiles can be supported. However, it may contain several supporting power vectors, from which we would like to select the best one in a sense defined below.

Definition 3.1 *For every two power vectors* $\mathbf{p}^1$ *and* $\mathbf{p}^2$*, we say that* $\mathbf{p}^1$ *is* **better** *than* $\mathbf{p}^2$ *if:*

(i) *Every transmitter which is supported under* $\mathbf{p}^2$*, is also supported under* $\mathbf{p}^1$*.*

(ii) $\mathbf{p}^1 \leq \mathbf{p}^2$*, component-wise.*

A ceiling stationary power vector $\mathbf{p}^*$ *is* **optimal** *(if exists), if it is* **better** *than any other ceiling stationary power vector.*

Note that the relation **better** is not necessarily a complete ordering among the ceiling stationary power vectors. Nevertheless, we require from an optimal power vector $\boldsymbol{p}^*$, to be **better** than any other ceiling stationary power vector. Such an optimal power vector may not exist in general. However, in the sequel, we prove that in our system such an optimal power does exist, and derive an algorithm that finds it.

Although the counterexample above shows that convergence cannot be guaranteed, there are certain initial power vectors from which the Ceiling and Floor DDPC do converge. Two special initial points are the zero power vector $(0,0,\ldots,0)$, and the maximum power vector $\overline{\boldsymbol{p}} = (\bar{p},\ldots,\bar{p})$. They are special in the sense that their convergence points define an envelope which contains all other stationary power vectors. Moreover (as will become apparent), when the power vectors do not converge, they oscillate within this envelope.

Lemma 3.2 *Let $\boldsymbol{p}(0) = (0,0,\ldots,0)$ ($\boldsymbol{q}(0) = (0,0,\ldots,0)$) be an initial power vector. Then, the power vector sequence $\{\boldsymbol{p}(k),\ k \geq 0\}$ ($\{\boldsymbol{q}(k),\ k \geq 0\}$) converges to a ceiling (floor) stationary power vector $\underline{\boldsymbol{p}}^*$ ($\underline{\boldsymbol{q}}^*$).*

Moreover, $\underline{\boldsymbol{p}}^*$ ($\underline{\boldsymbol{q}}^*$) is the smallest component-wise ceiling (floor) stationary power vector, and

$$\underline{\boldsymbol{q}}^* \leq \tilde{\boldsymbol{p}}^* \leq \underline{\boldsymbol{p}}^* \ .$$

Lemma 3.3 *Let $\boldsymbol{p}(0) = (\bar{p},\ldots,\bar{p})$ ($\boldsymbol{q}(0) = (\bar{p},\ldots,\bar{p})$) be an initial power vector. Then, the power vector sequence $\{\boldsymbol{p}(k),\ k \geq 0\}$ ($\{\boldsymbol{q}(k),\ k \geq 0\}$) converges to a ceiling (floor) stationary power vector $\overline{\boldsymbol{p}}^*$ ($\overline{\boldsymbol{q}}^*$).*

Moreover, $\overline{\boldsymbol{p}}^*$ ($\overline{\boldsymbol{q}}^*$) is the largest component-wise ceiling (floor) stationary power vector, and

$$\overline{\boldsymbol{q}}^* \leq \tilde{\boldsymbol{p}}^* \leq \overline{\boldsymbol{p}}^* \ .$$

In Lemmas 3.2 and 3.3 we specified the largest and the smallest stationary power vectors under the Ceiling and the Floor DDPC algorithms. We use them to define the two following envelopes:

$$B_c = \{\boldsymbol{p} \mid \underline{\boldsymbol{p}}^* \leq \boldsymbol{p} \leq \overline{\boldsymbol{p}}^*\} \ ,$$

$$B_f = \{\boldsymbol{q} \mid \underline{\boldsymbol{q}}^* \leq \boldsymbol{q} \leq \overline{\boldsymbol{q}}^*\} \ .$$

Theorem 3.1 *Let $\boldsymbol{p}(0)$ ($\boldsymbol{q}(0)$) be any feasible initial power vector. Then, the power vector sequence $\{\boldsymbol{p}(k),\ k \geq 0\}$ ($\{\boldsymbol{q}(k),\ k \geq 0\}$) satisfies the following convergence properties:*

(a) *If* $(0,0,\ldots,0) \le \mathbf{p}(0) \le \underline{\mathbf{p}}^*$ *(* $(0,0,\ldots,0) \le \boldsymbol{q}(0) \le \underline{\boldsymbol{q}}^*$ *), then* $\{\mathbf{p}(k)\}$ *(* $\{\boldsymbol{q}(k)\}$ *) converges to* $\underline{\mathbf{p}}^*$ *(* $\underline{\boldsymbol{q}}^*$ *).*

(b) *If* $\overline{\mathbf{p}}^* \le \mathbf{p}(0) \le \overline{\mathbf{p}}$ *(* $\overline{\boldsymbol{q}}^* \le \boldsymbol{q}(0) \le \overline{\mathbf{p}}$ *), then* $\{\mathbf{p}(k)\}$ *(* $\{\boldsymbol{q}(k)\}$ *) converges to* $\overline{\mathbf{p}}^*$ *(* $\overline{\boldsymbol{q}}^*$ *).*

(c) *Otherwise,* $\{\mathbf{p}(k)\}$ *(* $\{\boldsymbol{q}(k)\}$ *) will either converge to a stationary power vector inside the envelope* B_c *(* B_f *), or will eventually oscillate inside the envelope* B_c *(* B_f *).*

Moreover, let $\mathbf{p}^*$ *(* $\boldsymbol{q}^*$ *) be any power vector which is reached under the Ceiling (Floor) DDPC, after entering the envelope* B_c *(* B_f *). Then, the following ordering holds true:*

$$\underline{\boldsymbol{q}}^* \le \boldsymbol{q}^* \le \overline{\boldsymbol{q}}^* \le \tilde{\boldsymbol{p}}^* \le \underline{\boldsymbol{p}}^* \le \mathbf{p}^* \le \overline{\mathbf{p}}^* \; . \tag{1.9}$$

Theorem 3.1 above states that in the discrete case, the power vectors under the Ceiling (Floor) DDPC are converging in a somewhat weaker sense. That is, into an envelope B_c (B_f). Thus, even when the power sequence oscillates, it will be within a bounding envelope. The size of the envelope however, depends on the power level grid and on the link gains. A large envelope may result in a large fluctuation in the CIR. To evaluate how bad the CIR could become under the Ceiling DDPC, we consider the worst case scenario. The following Lemma provides a method to evaluate the lowest CIR.

Lemma 3.4 *Let* M *be the first step where the power vector sequence under Ceiling DDPC enters the envelope* B_c*. Then, the CIR for transmitter* i *at step* k*,* $\gamma_i(k)$*,* $k \ge M$*, is at least as large as*

$$\underline{\gamma}_i \stackrel{\text{def}}{=} \frac{\underline{p}_i^*}{\sum_{j\in\mathcal{N}} a_{ij}\overline{p}_j^* + \eta_i} \; .$$

An oscillating power sequence has two main drawbacks. One is clearly the added variability (due to the power control algorithm) in the link quality (CIR fluctuation). The other is the reduced number of supported mobiles compared with a converging algorithm. These two drawbacks call for an algorithm which converges to a unique power vector.

The power vector to which the algorithm converges, could be important. As motivated by our optimality notion above, it is always better to converge to the optimal one (if such exists).

Theorem 3.2 *The ceiling power vector $\underline{\mathbf{p}}^*$ is optimal.*

Theorem 3.2 identifies the optimal power vector. A distributed algorithm which drives the powers to the optimum, is a combination of the Floor and the Ceiling DDPC algorithms.

DDPC Algorithm

> *From any initial power vector, start updating the powers according to the Floor DDPC until the first entry into envelope B_f. From then on, continue with Ceiling DDPC.*

Since the power values are discrete, Theorem 3.1 implies that the power sequence enters the envelope B_f after a finite number of steps. The entrance of an individual transmitter i, can locally be detected by looking for a cycle (either of length one, or larger) in the sequence of powers $\{p_i(k)\}$. Coordinating the moment where all transmitters have entered the envelope, can be done by a simple signaling procedure among the base stations. We end this section with the following Theorem.

Theorem 3.3 *For every initial power vector, the power vectors generated by the DDPC algorithm converge to $\underline{\mathbf{p}}^*$.*

4 A MICROCELLULAR EXAMPLE

In this section we evaluate the performance of our discrete power control algorithms in a Manhattan-like microcellular system. The cellular system under investigation is a typical metropolitan environment consisting of building blocks of a square shape. We assume that the channel assignment is fixed and divides the cells into 3 different channel groups (cochannel cells).

The link gain, g_{ij}, is modeled as, $g_{ij} = l_{ij} \cdot s_{ij}$, where l_{ij} describes the distance dependent path loss, and s_{ij} is a log-normally distributed shadow fading factor (mean 0 dB, standard deviation 4 dB). A detailed description of the microcellular system is given in [3].

We use a grid of power levels $\mathcal{D}_\Delta$, which are equally spaced at distance Δ dB, from $10 \log \underline{p}$ dB up to $10 \log \overline{p}$ dB. The median SNR at the border of a cell when using maximum transmitter power ($\overline{p} = 1$ W) is equal to 80 dB, i.e., we have a strongly interference limited system. Further, the dynamic range value is 30 dB. Borrowing from GSM and Qualcomm's CDMA systems, we take three values for the grid spacing Δ, 0.5, 1 and 2 dB.

To investigate the potential power oscillation under the Ceiling DDPC, we study the cumulative distribution function (cdf) of $\Psi_{max} \stackrel{\text{def}}{=} \max_{1 \leq i \leq N} \Psi_i$, where Ψ_i is defined as $\{\Psi_i \stackrel{\text{def}}{=} 10 \log \overline{p}_i^* - 10 \log \underline{p}_i^* \mid 1 \leq i \leq N\}$. The cdf of Ψ_{max} describes the difference in power for the worst situated mobile, or equivalently, the maximum size of B_c.

We also investigate the outage probability, defined as $P_{outage} = P(\gamma < \gamma^t)$, where γ is the CIR of a randomly selected link.

The convergence speed is another practical performance measure that we consider. The convergence speed under the DCPC depends on its convergence criterion. In our examples, we stop the DCPC power update when the maximum relative difference between two consecutive power vectors becomes smaller than ϵ.

To evaluate the potential degradation of the CIR under the Ceiling DDPC, at those instances where the CIR is above the target under the DDPC, we compute the probabilities

$$P(\underline{\gamma} < x, \underline{\gamma}^* \geq \gamma^t) .$$

Here, $\underline{\gamma}^*$ is the CIR of a random link under $\underline{\boldsymbol{p}}^*$ (i.e., under DDPC); and $\underline{\gamma}$ is its lower bound under Ceiling DDPC (see Lemma 3.4).

All the statistics above are evaluated via simulation. For every power control algorithm we sample 10,000 instances of mobile locations and shadow fading values. The power updates are made synchronously, and we further assume a uniform CIR target, γ^t, for all transmitters.

Numerical results

Figure 1 depicts the cdf of Ψ_{max} for various CIR targets $(8, 12, 16, 20$ $dB)$ and a grid size of $\Delta = 2$ dB. Observe that the difference in power is largest (and therefore the potential power oscillation), when the CIR target is around 16 dB. As can be seen from the figure, the power oscillation may be severe.

In Figure 2 we compare between the outage probabilities of the DDPC, and that of the continuous DCPC. Figure 2 illustrates how much is being sacrificed in mobile outage by spreading the power levels. This is particularly important in light of the fact that the convergence speed is improved by doing so (see Figure 3). As expected, refining the power level grid reduces the outage probability. However, a very small CIR gain is obtained by refining the grid from $\Delta = 2\ dB$ to the continuous case. The increase in the expected number of update steps, however, is tremendous (from 20 to 160 steps).

The gain in the outage probability that is obtained by the DDPC, compared to the Ceiling DDPC, is demonstrated by Figure 4. We observe that for a grid size of $\Delta = 2\ dB$, 38% of the mobiles that are supported under the DDPC will not be supported under the Ceiling DDPC. This certainly gives an advantage to the DDPC compared to the Ceiling DDPC.

5 CONCLUSIONS

We studied transmitter power control algorithms in a cellular PCS which uses only discrete power levels. It has been shown that by simply "discretizing" the continuous power control algorithm, the convergence and the uniqueness properties of the continuous power control, are lost. A special distributed discrete power control algorithm, the Ceiling DDPC algorithm, has been shown to "converge" in a weaker sense, i.e., into an envelope of power vectors rather than to a unique vector. The power vector under this algorithm may oscillate, which results in a poorer link quality, and a higher outage probability. To alleviate the oscillation problem, a Distributed Discrete Power Control (DDPC) algorithm has been proposed, which converges to an optimal power vector.

REFERENCES

[1] J. M. Aein. Power Balancing in Systems Employing Frequency Reuse. *Comsat Tech. Rev.*, Vol. 3, No. 2, 1973.

[2] M. Andersin, Z. Rosberg and J. Zander. Gradual Removals in Cellular PCS with Constrained Power Control and Noise. *IEEE/ACM Wireless*

Networks; to appear.

[3] M. Andersin, Z. Rosberg and J. Zander. Soft Admission in Cellular PCS with Constrained Power Control and Noise. Proc. 5th WINLAB Workshop, Apr. 1995.

[4] M. Andersin, Z. Rosberg and J. Zander. Distributed Discrete Power Control in Cellular PCS. TRITA-IT R 94-23, ISSN 1103-534X, ISRN KTH/IT/R-94/23-SE, Radio Commun. Systems Lab., KTH, Sweden, Dec. 1994.

[5] N. Bambos and G. J. Pottie. On Power Control in High Capacity Cellular Radio Networks. Proc. 3rd WINLAB Workshop, Oct. 19-20, 1992.

[6] G. J. Foschini. A Simple Distributed Autonomous Power Control Algorithm and its Convergence. *IEEE Trans. on Veh. Tech.*, Vol. 42, No. 4, 1993.

[7] S. A. Grandhi, R. Vijayan and D. J. Goodman. A Distributed Algorithm for Power Control in Cellular Radio Systems. *IEEE Trans. on Commun.*, Vol. 42, No. 2/3/4, February/March/April 1994.

[8] S. A. Grandhi, J. Zander and R. Yates. Constrained Power Control. *Wireless Personal Communications*, Vol. 1, No. 4, 1995.

[9] S. V. Hanly. Information Capacity of Radio Networks. Ph.D. Thesis, Cambridge University, Aug. 1993.

[10] J. C. Lin, T. H. Lee and Y. T. Su. Power Control Algorithm for Cellular Radio Systems. *IEEE Trans. on Veh. Tech.*, Vol. 44, No. 1, 1995.

[11] D. Mitra. An Asynchronous Distributed Algorithm for Power Control in Cellular Radio Systems. Proc. 4th WINLAB Workshop, Oct. 19-20, 1993.

[12] An Overview of the Application of Code Division Multiple Access (CDMA) to Digital Cellular Systems and Personal Cellular Networks. Document Number EX60-10010, QUALCOMM Inc., May 1992.

[13] J. Zander. Transmitter Power Control for Co-channel Interference Management in Cellular Radio Systems. Proc. 4th WINLAB Workshop, Oct. 19-20, 1993.

[14] J. Zander. Performance of Optimum Transmitter Power Control in Cellular Radio Systems. *IEEE Trans. on Veh. Tech.*, Vol. 41, No. 1, 1992.

[15] J. Zander. Distributed Cochannel Interference Control in Cellular Radio Systems. *IEEE Trans. on Veh. Tech.*, Vol. 41, No. 3, 1992.

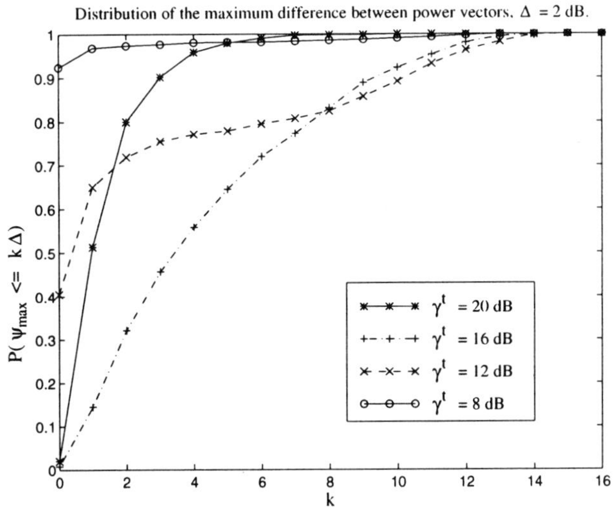

Figure 1 The cumulative distribution function of the maximum difference between the two stationary power vectors $\underline{p}^*$ and $\overline{p}^*$. The grid spacing $\Delta = 2\ dB$.

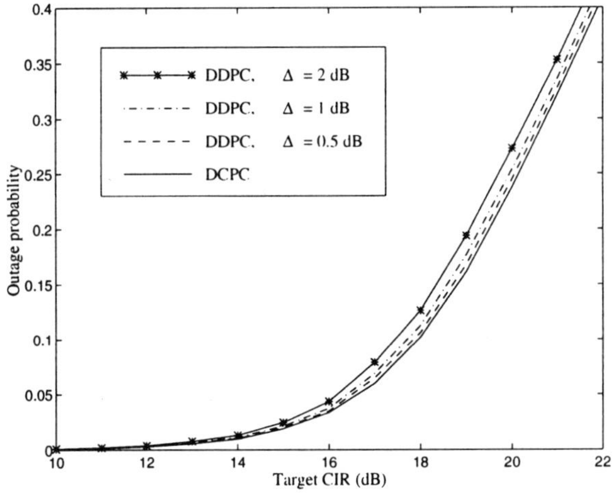

Figure 2 Outage probabilities for the DCPC and the DDPC algorithms. The outage probability for DDPC is obtained under the optimal stationary power vector $\underline{p}^*$. The outage probability for DCPC is obtained using $\epsilon = 10^{-8}$.

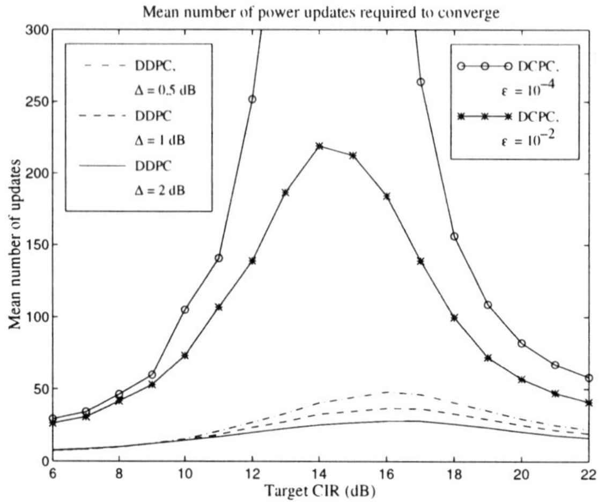

Figure 3 The mean number of power updates that is required to converge for the DCPC algorithm and the Ceiling DDPC algorithm.

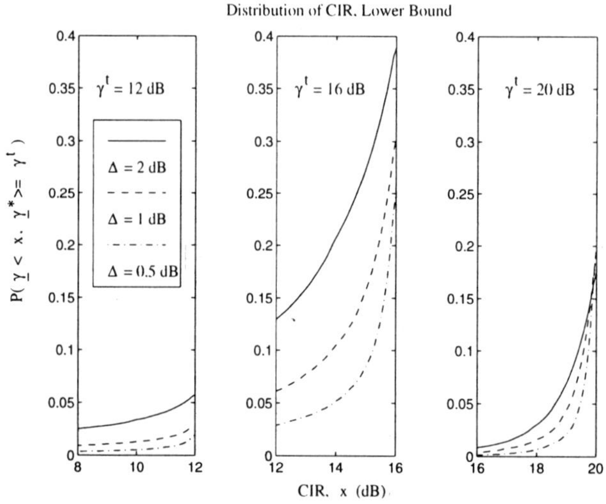

Figure 4 The distribution of the CIR according to the lower bound for target CIR $\gamma^t = 12, 16$ and $20\ dB$, at the instances where the communication link is supported under $\underline{\boldsymbol{p}}^*$.

CONGESTION MEASURES AND CAPACITY CONSTRAINTS IN SPREAD SPECTRUM NETWORKS

Stephen HANLY

Dept. Electrical and Electronic Engineering

University of Melbourne, Parkville Vic. 3052 Australia

1 INTRODUCTION

When can a call be admitted to a radio network and when should it be blocked due to congestion in the network? In order to answer such a question, we need to be able to compute traffic constraints that are imposed by quality of service considerations. This is a nontrivial problem with an interesting spatial aspect, as users are distributed amongst the receivers, and may be in motion. The great benefit of spread spectrum is that the traffic constraints turn out to be "macroscopic", in the sense that the constraints are only imposed on the overall traffic in relatively large geographical areas. This averaging over space is a form of statistical multiplexing, and simplifies the analysis of mobility, fading, and burstiness, although we do not directly consider these issues in the present paper. It implies that we can dimension a network, and control the admission of calls, without needing to predict at call admission the future fading levels of individual mobiles throughout the lifetimes of their calls.

We assume users place "demands" on the network in accordance with their quality of service requirements. For voice (or video) in CDMA, quality of service (Q o S) can be translated into C/I constraints; for a particular bit rate and bit error rate, there is an associated C/I level required. These C/I constraints give rise to a set of equations in transmitter power levels that must be solved in order to meet the Q o S constraints. Further, it is necessary to place constraints on the transmit power levels of users. In fact, the aim is to provide communications with as low a set of transmit power levels as possible. We emphasize that a CDMA network is a large collection of "microscopic elements", the mobiles, and these elements are highly random (in their fadings and burstiness). It is just too complex to keep track of the network at the microscopic level of detail. In particular, it does not appear to be tractable to predict future evolution of a network at the microscopic level of detail. Rather, we take a "macroscopic approach"; we identify macroscopic measures of congestion and relate these to the traffic in the network. These measures relate to the solubility of the equations and also to the transmit power requirements on the network as a whole.

When the network becomes congested in these terms, it means that the probability that a user is required to use more than its maximum transmit power capability has become unacceptably large. These measures of congestion can be used to derive traffic constraints, consisting of bounds on traffic in geographical regions, ranging from local constraints on the traffic within a cell, to constraints on the traffic in several cells, to an overall constraint on the traffic in the whole network.

The format of the paper is as follows. We begin by considering a single cell and derive a measure λ of congestion, which we relate to both the traffic in the cell, and "power warfare" in the cell. We consider this power warfare measurement as fundamental; it is ultimately transmit power level constraints that limit the capacity of CDMA systems. We then consider a macrodiversity model, in which no cellular structure is assumed; effectively we have a large antenna array. We derive an analogous measure, λ, for this model, measuring local power warfare in the network. Finally, we consider a cellular system, and by providing bounds on λ, we obtain explicit traffic constraints on the network.

2 A SINGLE RECEIVER CDMA NETWORK

We begin this section with an assumption to the effect that a bit rate and quality of service requirement can be expressed as a carrier to interference ratio requirement (C/I). Implicit in this is the assumption that all other users' signals are treated as *noise* by the decoder of a given user; thus, we do not consider joint multi-user decoding here.

Suppose that all users have the same bit rate and quality of service requirement. Then it follows that all users must be received at the same power level Q, and all users attain a C/I given by $\frac{C}{I} = \frac{Q}{(M-1)Q + \eta W}$, where we denote the spread spectrum bandwidth by W Hz, and assume a Gaussian external white noise of spectral density $\frac{\eta}{2}$. In spread spectrum, it makes sense to take into account the processing gain, so we multiply this by the bandwidth, W, to get the $CIDR$ (carrier to interference spectral density ratio) $CIDR \equiv \frac{WQ}{(M-1)Q + \eta W}$. Let us suppose that each user requires $CIDR = \alpha$ Hz. Then, it is required that

$$(W - (M-1)\alpha)Q = \eta W \alpha \tag{1.1}$$

and hence, for feasibility, we require

$$\lambda \equiv \frac{(M-1)\alpha}{W} < 1 \tag{1.2}$$

Note that the solution $Q^\star$ to (1.1) satisfies

$$Q^\star = \frac{\eta\alpha}{1-\lambda}. \tag{1.3}$$

Thus, as $\lambda \uparrow 1$, $Q^\star(\lambda) \uparrow \infty$. We can readily see that a "power war" is being fought as users are added to the system, and that as the limit of $\frac{W}{\alpha}+1$ users is approached, the required received power goes to infinity, as users fight to be heard. Without external noise, $\frac{W}{\alpha}+1$ users can achieve a common $CIDR$ of α Hz. The observed power escalation arises in order to make the thermal noise less and less significant. In Section 4 we shall see that it also arises to overcome other-user interference. We can also measure the rate of increase of the received power:

$$\frac{d}{d\lambda}Q^\star(\lambda) = \frac{1}{1-\lambda}Q^\star(\lambda) \tag{1.4}$$

We interpret λ as the normalized traffic, a measure of congestion, and note that it is also a measure of power warfare.

We shall find it a simplification to include a user's own signal power as part of the interference power. This results in a measure of congestion that is slightly conservative in that it overestimates the level of congestion in the network. Thus, we take $\lambda \equiv \frac{M\alpha}{W}$ in the present case. For a network with large W, this has negligible effect. Formally, if $\alpha \asymp O(1)$ as $W \uparrow \infty$, then the discrepency is $O(1/W)$.

3 A MULTI-RECEIVER MACRODIVERSITY MODEL

A macrodiversity model for a CDMA network has recently been proposed [3] in which each user is decoded by a "super base station" which combines the signals from all the receivers in the network. Denote the number of users by M and the number of receivers by K. We begin by obtaining a condition for feasibility when no transmit power constraints are imposed. This condition is analogous to $\lambda < 1$ from the preceding section.

As in Section 2, we can consider the power control problem of finding transmit powers to satisfy C/I constraints. In the present model, user i transmits with power q_i and is received at receiver k at power $\Gamma[i,k]q_i$. We call $\Gamma[i,k]$ the "path gain" from mobile i to receiver k. It follows that the total interference at receiver k contributed by *users* is given by $Q_k \equiv \sum_{i=1}^{M}\Gamma[i,k]q_i$, and the CIDR of user i at receiver k is given by $\frac{Wq_i\Gamma[i,k]}{Q_k+\eta_k W}$, where η_k is the thermal noise spectral density at receiver k. It is shown in [3] that the appropriate quality of service measure for user i, when the decoder combines the signals from all K receivers, is given by $\mathrm{CIDR}_i \equiv \sum_{k=1}^{K}\frac{Wq_i\Gamma[i,k]}{Q_k+\eta_k W}$.

Let us assume that user i requires

$$\mathrm{CIDR}_i = \alpha_i \quad i = 1, 2, \ldots, M \tag{1.5}$$

which gives M nonlinear equations in M unknowns $(q_1, q_2, \ldots, q_M)$. Pick a receiver ℓ and rewrite (1.5) as

$$q_i \Gamma[i, \ell] = \frac{\alpha_i}{W} \left(\sum_{k=1}^{K} \frac{\Gamma[i, k]}{\Gamma[i, \ell]} \frac{1}{Q_k + \eta_k W} \right)^{-1}$$

which is precisely the interference created by user i at receiver ℓ. Summing over the users, we obtain the fixed point equation

$$Q_\ell = \sum_{i=1}^{M} \frac{\alpha_i}{W} \left(\sum_{k=1}^{K} \frac{\Gamma[i, k]}{\Gamma[i, \ell]} \frac{1}{Q_k + \eta_k W} \right)^{-1} \tag{1.6}$$

The task of solving (1.5) is therefore that of finding the fixed point Q^* of the mapping

$$T: \quad \mathbb{R}_+^K \to \mathbb{R}_+^K \tag{1.7}$$

$$(Q_\ell)_{\ell=1}^K \rightsquigarrow \left(\sum_{i=1}^{M} \frac{\alpha_i}{W} \left(\sum_{k=1}^{K} \frac{\Gamma[i, k]}{\Gamma[i, \ell]} \frac{1}{Q_k + \eta_k W} \right)^{-1} \right)_{\ell=1}^{K}$$

Theorem 1 gives a necessary and sufficient condition for the existence of Q^*:

Theorem 1 *The M CIDR equations (1.5) have a solution Q^* if and only if $x < 1$ where $x \equiv \frac{\sum_{i=1}^{M} \alpha_i}{KW}$. If $x < 1$ then Q^* is unique and*

$$x = \frac{1}{K} \sum_{k=1}^{K} \frac{Q_k^*}{Q_k^* + \eta_k W} \tag{1.8}$$

Proof: See [3]. □

The important point to note is that we can interpret x as the total traffic in the network, if we interpret $\frac{\alpha_i}{KW}$ as the bandwidth required by user i. We have normalized the traffic by the spread spectrum bandwidth, and the number of receivers in the network, and have obtained the constraint:

$$\sum \text{traffic in network} \; < 1 \tag{1.9}$$

4 POWER LEVELS AND POWER WARFARE

4.1 Average congestion

In Section 2, we related the traffic measure λ to power levels and power warfare; see equations (1.3) and (1.4) respectively. In the macrodiversity model, x is related to average power levels by (1.8). Note that we can interpret $\frac{Q_k^\star}{Q_k^\star+\eta_k W}$ as a normalized measure of interference at receiver k, and x is then the average of these measurements over the whole network.

4.2 Localized congestion and power warfare

Theorem 1 implies that $\frac{1}{K}\sum_{k=1}^{K}\frac{Q_k^\star}{Q_k^\star+\eta_k W}$ does not depend on the users' locations in the network. However, it is clear that an individual $\frac{Q_k^\star}{Q_k^\star+\eta_k W}$ does depend on these locations. If all users are located near a receiver, then the corresponding $Q_k^\star$ can be extremely high, and indeed the individual transmit powers of many users may be excessively high. The configuration of users may be feasible, given unlimited transmit power, but the users may need sufficient transmit power to reach receivers that are far away. It is clear that the macrodiversity network can suffer local congestion in a way that is not reflected in the global traffic measure x. Can we provide a macroscopic measure of local congestion?

We have seen that "power warfare" occurs as congestion occurs. This suggests that in a network local congestion should give rise to "local power warfare". We therefore consider the effect of increasing the level of traffic in the network.

We assume that $Q^* = (Q_1^*, Q_2^*, \ldots, Q_K^*)$ is the unique fixed point of the mapping T defined in (1.7), for the CIDR requirements in the vector α. If we try and increase the traffic in the network, we face the problem of where to put the new users, so instead we consider scaling up the CIDR of the users that we already have. For $\mu > 1$, we let $Q^*[\mu]$ be the new fixed point, given that the CIDR requirements of the users are now $\mu\alpha$. By definition, $Q^* \equiv Q^*[1]$. Note that by Theorem 1, $Q^*[\mu]$ exists provided $\mu < \frac{KW}{\sum_{i=1}^{M}\alpha_i}$.

The matrix $\left.\frac{\delta T}{\delta Q}\right|_{Q^*}$ is a positive matrix and we let λ be its Perron-Frobenius eigenvalue. As before $x \equiv \frac{\sum_{i=1}^{M}\alpha_i}{KW}$. Then we have the following result:

Theorem 2 *If $x < 1$ then $x \leq \lambda < 1$, and there exists a receiver k for which*

$$\left.\frac{dQ_k^*}{d\mu}\right|_{\mu=1} \geq \frac{1}{1-\lambda} Q_k^* \tag{1.10}$$

Proof: See [4] □

This result is a generalization to networks of (1.4). A single cell has a capacity bound and as this bound is approached, the required transmitter powers of the users increase without bound. In a *network*, local congestion will manifest itself as a localized "power war" between the users. Theorem 2 shows that λ, the Perron-Frobenius eigenvalue of $\left.\frac{\delta T}{\delta Q}\right|_{Q^*}$, has a role to play in characterizing this congestion. We shall provide some examples to illustrate Theorem 2 in Section 5. Before doing this, it is useful to write down explicitly the derivative matrix $\left.\frac{\delta T}{\delta Q}\right|_{Q^*}$, and obtain some bounds on λ. These bounds provide further insight into the relationship between λ and "localized congestion".

4.3 The matrix $\frac{\delta T}{\delta Q}$ and bounds on λ

For notational convenience, let us denote the $\frac{C}{I}$ of user i at receiver k by s_{ik}^*. Then

$$\left.\frac{\delta T_k}{\delta Q_j}\right|_{Q^*} = \sum_{i=1}^{M} \frac{W}{\alpha_i} s_{ik}^* s_{ij}^* \frac{Q_k^* + \eta_k W}{Q_j^* + \eta_j W}$$

Let us also define the $K \times K$ *symmetric* matrix $\mathcal{M}$ by

$$\mathcal{M}_{k,j} = \sum_{i=1}^{M} \frac{W}{\alpha_i} s_{ik}^* s_{ij}^*$$

and note that $\left.\frac{\delta T}{\delta Q}\right|_{Q^*}$ is *similar* to $\mathcal{M}$, in the sense that

$$\left.\frac{\delta T}{\delta Q}\right|_{Q^*} = D^{-1}\mathcal{M}D$$

where $D = \text{diag}\ (Q_1^* + \eta_1 W, Q_2^* + \eta_2 W, \ldots, Q_K^* + \eta_K W)$. It follows that $\mathcal{M}$ has the same eigenvalues as $\left.\frac{\delta T}{\delta Q}\right|_{Q^*}$. An interesting and useful feature of $\mathcal{M}$ is given in the following lemma.

Lemma 1 *The jth row sum of $\mathcal{M}$ is $\frac{Q_j^*}{Q_j^* + \eta_j W}$.*

Proof: See [4]. □

Corollary 1 $\lambda \leq \max_j \dfrac{Q_j^*}{Q_j^* + \eta_j W}$.

Proof: The Perron-Frobenius eigenvalue of $\mathcal{M}$ is dominated by the maximum row sum of $\mathcal{M}$ (see [5]) and, by similarity, this eigenvalue is λ. □

Recall that we have interpreted $\frac{Q_j^*}{Q_j^*+\eta_j W}$ as the normalized interference at receiver j. The next result shows that λ is *lower* bounded by the average interference.

Lemma 2 $\lambda \geq x \equiv \dfrac{1}{K} \sum_{k=1}^{K} \dfrac{Q_k^*}{Q_k^* + \eta_k W}$.

Proof: See [4]. □

In Section 3, we interpreted x as the average congestion in the network. Theorem 2 provides an interpretation of λ as a measure of "worst-case power warfare" in the network. In this section, we see that λ lies between the average congestion, and the worst case local interference at any receiver.

5 EXAMPLES TO ILLUSTRATE THEOREM 2.

5.1 Uniform interference

In this section we consider any configuration of users for which there exists Q^*, η such that

$$Q_k^* \equiv Q^*, \qquad \eta_k \equiv \eta, \qquad k = 1, 2, \ldots, K$$

We call this condition "uniform interference"; the congestion is truly global in this example.

This example is one of the few in which we can make explicit computations. It follows from Theorem 1 that $\dfrac{Q^*}{Q^* + \eta W} = x \equiv \dfrac{\sum_{i=1}^{M} \alpha_i}{KW}$ and hence $Q^* = \dfrac{x}{1-x}\eta W$. We can also compute λ and its associated eigenvector e:

$$e = \frac{1}{\sqrt{K}}(1, 1, \ldots, 1)$$

$$\lambda = x$$

This is the only case for which $x = \lambda = \max_k \dfrac{Q_k^*}{Q_k^* + \eta_k W}$.

Let us now consider the effect of increasing traffic, by taking the vector of required CIDRs to be $\mu\alpha$. Then $Q^*[\mu\alpha] = \dfrac{\mu x}{1 - \mu x}\,\eta W$ and hence

$$\left.\frac{dQ^*}{d\mu}\right|_{\mu=1} = \frac{1}{1-x}Q^*$$

$$= \frac{1}{1-\lambda}Q^*$$

In this example only, all receiver interference levels increase at rate $\frac{1}{1-\lambda}Q^*$, and equality holds in (1.10) for all receivers $k = 1, 2, \ldots, K$.

5.2 Non-uniform interference in a 3 receiver network

Consider the network depicted in Figure 1, (a). There are 3 receivers labelled 1, 2 and 3, and $M = 32$. Each user requires $\alpha = 0.038$ MHz and is located randomly within a square of side 0.1 around receiver 2. For simplicity, we assume that power degrades as the 4th power of distance. In this case, the total interference at receiver 2 is 44, in units of thermal noise (we have scaled power so that $\eta W = 1$). The eigenvalue for this configuration is $\lambda = 0.972$.

In Figure 1, (c), $M = 92$ users are located at random in a square of side 1 around receiver 2. Here, the interference at receiver 2 is 382, which is much higher than in (a). This is not surprising, as there are many more users in this network. The eigenvalue in this case is $\lambda = 0.938$, which we note to be *lower* than in (a).

In (b), two more users have been added randomly to the congested square in (a), so now $M = 34$. The interference at receiver 2 has now increased drastically to $Q_2 = 6760$. In (d), six users have been added to the square in (c) for a total of $M = 98$ users. The interference at receiver 2 also increases substantially, but we note that the final interference is *less* than in (b); here, $Q_2 = 5147$. This example illustrates Theorem 2, by showing that when λ is close to unity, there is great sensitivity in Q^* to increases in traffic.

From (c) to (d), λ increases from 0.938 to 0.993. However, from (a) to (b), λ *decreases* from 0.972 to 0.969. We shall see that this is because users in (b) have greater diversity than in (a); there is a decrease in the level of power warfare as a consequence of an increased number of "local resources". We shall interpret this example by saying that in (b), the middle receiver has "overloaded" so the users have increased transmit power drastically to reach the other two receivers. The users then compete for the middle receiver *less* than in (a), because they have access now to the other two receivers. This example shows that although local power warfare has decreased, interference levels have increased substantially. We claim that λ measures an important aspect of local congestion, as quantified in Theorem 2; it is clear from this example that other measures, such as the values $\left(\frac{Q_k^*}{Q_k^*+\eta_k W}\right)_{k=1}^{K}$, are required as well.

We shall see in Section 6 that the phenomenon of λ decreasing with increasing traffic does *not* occur when we limit the diversity that the users are allowed. However, we also show that λ can be decreased by relaxing the diversity constraints (see also [2]). Thus the network controller can decrease local power warfare by increasing the number of users in cell-site diversity mode ([6]) or "soft handover".

6 A CELLULAR MODEL WITH MACRODIVERSITY

The full macrodiversity model of Section 3 is a useful abstraction, but in practice it is perhaps unreasonable to expect that all receivers in a network be connected to form one "super cell". However, we can incorporate limited macrodiversity into a cellular model. Figure 2 (a), depicts a cellular network in which the cells are squares, with centrally located cell sites, but with antennas located at the cell corners. In this way, each cell has 4 widely separated antennas. A user in a cell is only decoded by a cell-site associated with the four adjacent receivers. Thus cell C_5 is decoded by combining signals from receivers $6, 7, 10, 11$. The motivation here is to impose on the network a degree of localization; an attempt to limit transmitter power, without actually imposing an explicit power constraint. We also wish to limit the complexity in the network. For cell C users, let $R(C)$ be the set of receivers that are used to decode the users in that group. Thus $R(C_5) = \{6, 7, 10, 11\}$.

As in Section 3, the total interference at receiver k is given by $Q_k = \sum_{i=1}^{M} \Gamma[i,k]q_i$. The only difference between the present model and full macrodiversity is that we now require

$$\sum_{k \in R(C)} \frac{W q_i \Gamma[i,k]}{Q_k + \eta_k W} = \alpha_i \quad \forall C \ \ \forall i \in C$$

In this case the mapping T is given by

$$T(Q)_j = \sum_{i=1}^{M} \frac{\alpha_i}{W} \left(\sum_{k \in R(C_i)} \frac{\Gamma[i,k]}{\Gamma[i,j]} \frac{1}{Q_k + \eta_k W} \right)^{-1}$$

where C_i is the cell of user i, and hence

$$\frac{\delta T_j}{\delta Q_k} = \sum_{i=1}^{M} \frac{W}{\alpha_i} s_{ij} s_{ik} \frac{Q_j + \eta_j W}{Q_k + \eta_k W} I[k \in R(C_i)]$$

We let λ be the Perron-Frobenius eigenvalue of $\frac{\delta T}{\delta Q}\Big|_{Q^\star}$, where $Q^\star$ is the unique fixed point of T (we assume this exists). Then Theorem 3 can be proven in a similar manner to Lemma 2.

Theorem 3 $\lambda \geq \max_G \frac{\sum_{C \in G} \sum_{i \in C} \alpha_i}{|L(G)|W}$, *where G is a set of contiguous cells, and $L(G)$ is the set of receivers in the cells in G.*

Proof: See [4]. □

A special case of Theorem 3 is the bound $\lambda \geq \frac{\sum_{i=1}^{M} \alpha_i}{KW}$, where we are taking G to be the set of all cells in the network. This bound also held in the full macrodiversity model (Lemma 2). However, it is now possible for the bound $\lambda \geq \frac{\sum_{i \in C} \alpha_i}{4W}$ to be the tightest of the considered bounds, if a single cell C is highly congested with the rest of the network lightly loaded. In general, we must consider all groups of contiguous cells. When combined with the feasibility condition $\lambda < 1$, the bounds of Theorem 3 imply traffic constraints, of the form (1.9), but one for each group of contiguous cells.

The bounds in Theorem 3 are not tight in general. For example, unless cell C is the only cell with any traffic, then we will have strict inequality in $\lambda \geq \frac{\sum_{i \in C} \alpha_i}{4W}$. The traffic constraints are necessary but not sufficient for network feasibility. However, in a 3 receiver network studied in [1], in which two 2-receiver cells share a common receiver, the constraints can be sufficient. The condition for sufficiency is that users only create interference at the receivers in their own cell. This result suggests that the constraints that we have identified may be good approximations to the capacity region in general, provided path gains from users to receivers in other cells are small.

The following numerical work also includes the effect of having directional antennas. It is shown in [4] that similar bounds on λ to those derived above can be derived

receiver	cell, dir.	cell, omni.	macro., omni
15	6.3	386.8	22.2
16	4.8	183.5	12.0
19	5.8	330.7	20.4
20	3.9	172.3	12.0
eigenvalue	0.803	0.98	0.77

Table 1 Interference levels and eigenvalue

for the network with directional antennas, in which each antenna has four directional beams, one for each quadrant.

Consider the network of 36 cells depicted in Figure 2 (b). The bandwidth $W = 1.25$ MHz, each user requires $\alpha = 0.04$ MHz and there are 574 such users. There is a receiver at each integer-valued point in $[1,6] \times [1,6]$. The users are distributed over $[1,6] \times [1,6]$ and it is apparent that there is a hotspot in $[3,4] \times [3,4]$.

Let us begin with the assumption that directional antennas are used, for a total of 100 directional antennas. The "limited directional" column in Table 6 plots the total interference at the four receivers in cell $[3,4] \times [3,4]$ and we note that the maximum interference is 6.3 units (1 unit = thermal noise power, assumed to be the same at each receiver). The Perron-Frobenius eigenvalue of $\left.\frac{\delta T}{\delta Q}\right|_{Q^*}$ is 0.803. The lower bound of Section 6 corresponding to the congested central cell is

$$\frac{100(0.04)}{4W} = 0.8$$

since there are 100 users in the central congested cell. Note that the single cell constraint is quite tight in this case.

Now let us suppose that the receivers are omnidirectional. Then instead of 100 receivers, we have only 36 and the interference levels at the four receivers in cell $[3,4] \times [3,4]$ increase dramatically. The "limited diversity, omnidirectional" column in Table 6 shows the new interference values, and we note that the system is almost overloaded: $\lambda = 0.98$. The maximum interference level is now 387. In this case, the single cell constraint is far from tight, because the other-cell interference is now significant.

Suppose the network controller decides to relieve this congestion by increasing the level of diversity in the network. In practice, soft handover between cells as in [6] may be appropriate, but here we shall consider the extreme case of full macrodiversity. We see from the "full macrodiversity" column that the maximum interference has dropped to 22 under full macrodiversity. The Perron-Frobenius eigenvalue, 0.77, is now lower than in the directional antennas case, even though there are only 36 available receivers here, compared to 100 in the directional case. This example indicates that increasing the diversity in the network can have a significant impact on network performance. In [2], a similar set of results is provided for soft handover for the traditional cellular model with one receiver in the center of each cell.

Acknowledgements

Research undertaken at AT & T Bell Labs, and completed at Melbourne University. Special thanks to Debasis Mitra and Emre Telatar for useful discussions.

REFERENCES

[1] Hanly, S.V. (1993) Information capacity of radio networks *Ph. D. Thesis, Cambridge University* August.

[2] Hanly S. V. (1995) An algorithm for combined cell-site selection and power control to maximize cellular spread spectrum capacity *IEEE Journal on Selected Areas, special issue on the fundamentals of networking, Vol. 13, No. 7* September.

[3] Hanly, S.V. (1996) Capacity and power control in spread spectrum macrodiversity radio networks. *IEEE Trans. on Communications, Vol. 44, No. 2* Feb.

[4] Hanly, S.V. Congestion measures and traffic constraints in spread spectrum networks *submitted for publication*

[5] Seneta E. (1981) Non-negative matrices and markov chains *Springer-Verlag.*

[6] Mobile station - base station compatibility standard for dual-mode wide-band spread spectrum cellular systems (IS95 standard).

(a) 32 users

$Q_2 = 44$

$\lambda = 0.972$

(b) 34 users

$Q_2 = 6760$

$\lambda = 0.969$

(c) 92 users

$Q_2 = 382$

$\lambda = 0.938$

(d) 98 users

$Q_2 = 5147$

$\lambda = 0.993$

Figure 1 Sensitivity to increases in traffic

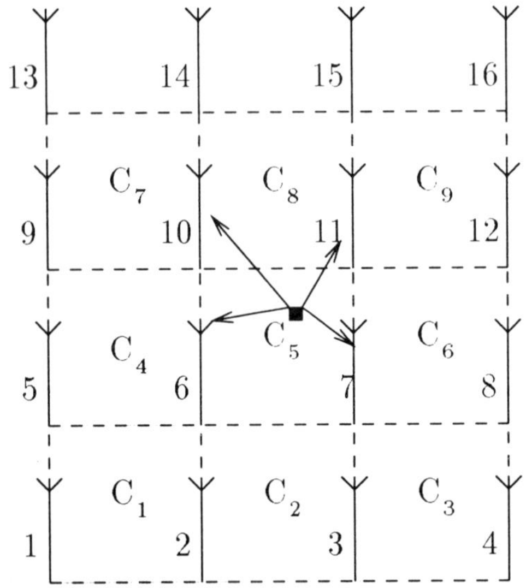

(a) Limited macrodiversity

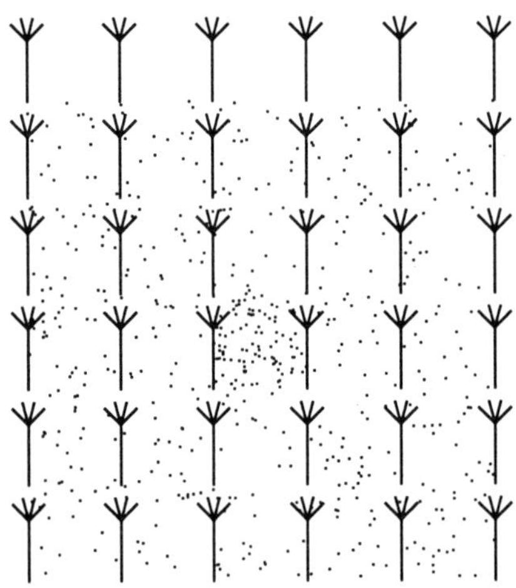

(b) Random locations

Figure 2 A cellular network

CHANNEL CONTROL AND MULTIPLE–ACCESS

R. Knopp, P.A. Humblet

Mobile Communications Dept.,
Institut Eurécom,
B.P. 193, Sophia Antipolis,
06904 FRANCE

ABSTRACT

This work considers fundamental performance limits for single and multiuser channels on time–varying (fading) channels. We consider systems which are burst–oriented along the lines of GSM and IS-54. We examine the benefits of exploiting channel feedback to vary transmit information rate/power in a single–user (or orthogonal multiuser) scenario. It is shown that significant increases in spectral efficiency can be expected over constant–rate systems. Even in the latter case, however, higher spectral efficiencies can be expected if high outage probabilities are tolerated. For the general multiuser case, channel feedback can be used to exploit *multiuser diversity*, which can be seen as a form of selection diversity. We show that the optimal multiple–access method in terms of spectral efficiency is channel–controlled TDMA/FDMA. We discuss some practical issues concerning channel–controlled multiple–access.

1 INTRODUCTION

In this work we examine the role of channel state feedback in mobile radio systems from the point of view of increasing spectral efficiency. We restrict our treatment here to a single–cell model, since there is significant practical insight to be gained without complicating the model to include inter–cell interference. Moreover, when considering a multi–cell environment, we must consider two possible scenarios. Firstly, we can consider the case when there is joint processing between receivers, which, in effect, reduces the problem to a single–cell situation with multiple independent receivers. This is treated in [1][2] for non–fading Gaussian channels. The other case of interest is when processing at the receivers is restricted to the signals of the users within the cell. This calls for the inter–cell interference to be modeled.

We are mostly interested in burst–oriented systems along the lines of GSM [3] or IS–54. Here, data is transmitted in bursts and processed respecting some time–constraint. We consider both fixed and variable–rate systems. Such systems were modeled and analysed by Ozarow *et al.* in [5]. We review the techniques from [5] necessary to demonstrate our results.

In mobile radio systems, the received signal strength is dependent on the physical locations of the receiver(s) and transmitter(s), and includes both short and long–term attenuation effects. These effects become time–varying when the receiver/transmitter are in motion, and result in signal fading. Long–term or slow–fading effects are attributed mostly to the path loss due to the separation between transmitters and receivers. Another long–term fading effect is shadowing [4]. The short–term or fast–fading effects are do to multipath propagation, which is arguably the most difficult aspect to handle in mobile radio systems.

Power control is necessary for combating these time–varying signal attenuations. By power control we mean the exploitation of channel knowledge at the transmission end in any form. We will show that such channel information can be used for variable–rate coding and dynamic channel–access.

2 SINGLE–USER BURST–ORIENTED COMMUNICATIONS

Consider the single–user information signal $x(t)$ transmitted over an arbitrary time–varying channel $h(t,\tau)$ and subject to additive white Gaussian noise, $z(t)$, with power spectral density $N_0/2$ upon reception. The received signal is given by

$$y(t) = \int_{-\infty}^{\infty} x(\tau)h(t,\tau)d\tau + z(t). \tag{1.1}$$

This can also represent an orthogonal multiuser system. Following the approach in [5] we divide the information signal into bursts $x_n(t)$ which are time–limited to $[(n-1)(T+T_G)/2, (n+1)(T+T_G)/2]$, where $T_G > T_o - T$ is a guard–time between bursts during which $x_n(t) = 0$. It is inserted so that the received signal in a given burst does not contain information from previous bursts (i.e. T_G is chosen to be slightly larger than the maximum time–delay spread experienced in the system.) We also place the following average energy constraint on the information signal

$$\lim_{N\to\infty} \frac{1}{N(T+T_G)} \sum_{i=-N/2}^{N/2} \int_{(i-1)T}^{iT} \overline{x_i(t)^2} dt = S. \tag{1.2}$$

When processing is performed over a finite number, N, of such bursts and the receiver has complete knowledge of the channel, the maximum attainable information rate is given by

$$\begin{aligned} I_N &= \frac{1}{N(T+T_G)} \sum_{n=0}^{N-1} I(Y_n(t), H_n(t,\tau); X_n(t)) \\ &= \frac{1}{N(T+T_G)} \sum_{n=0}^{N-1} I(Y_n(t); X_n(t,\tau)|H_n(t)) + I(H_n(t); X_n(t)) \\ &= \frac{1}{N(T+T_G)} \sum_{n=0}^{N-1} I(Y_n(t); X_n(t)|H_n(t,\tau)) \end{aligned} \tag{1.3}$$

where $y_n(t)$ and $h_n(t,\tau)$ are the received signal and channel response for the nth burst. Equation (1.3) follows from the assumption that the information process is statistically independent of the channel and that the received signal in a given burst is dependent only on the signal transmitted during that burst. Processing over several bursts is done to some extent in GSM and IS-54, where the data are interleaved over F bursts transmitted over F independent fading channels (via frequency–hopping in GSM and time–separation in IS-54). It is for this reason that we have included the subscript n in $h_n(t,\tau)$, since it may represent a different realization of a fading channel. Assuming an L–path fading channel, $h_n(t,\tau)$ is given by

$$h_n(t,\tau) = \sum_{l=0}^{L-1} c_{n,l}(t)\delta(t-\tau-d_{n,l}(t)), \tag{1.4}$$

where $c_{n,l}(t)$ and $d_{n,l}(t)$ are the gain and propagation delay for the lth path in the nth burst respectively. If the rate of channel variation is slow enough so that $h_n(t,\tau) \approx h_n(t-\tau)$, $t \in [(n-1)(T+T_G)/2, (n+1)(T+T_G/2)]$ (i.e. the $c_{n,l}(t)$ and $d_{n,l}(t)$ are constant during the burst), then we have that [5]

$$I_N \geq \frac{1}{N} \sum_{n=0}^{N-1} \int_{-W/2}^{W/2} \log\left(1 + \frac{S}{WN_0} S_n(f) \left|H_n(f)\right|^2\right) df, \tag{1.5}$$

with equality as $T \to \infty$. $H_n(f)$ is the Fourier transform of $h_n(t-\tau)$ and $S_n(f)$ is the power spectral density of the Gaussian information process during the nth burst.

We restrict $|H_n(f)|^2$ to be an ergodic process for all f, which means that the slowly–varying statistics (i.e. path loss and shadowing) are made transparent to the receiver. This is made possible by multiplying the information signal by a slow power control signal inversely proportional to the path loss and shadowing prior to transmission. In other words, the *average* received signal power is constant for all time. This is quite different from perfect fast power–control which keeps the ***instantaneous*** received signal power constant. We also assume that $H_n(f)$ is identically distributed at all frequencies and that $\overline{H_n(f)} = 1$. Because of the ergodicity of $H_n(f)$, if we can process over infinitely many bursts the maximum achievable information rate is bounded below by

$$I_\infty = \overline{\int_{-W/2}^{W/2} \log\left(1 + \frac{S}{WN_0} S_n(f)|H_n(f)|^2\right) df}. \tag{1.6}$$

Considering first the case when $H_n(f)$ is not known to the transmitter, it follows straightforwardly from (1.6) and the fact that the statistics of $|H(f)|^2$ do not vary with f that $S_n(f) = 1$ maximizes I_∞. When $H_n(f)$ is known to the transmitter and processing over multiple bursts is not possible, I_∞ can be achieved by variable–rate coding. Over each burst, the transmitter would choose a rate R_n satisfying

$$R_n < \int_{-W/2}^{W/2} \log\left(1 + \frac{S}{WN_0} S_n(f)|H_n(f)|^2\right) df, \tag{1.7}$$

so that the average information rate would also be I_∞. If the transmitter wishes to optimize power allocation we may choose $S_n(f)$ to maximize I_∞ subject to the power constraint in (1.2), resulting in the capacity C_∞. Applying the Kuhn–Tucker conditions [6], the maximizing $S_n(f)$ is given by

$$S_n(f) = \left[B\frac{S}{WN_0} - \frac{1}{|H_n(f)|^2}\right] ind\left(|H_n(f)|^2 > \frac{WN_0}{SB}\right), \tag{1.8}$$

so that

$$C_\infty = \overline{W \log\left(B\frac{S}{WN_0}|H_n(f)|^2\right) ind\left(|H_n(f)|^2 > \frac{WN_0}{SB}\right)}, \tag{1.9}$$

where B is a constant chosen to satisfy the power constraint, and $\text{ind}(\cdot)$ is the indicator function. This is water–filling both in time(block) and frequency and suggests a time–varying OFDM scheme. In a given block, high–rate/high–power codes would be used on the strong carriers, and low–rate(or zero)/low–power codes on the weak carriers. Looking at a particular frequency, this is simply a fast power–control law which works in an opposite fashion to conventional power control, since more power is allocated when the channel is strong and less when it is weak.

In existing systems such as GSM or IS-54 which use a fixed information rate, the ergodicity of the channel cannot be exploited fully because of time–delay constraints. In GSM, for example, data is interleaved over 4 or 8 bursts which are transmitted on a different carrier (ideally) which, when combined with error–control coding, provides a certain amount of diversity. Nevertheless, the short decoding interval and small number of frequencies do not provide enough realizations of the fading channel to achieve information rates close to I_∞. As a result, a more useful benchmark for system performance limits is the information outage probability

$$P_{\text{out}}(R) = \Pr(W^{-1} I_N \leq R). \tag{1.10}$$

This indicates the probability that reliable communication of the N bursts is impossible at rate R. For frequency–flat or single–path systems, where the channel strength is a constant H for all f, we can write this as

$$P_{\text{out}}(R) = \Pr\left(H \leq \left(\frac{WN_0}{S}\right)\left(e^R - 1\right)\right), \tag{1.11}$$

by inverting the logarithm in (1.7). For systems employing variable–rate coding and possibly fast power control, it makes little sense to consider information outage probabilities. This is because the rate R_n will always be chosen such that reliable communication is possible. Practically speaking, it may be useless to code at an achievable rate during a deep fade, and communication could be stopped. The optimal power control law in (1.8) indicates this as well. This would be the only outage probability to consider but would not be a fair means for comparison. We can, however, compare the systems based on average throughput. For the variable–rate coding case, the average throughput is simply I_∞ or C_∞ depending whether or not fast power–control is performed. For the fixed–rate case, we define the average throughput as

$$T(R) = R\left(1 - P_{\text{out}}(R)\right), \tag{1.12}$$

which assumes that no information is conveyed when the achievable information rate falls below R. This is the case in GSM, for instance, where an error–detecting code is used to determine whether or not the burst has been received reliably. If not, the burst is rejected and speech information is obtained by extrapolating information from previous bursts. Under this assumption, the maximum throughput, $T_{\max}$ is attained by choosing the operating information rate R_{opt}, which maximizes $T(R)$ for a given SNR.

In order to compare different systems, we choose a Rayleigh fading model for the path gains in (1.4). In this case the $|H(f)|^2$ are exponential with $p_H(\alpha) = e^{-\alpha}, \quad \alpha > 0$. This is, in some sense, a worst–case model for what is experienced in real systems. It is also convenient for analysis purposes. The corresponding mutual informations are given by

$$I_\infty = We^{\frac{WN_0}{S}} \operatorname{Ei}\left(\frac{WN_0}{S}\right), \tag{1.13}$$

$$C_\infty = W\operatorname{Ei}\left(\frac{WN_0}{BS}\right), \tag{1.14}$$

and B is the solution to

$$Be^{-\frac{WN_0}{BS}} - \frac{WN_0}{S}\operatorname{Ei}\left(\frac{WN_0}{BS}\right) = 1, \tag{1.15}$$

where $\operatorname{Ei}(\cdot)$ is the first–order exponential integral. We note for large S/WN_0 that $B \approx 1$ and therefore that $I_\infty \approx C_\infty$. Thus, when ergodicity can be exploited, fast power–control, and transmitting only when channel conditions are favourable, have no advantage. We plot $W^{-1}I_\infty$ and $W^{-1}C_\infty$ in Fig. 1. We also show the spectral efficiency of a non–fading channel,

$$W^{-1}C_G = \log\left(1 + \frac{S}{WN_0}\right). \tag{1.16}$$

It is worthwhile noting that C_G is the capacity with perfect fast power–control (i.e. when the instantaneous received SNR is kept constant) and that it is not significantly higher than systems with a variable received SNR. Moreover, in Rayleigh fading, a perfect power–control law cannot satisfy the average–power constraint since the average transmit power is infinite. In practice, the transmit power is limited to some peak level, and when this is considered, it turns out that the capacity is significantly less than C_∞ [7].

When $N = 1$, the information outage probability for frequency–flat fading in (1.11) is given by

$$P_{\text{out}}(R) = 1 - \exp\left(-\frac{WN_0}{S}\left(e^R - 1\right)\right). \tag{1.17}$$

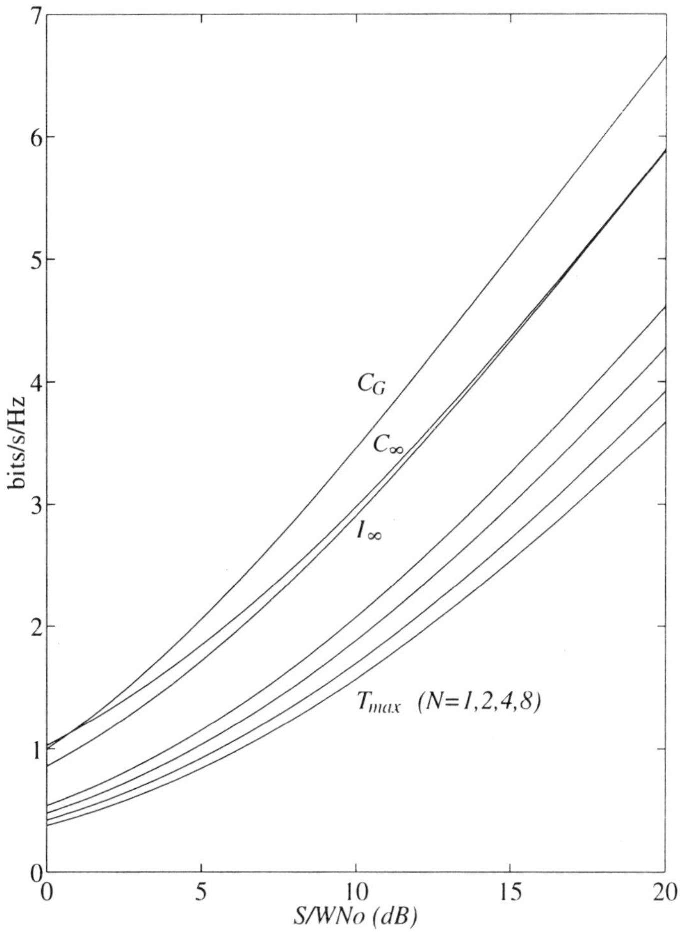

Figure 1: Information Capacity Comparison

It is easily shown that the optimal operating information rate is the solution to the following non–linear equation

$$R_{\text{opt}} e^{R_{\text{opt}}} = \frac{S}{W N_0}. \tag{1.18}$$

The maximum throughput, $T_{\max}$, when R_{opt} is chosen for each SNR, is also shown in Fig. 1. For the case when processing is performed over several bursts, the information outage probability can be calculated numerically. We have

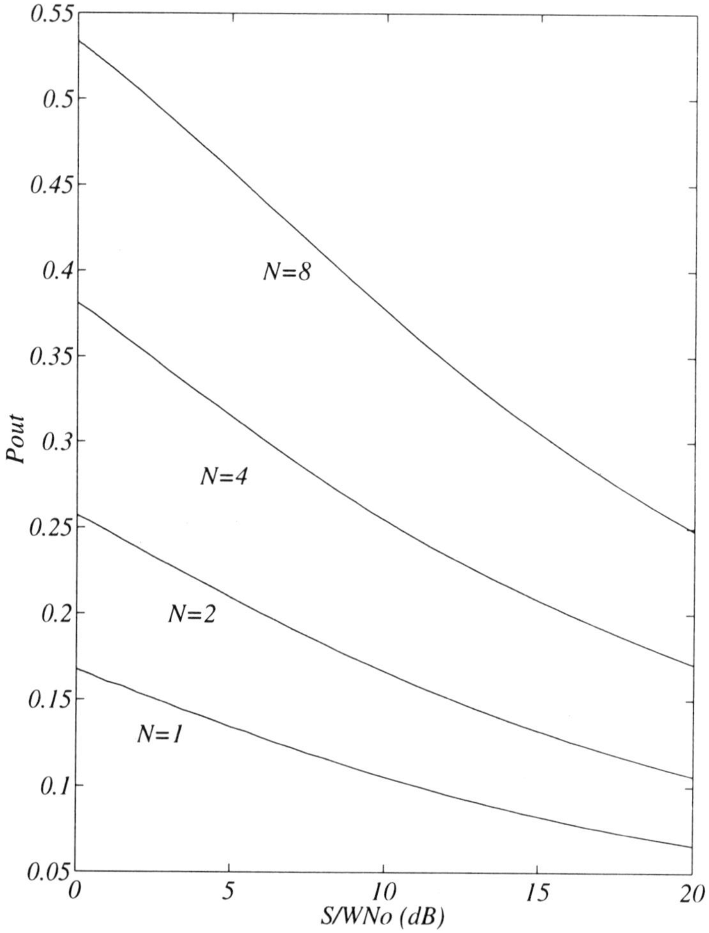

Figure 2: Optimal operating outage rates

performed the calculations for $N = 2, 4, 8$ which corresponds to ideal conditions in the IS–54, half–rate GSM, and full–rate GSM systems respectively. Although the spectral efficiency increases towards I_∞, it is clear that variable–rate coding with and without fast power–control, which achieve either I_∞ or C_∞, can have a significant advantage for practical SNR. In Fig. 2 we show the operating outage rates which achieve maximum throughput in constant–rate systems. Even here, we see that for maximum throughput, the channel is used effectively only when not in a deep fade since the outage probabilities are high (i.e. many bursts will be received in error).

We have not considered the information outage probability for frequency-selective channels. In [5] expressions for P_{out} on two-path Rayleigh fading channels are given. Although the analysis is more complicated than the single-path case, it is shown that at a given SNR P_{out} is significantly lower when multipath diversity can be exploited. This means that the average throughput also increases towards $W^{-1}I_\infty$.

3 MULTIUSER DIVERSITY

In this section we generalize the ideas of Section 2 for a single-cell multiuser system. Let us assume that we have K users sharing a fixed bandwidth KW, such that the bandwidth per user is W. Each user's signal, $x^{(i)}(t)$, is transmitted over a different channel with time-varying impulse response $h^{(i)}(t,\tau)$. The information signals are average-power limited as in (1.2). We have the following composite complex baseband signal at the basestation,

$$y(t) = \sum_{i=0}^{K-1} \int_{-T/2}^{T/2} x^{(i)}(\tau) h^{(i)}(t,\tau) d\tau + z(t). \tag{1.19}$$

We assume that the attenuations on different channels are independent processes. As in the single-user case, we consider the case when the channel attenuations are ergodic processes by assuming perfect slow power control. The achievable data rates in the nth burst belong to the region inner-bounded by

$$\sum_{i\in\mathcal{U}} R_i^{(n)} < \int_{-KW/2}^{KW/2} \log\left(1 + \frac{2}{N_0} \sum_{i\in\mathcal{U}} S_{in}(f) \left|H_{in}(f)\right|^2\right) df, \tag{1.20}$$

where $S_{in}(f)$ and $H_{in}(f)$ are the information power spectrum and channel response for user i over the nth burst respectively, and $\mathcal{U} \in \{0, 1, \cdots, K-1\}$. Generalizing the results of the single-user case, the achievable rate region for systems where processing is carried out over an infinite number of bursts

$$\sum_{i\in\mathcal{U}} R_{i,\infty} < \overline{\int_{-KW/2}^{KW/2} \log\left(1 + \frac{2}{N_0} \sum_{i\in\mathcal{U}} S_{in}(f) \left|H_{in}(f)\right|^2\right) df}. \tag{1.21}$$

This region is that of SSMA (Spread-Spectrum Multiple Access) systems, since all the users transmit over the entire available bandwidth. It is shown in [8] that the achievable rate region for FDMA (or TDMA) is strictly enclosed by the region of an SSMA system if the channels are all statistically identical. Moreover, diversity techniques such as frequency–hopping or time–separation cannot enlarge the average achievable rate region. This implies that higher data rates are theoretically possible with SSMA than with systems like GSM or IS-54.

In the case of equal average received powers, the equal–rate point on the achievable rate region is achieved by the total sum–of–rates (i.e. the inequality in (1.21) corresponding to the case when $\mathcal{U} = \{0, 1, \cdots, K-1\}$). From this point we define the per user average achievable spectral efficiency as $W^{-1} I_\infty^K = (1/K) \sum_{i=0}^{K-1} R_{i,\infty}$. In [9], it is shown that the gain in $W^{-1} I_\infty^K$ for SSMA over FDMA/TDMA in Rayleigh fading tends to $\gamma / \ln 2 = .8327$ bits/s/Hz, where γ is Euler's constant, for a large number of users and high SNR. This means that K users transmitting over K independent fading channels can transmit more total information than a single user with K times as much power. Furthermore, as the number of users increases, $W^{-1} I_\infty^K$ increases towards $W^{-1} C_G$, the spectral efficiency of a non–fading Gaussian channel.

In order to operate on the boundary of the achievable rate region, we must employ a successive decoding strategy [10]. For each group of N bursts (N large) over which decoding is performed, we choose an order in which to decode the users' signals. Each $x^{(i)}(t)$ is decoded, treating the others as noise, regenerated and subtracted from $y(t)$. To operate at the equal–rate point, we must choose the ordering uniformly at random each time decoding is performed. This implies that the transmitters must adjust their code rates according to the ordering to be used by the decoder.

Unlike single–user channels, channel state feedback may be used to allocate users in time and frequency dynamically depending on the states of *all* the channels. This assumes, of course, a closed–loop system where the basestation controls all the users. We illustrate this with a simple example. Consider a two-user system in a flat fading environment using modified TDMA. As in ordinary TDMA only one users transmits at any given time, but now we use channel knowledge to allocate the channel to the user who has the stronger instantaneous received signal. In doing so, we have effectively made this equivalent to a single-user system with selection diversity, except that the diversity is not due to multiple antennas but to multiple users. In the remainder of this section we investigate the effect of channel knowledge at the transmission end more closely, and show that a generalization of this simple idea is optimal under certain conditions.

The capacity region for burst–oriented systems is achieved by a multiuser water–filling solution similar to that described in [11] for time–invariant Gaussian channels. We do not discuss this at any length here, except that, as in the single–user case, it is water–filling both in time(block) and frequency. We consider only the equal–rate point on the capacity region which, with equal average received powers, is achieved by the total sum–of–rates.

For a system with K users, the maximum sum-of-rates is achieved by having only the *strongest* user transmitting at any given frequency and time [9], and the corresponding power controllers are given by

$$S_{in}(f) = \left[B\frac{S}{KWN_0} - \frac{1}{|H_{in}(f)|^2}\right]^+, \quad |H_{in}(f)| \geq |H_{jn}(f)| \forall j \neq i \tag{1.22}$$

where B is chosen to satisfy the average–power constraint. This is a result of applying the Kuhn–Tucker conditions to the sum–of–rates inequality in (1.21).

We now give numerical results for the per–user spectral efficiencies for the various multiple–access schemes in Rayleigh fading assuming equal average received powers. The spectral efficiency per user for FDMA is simply $W^{-1}I_\infty$ in (1.13) (or $W^{-1}C_\infty$ if the users are aware of the states of only their own channels). For SSMA, it can be calculated numerically. For the optimal scheme, the spectral efficiency per user is given by

$$W^{-1}C_\infty^K = \sum_{i=1}^{K}(-1)^{i-1}\binom{K}{i}\mathrm{Ei}\left(\frac{iK}{B}\frac{WN_0}{S}\right) \quad \text{nats/s/Hz} \tag{1.23}$$

where B is the solution to

$$\sum_{i=1}^{K}(-1)^{i-1}\binom{K}{i}\left[B\exp\left(-\frac{iK}{B}\frac{WN_0}{S}\right) - \left(iK\frac{WN_0}{S}\right)\mathrm{Ei}\left(\frac{iK}{B}\frac{WN_0}{S}\right)\right] = K. \tag{1.24}$$

These are shown in Fig. 3.

As mentioned earlier, we see that spectral efficiency per user increases slightly with the number of users for SSMA. For the optimal scheme with multiuser diversity, a significant increase is achievable, which is not the case with the non-fading Gaussian channel or FDMA on a fading channel. This is not surprising, since with many users, the probability that one of the channels is good is high which means that the corresponding user can transmit at a high rate. Moreover, even in the two-user case a system exploiting multiuser diversity on Rayleigh

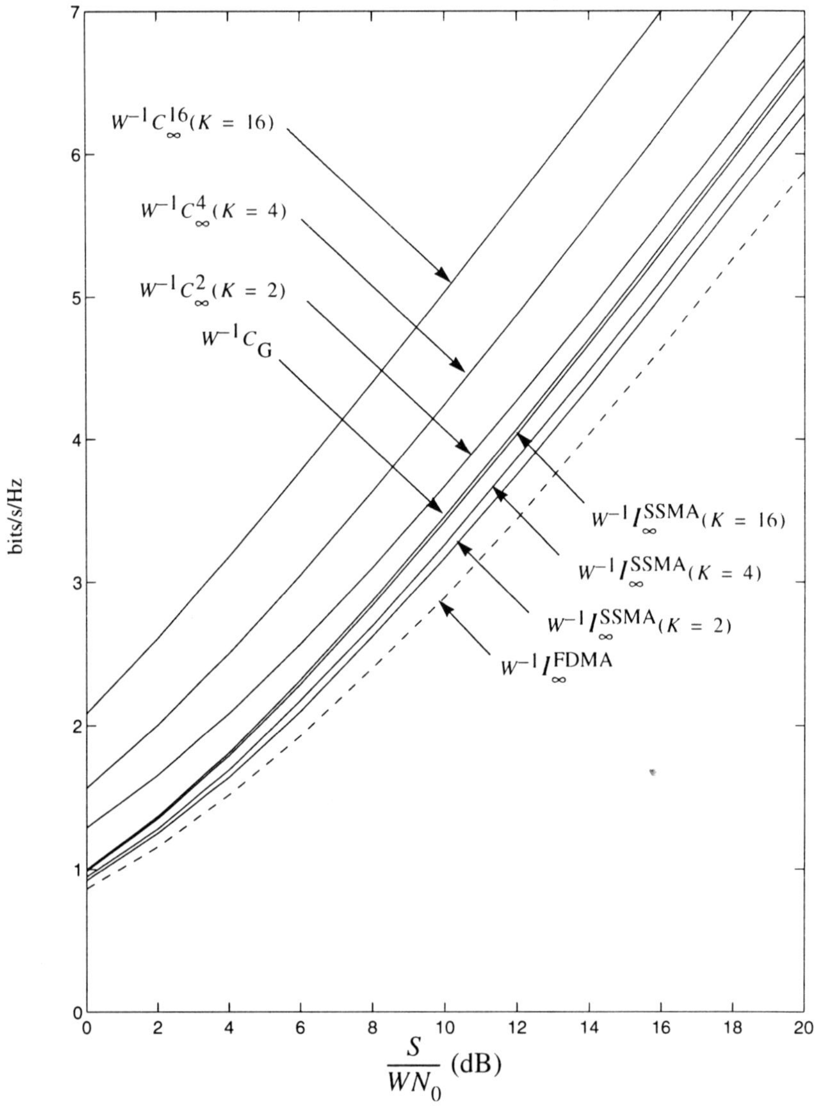

Figure 3: Spectral Efficiency Comparison

fading channel can theoretically achieve higher spectral efficiency than a system transmitting over the non-fading Gaussian channel.

Although not shown here, if we compute the maximum average throughput (T_{max}) when the transmitters use a constant–code rate with multiuser diversity, we find that high spectral efficiency can be obtained without having to decode over many bursts. The calculation of the information outage probability is identical to the case of selection diversity discussed in [5].

4 DISCUSSION

Optimal bandwidth partitioning based on channel knowledge at the user terminals would be hard to achieve practically. A more realistic alternative would be to divide the entire bandwidth into equal size subbands and allocate a single user to each these subbands based on the instantaneous frequency responses of all the users. The DECT system [12, chap. 30] employs a technique along these lines. In this system, the available bandwidth is divided into several subbands, which we assume to be frequency flat. The users measure the strength of each subband (from the downlink which operates in time–division duplex with the uplink), and choose the best available subband in which to transmit. Our results suggest that the opposite should be done, namely that the basestation would instruct the user with the best channel response in a particular subband to use that subband alone. The subbands need not be adjacent, and, in fact, should not be. This would assure independence of the subband channel gains. We could envisage a variable–rate system where the total bandwidth per cell is divided into L independent subbands (i.e, with carrier spacings greater than the coherence bandwidth) of bandwidth KW/L where the basestation instructs the strongest user to transmit at a constant rate R in each subband. The instantaneous data rate per user would be R times an $(L,1/K)$ binomial random variable, so that the average data rate would be RL/K. The probability that a user transmits no information in any given burst would be $(\frac{K-1}{K})^L \approx e^{-L/K}$, which can be made arbitrarily small by increasing L/K. We can interpret L as the fineness control of the variable–rate system.

In this paper, we have ignored the problem of how channel measurements can be made available to the users. In two–way systems the measurements could be made at the basestation and passed to the users via the downlink. The main problem, however, lies in obtaining the state of the channels when the users are not transmitting in certain parts of the spectrum. This would call for wideband pilot signals to be transmitted periodically if the channel changes slowly. For quickly–varying channels, low–power measurement signals (several dB below the information signals) could be overlaid at all times so that sudden changes could be tracked and accounted for.

It would also be interesting to investigate the effect that channel–controlled multiple–access schemes and variable–rate coding would have on cellular system capacity by considering adjacent–cell interference.

REFERENCES

[1] Wyner A.D., "Shannon–Theoretic Approach to a Gaussian Cellular Multiple Access Channel," IEEE Trans. on Information Theory, vol. 40, no. 6, pp. 1713–1727.

[2] Hanly S.H., *Information Capacity of Radio Networks*, PhD Thesis, Cambridge, UK, 1993.

[3] Steele R., *Mobile Radio Communications*, Pentech Press 1992, chap 8.

[4] Lee W.C.Y., *Mobile Communications Engineering*, McGraw-Hill, 1982.

[5] Ozarow L.H., Shamai S., Wyner A.D., "Information Theoretic Considerations for Cellular Mobile Radio," IEEE Trans. on Vehic. Tech., vol. 43, no. 2, May 1994, pp. 359-378.

[6] Gallager, R.G., *Information Theory and Reliable Communication*, Wiley & Sons, 1968.

[7] Goldsmith A., Varaiya P., "Increasing Spectral Efficiency Through Power Control," Proc. ICC '93, Geneva, Switzerland, pp. 600–604.

[8] Gallager R.G., "An Inequality on the Capacity Region of Multiaccess Multipath Channels," in Communica tions and Cryptography - Two Sides of One Tapestry -, Kluwer 1994, pp. 129-139.

[9] Knopp R., Humblet P.A., "Multiple Accessing over Frequency–Selective Fading Channels," Proc. PIMRC '95, Toronto, Canada, pp. 1326–1330.

[10] Cover T., Thomas J., *Elements of Information Theory,*, Wiley & Sons, 1991.

[11] Cheng R.S., Verdú S.,"Gaussian Multiaccess Channels with Capacity Region and Multiuser Water–Filling", IEEE Trans. on Information Theory, vol. 39, no. 3, May 1993.

[12] *The Mobile Communications Handbook,*, ed. J.D. Gibson, CRC Press, 1996.

ON THE EFFECT OF CHIP SYNCHRONIZATION ERROR IN MC-CDMA SYSTEMS

Luciano Tomba*, Witold A. Krzymien**

**University of Padova/TRLabs,*
***University of Alberta/TRLabs*

ABSTRACT

In this paper we investigate the effect of chip synchronization error (or chip timing jitter) in multi-carrier code division multiple access (MC-CDMA) schemes for indoor environments. We model the chip timing jitter as a stationary random process with a known statistic and the error produced by chip timing jitter as an additive noise source. We evaluate the spectral density of this noise and we show that it can be split into two uncorrelated components. These results are important because they allow us to use an analytical or semi-analytical formula for the error probability evaluation. This is helpful, because the simulation of a MC-CDMA system is complex and time consuming. To mitigate the effect of chip synchronization error, a time diversity technique (*average diversity combining*) is suggested.

1 INTRODUCTION

The increasing demand for mobile terminals, especially for indoor applications (e.g. offices, shopping malls, factories), is one of the major motivations to develop new, highly efficient, wireless networks. Beside the TDMA scheme and the CDMA one, a scheme based on the joint use of CDMA and OFDM [1, 2] has been recently proposed for high data rate transmission (e.g. images, ATM packets at 155 Mbit/s [3]). At such high bit rate DS-CDMA is impractical, because of the very short chip time necessary when a reasonably high processing gain is used. At the present state of the art such very short chips are impossible (or very difficult and expensive) to implement. The multicarrier (MC) CDMA technique, instead, permits lower chip rates at high processing gains because of the increased symbol duration. Moreover, in principle, this technique is able to combat the frequency selectivity of the channel by using a simple one-tap equalizer. Since the OFDM symbol duration is usually very large compared to the one of single carrier modulation, the radio channel can be considered frequency non-selective. However, to get the same data rate with a larger symbol duration, many symbols must be transmitted in parallel without interference and this is possible by modulating carriers which are orthogonal to each other over the symbol duration. The resulting signal is given by the summation of all modulated subcarriers. The subchannels corresponding to each subcarrier spectrally overlap but, if the channel distorsion is sufficiently mild, the signals of each subchannel can be demodulated with negligible penalty because of the orthogonality property.

The performance of a MC-CDMA system is good enough if we assume a nearly ideal system, i.e. with a linear power amplifier at the transmitter side, a linear channel, ideal carrier frequency and sampling instant regeneration. When these assumptions are relaxed, as it happens in practical systems, performance degrades severely. This is the case for the presence of an offset and/or noise in the carrier frequency recovery loop [4], or for radio channels with delay spread longer than an estimated maximum value [5]. Moreover, this performance loss increases with the number of subcarriers.

Here, our attention is devoted to the effect of *chip synchronization error* or *chip timing jitter* in the downlink receiver; the same ideas may be applied to the uplink, too. We model the chip timing jitter as a stationary random process with a known statistic. The error produced by chip timing jitter is modelled as an additive noise source. We evaluate the spectral density of this noise and we show that it can be split into two uncorrelated components. These results are important because they allow us to use analytical or semi-analytical formulas for the error probability evaluation. This is helpful, because the simulation of a MC-CDMA system is complex and time consuming. Lastly, to mitigate the effect of chip timing jitter, a time diversity technique (*average diversity combining*) is considered.

This paper is organized as follows. In Section 2, a brief overview of the MC-CDMA transmission scheme is reported. Section 3 is devoted to the spectral analysis of the chip timing jitter error and performance evaluation is carried out in Section 4. Finally, in Section 5 the numerical and simulation results are discussed.

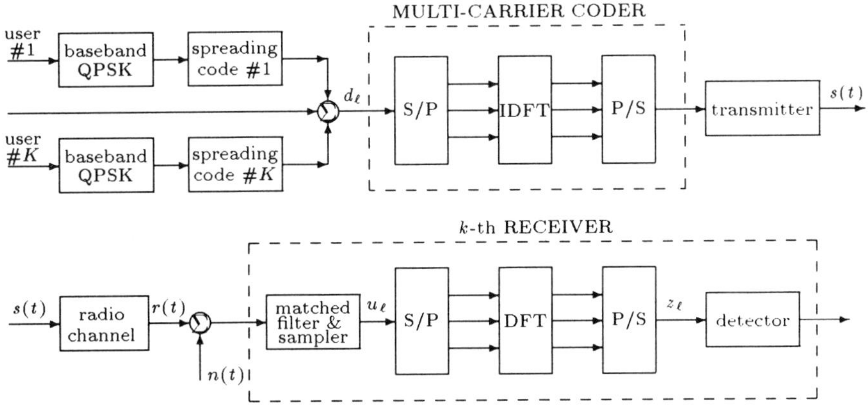

Figure 1 Simplified block diagram for the downlink of the MC-CDMA system.

2 OVERVIEW OF MC-CDMA

The transmitter and receiver block diagrams are shown in Fig. 1. The k-th user, $1 \leq k \leq K$, transmits QPSK symbols which are spread by a PN sequence of length L. In order to reduce the multiple access interference among different users, the sequences are required to have very low cross-correlation functions, while their auto-correlation function must be very sharp around the proper alignment instant to guarantee proper acquisition performance [6]. Usually, Gold, Kasami or Walsh sequences are considered in CDMA systems; here Walsh sequences are used.

The spread sequence, d_ℓ, is transmitted by a MC-CDMA system capable of transmitting N complex numbers at the same time. Basicly, a multi-carrier signal [1, 2] consists of the superposition of N sinusoidal carriers with frequency spacing F/N, $F \triangleq 1/T$; they are modulated by QAM symbols with a duration NT. The modulated carriers overlap spectrally but, since they are orthogonal within the symbol duration (the ℓ-th carrier's frequency is $f_\ell = f_0 + \ell F/N$, where f_0 is some reference frequency), the signal associated with each tone can be recovered as long as the channel does not destroy their orthogonality. In practice, the samples of the OFDM signal are generated by taking the IDFT of the discrete-time input sequence and passing the transform samples through a pulse shaping filter. As it is well known [6, 7], in order to (ideally) eliminate intersymbol and interchannel interference caused by channels with non-zero delay spread, a guard time including a *cyclic prefix* is inserted. For non-causal filters, both *prefix* and *postfix* are necessary (we call their lengths p_1 and p_2, respectively). Let c_{mi} be the m-th sample of the i-th OFDM symbol at the output of the IDFT processor, namely

$$c_{mi} = \frac{1}{N} \sum_{n=0}^{N-1} d_{ni} e^{j2\pi n \frac{m}{N}} \quad , \quad j \triangleq \sqrt{-1} \quad , \quad 0 \leq m < N \tag{1}$$

where $d_{ni} \triangleq d_{n+iN}$ is the data at the input of the IDFT processor. Then, the complex envelope of the transmitted signal is given by

$$s(t) = \sum_{i=-\infty}^{+\infty} \sum_{m=-p_1}^{N+p_2-1} \acute{c}_{mi} q \left[t - mT - i(N + p_1 + p_2)T\right] \quad , \quad t \in \mathcal{R} \quad , \tag{2}$$

where $\acute{c}_{mi}$ corresponds to the sequence c_{mi} completed with the prefix and postfix information and $\mathcal{R}$ is the set of real numbers. We assume $q(\cdot)$ to have a square-root raised cosine spectral characteristic with bandwidth $(1 + \beta)/2T$, where β is the roll-off factor [6]. At the receiver end, a filter with impulse response $q^*(-t)$ is used, i.e. matched to the transmitted pulse (the symbol * denotes complex conjugate).

The channel is modelled as a multipath tapped delay line [8], namely

$$h(\tau) = \sum_{p=0}^{P-1} \varrho_p \delta(\tau - \tau_p) \tag{3}$$

where ϱ_p is the complex tap weight (approximately time-invariant over, at least, the duration of one OFDM symbol), τ_p is the relative time delay of the p-th path, P is the total number of taps and $\delta(\cdot)$ is the Dirac delta function. The received signal, $r(t)$, is corrupted by additive white Gaussian noise (AWGN), $n(t)$, and the noisy composite signal is low-pass filtered, sampled and processed by the multi-carrier decoder. Finally, the transmitted sequence is detected by a correlation procedure of the received samples with the spreading sequences.

3 CHIP TIMING JITTER ANALYSIS

3.1 Jitter model

Let us refer to Fig. 2. It represents the MC-CDMA scheme from the input of

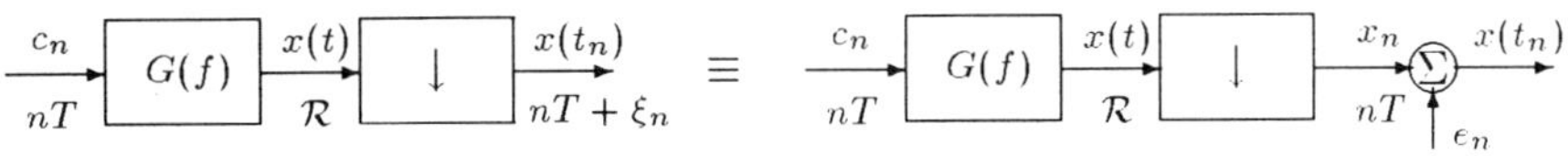

Figure 2 Equivalent scheme for the analysis of the chip timing jitter error ($G(f)$ is the frequency response of the interpolating filter and the block marked with a down-arrow is a sampler).

the transmitter filter to the output of the jittered sampler at the receiver side (the AWGN has been omitted). The discrete-time sequence, c_n, is ideally interpolated and filtered by a filter with frequency response $G(f) = H(f)|Q(f)|^2$, where $H(f)$ ($Q(f)$) is the Fourier transform of $h(t)$ ($q(t)$). We have implicitly assumed that all devices, i.e. the transmitter filter, the channel and the receiver filter could be modelled as linear filters. The signal $x(t)$ is sampled by the jittered sampler every T seconds. This means that the sampling instants, t_n, are not nT but

$$t_n = nT + \xi_n \tag{4}$$

where ξ_n is the chip timing jitter of the n-th sampling instant; therefore, the sampled signal is $x(nT + \xi_n)$ instead of $x(nT)$. Since the jitter sequence, $\xi_n \triangleq \xi(nT)$, is random, the error caused by the chip timing jitter on the sampled values has a noise-like nature. This error propagates through the subsequent operations and it causes degradation of the system.

Using the decomposition $x(nT + \xi_n) = x(nT) + e(nT)$, the jittered sampler can be replaced by an ideal sampler and an additive noise source (see Fig. 2). We assume that the jitter sequence, ξ_n, is a stationary random process (r.p.), statistically independent of the input signal and with a known statistic. We denote by m_ξ the mean value of ξ_n (it represents the error in the time delay estimate provided by the chip synchronizer), and by σ_ξ its standard deviation. Usually σ_ξ is very small for PLL synchronization circuits, while m_ξ may assume significant values (e.g., from 5 to 20 percent of the sampling period).

Since we are interested in the spectral density, the mean value and the power of $e(nT)$, an average spectral analysis of the system must be provided. The derivation of the spectral density requires knowledge of the first and second order characteristic functions of the jitter. A reasonably simple model for the jitter is the *white jitter*, which means that the jitter values ξ_n and ξ_{n+m} are statistically independent for $m \neq 0$. In this case, the second order characteristic function is

$$\Psi_\xi(f_1, f_2; mT) = \begin{cases} \Psi_\xi(f_1 + f_2) \quad , & m = 0 \\ \Psi_\xi(f_1)\Psi_\xi(f_2) \quad , & m \neq 0 \end{cases} \tag{5}$$

This model is acceptable for performance comparison between different system architectures.

3.2 Jitter spectral analysis

We assume to know the average spectral density, $R_{c_n}(f)$, of c_n, the characteristic function, $\Psi_\xi(f)$, of the chip timing jitter and the frequency response $G(f)$. Previous assumptions are sufficient to get the spectral density of $e(nT)$ (see the Appendix for the mathematical details):

$$\begin{aligned} R_{e_n}(f) =& R_{c_n}(f) \sum_{k=-\infty}^{+\infty} F \left| G(f+kF)[1-\Psi_\xi(f+kF)] \right|^2 \\ &+ \int_{-\infty}^{+\infty} R_{c_n}(\lambda) |G(\lambda)|^2 \left[1 - |\Psi_\xi(\lambda)|^2\right] d\lambda \end{aligned} \tag{6}$$

From (6), we see that $R_{e_n}(f)$ is not constant, but a coloured component is superimposed on to a white one. However, to get analytical results for the bit error rate (BER), we treat $e(nT)$ as a white noise. In addition, because it is difficult to evaluate the probability density function of $e(nT)$, we assume it is a stationary Gaussian r.p.. The simulation results will give us an idea of the roughness of our approximations. With this model, we are able to evaluate the signal-to-noise ratio (SNR) loss [4] produced by the jitter. Hence, we have a guideline in choosing a proper error correction code to mitigate the impairment produced by the jitter error and an estimate of the precision required of the chip timing recovery circuit.

To reduce the error rate produced by chip timing jitter and additive noise, we suggest the scheme presented in [9] for DS/CDMA. It consists of an error detection encoder/decoder and an error correction encoder/decoder (convolutional coding with Viterbi decoding). In addition, to increase the number of correctly decoded packets, an average diversity combining (ADC) technique is also used. The basic idea is that, while in a classical ARQ scheme the damaged packets are discarded before a new transmission is requested, in this scheme the damaged packets are kept and combined together with the newer ones to form more reliable packets.

4 PERFORMANCE EVALUATION

Performance of the uncoded system can be determined analytically. However, since some approximations must be made in the analysis, the analytical results are compared to the ones produced by simulation (at least for BER values greater than 10^{-4}) to verify their accuracy. The most significant approximation is to consider the jitter error to be an AWGN. In fact, while it is reasonable to assume that it is Gaussian, in general it is less reasonable to assume that it is white (see (6)). If the cyclic prefix-postfix is long enough to approximate each subchannel of the MC system with a flat-fading one, at the output of the DFT processor, the samples, z_ℓ, are given by

$$z_\ell = d_\ell G_\ell + \eta_\ell \quad , \quad 0 \le \ell < N \tag{7}$$

where G_ℓ is the ℓ-th DFT sample of the impulse response of the filter equivalent to the cascade of the transmitter, the channel and the receiver filter and η_ℓ are samples of the additive noise composed of two components: the first one is the filtered thermal noise, while the second one is produced by the sampling error. Each r.v. η_ℓ is assumed to be Gaussian with zero mean and equal variance σ_η^2. From (7), assuming the channel is known, we can retrieve the symbol values d_ℓ using the orthogonality-restoring procedure [1], namely

$$\tilde{d}_\ell = z_\ell \tilde{G}_\ell^{-1} \tag{8}$$

where $\tilde{d}_\ell$ ($\tilde{G}_\ell$) is an estimate of d_ℓ (G_ℓ). As observed in [10] and references therein, this technique is improved if an appropriate threshold is used, and the samples $\tilde{d}_\ell$, for which $|\tilde{G}_\ell|$ falls below the threshold, are discarded. Actually, as noted in [11], an equalization procedure based on the MMSE criterion [6] gives the same or better performance and it avoids the computation of the optimal threshold. However, in this work results are given for the simple orthogonality restoring procedure given by (8).

It can be shown [12] that for a given channel realization the conditional symbol error probability is

$$P_e^c = \mathrm{erfc}(\sqrt{\gamma_0}\,)\left[1 - \frac{1}{4}\,\mathrm{erfc}(\sqrt{\gamma_0}\,)\right] \quad , \quad \gamma_0 \triangleq \frac{1}{2}\frac{\mathrm{E}[|d_0|^2]}{\mathrm{E}[|\zeta_0|^2]} = \frac{1}{\mathrm{E}[|\zeta_0|^2]} \quad , \tag{9}$$

where $\mathrm{erfc}(\cdot)$ is the complementary error function [6] and $\mathrm{E}[\cdot]$ is the expectation operator. In (9) ζ_0 is a zero-mean r.v. with variance

$$\sigma_{\zeta_0}^2 = \frac{\sigma_\eta^2}{L}\sum_{\ell=0}^{L-1} |\tilde{G}_\ell^{-1}|^2 \quad . \tag{10}$$

Unfortunately, since $\{\tilde{G}_\ell\}$ are correlated r.v.'s, a closed form solution, even if theoretically possible, is very cumbersome and its numerical computation is not feasible for reasonable values of L. An accurate procedure to get the unconditional probability involves averaging the conditional error probability using numerical integration techniques. A semi-analytical procedure can also be used. It involves evaluating $\sigma^2_{\zeta_0}$ by simulation, computing the conditional error probability according to (9) and taking the arithmetic mean of these probability values to get the unconditional error probability.

5 RESULTS

If the data stream of each user is composed of independent and equiprobable symbols, it can be shown that due to the orthogonality of the Walsh sequences the spectral density of c_n is constant, i.e. $R_{c_n}(f) = \frac{2N_a}{NL}$, where N_a is the number of active users. Hence, using (6), we get the spectral density of the jitter error. Fig. 3 shows the spectral density of the jitter error for $m_\xi = 0$ and $\beta = 0$. From the plots we observe that the coloured component of the spectral density of e_n, $R_{e_n}(f)$, is almost negligible (with respect to the white component) when the normalized jitter standard deviation, $\sigma_\xi F$, is less than about 0.1, as it happens in practical systems. However, when $m_\xi \neq 0$ and $\sigma_\xi = 0$, the white component is quite negligible (exactly zero for Gaussian jitter, see (6)) and only a coloured component exists. In this case it is not correct to view the jitter error as an AWGN. However, to get very small values of BER, it is impossible to use simulation and we accept this approximation (this approximation should be optimistic, because we have approximated a coloured noise with a white noise). Because of the space limitations the spectral densities for $\beta > 0$ are not shown; however, we have determined that the power of e_n for $\beta > 0$ is less than the one for $\beta = 0$.

Fig. 4a shows a comparison between numerical and simulation results assuming $m_\xi = 0$. The agreement between them is very good, especially for $\sigma_\xi F \leq 0.1$. These results have been obtained neglecting fading in order to know the impact of the jitter error only. When just one user is active, the jitter error is almost negligible, at least for $\sigma_\xi F < 0.1$. In the worst case condition (all users are active), a value $\sigma_\xi F = 0.1$ gives about 3 dB of loss in the SNR at BER$= 10^{-4}$ while the performance loss is almost negligible for $\sigma_\xi F < 0.05$. Fig. 4b shows a comparison between numerical and simulation results assuming $\sigma_\xi = 0$. In this case the approximation is less accurate, bust still acceptable, especially for $m_\xi F \leq 0.1$, which are values of practical interest.

Fig. 5 shows the BER for different values of β and $m_\xi F$ for the AWGN channel (the curves are obtained using the analytical approximation). We observe that larger values of β are helpful to mitigate the jitter error.

Fig. 6 shows the error probability in the presence of both jitter and fading for different values of $\sigma_\xi F$ and $m_\xi F$. A three-path slowly time-varying Rayleigh channel has been assumed [8]. The paths' arrival time has been assumed uniformly distributed on the channel impulse response length. The BER has been evaluated by the semi-analytical approach. The error probability is very high if compared with previous results for the AWGN, so proper measures (e.g. diversity combining) must be applied to improve performance. We see that a chip timing jitter with $\sigma_\xi F > 0.05$ significantly degrades performance, especially when all users are active. Hence, the effect of jitter is more significant in the presence of fading rather than on an AWGN channel. Less severe is the effect of a fixed chip timing error.

We have also evaluated the error probability in the presence of both jitter and fading vs. the number of combined packets, R, in the ADC for different values of $\sigma_\xi F$ and $m_\xi F$. The gain of the ADC is about one order of magnitude in the BER (for $R > 10$ the BER lines show a floor), and combining 2 or 3 packets gives significant gain. However, to further improve performance, use of a more effective detection scheme combined with error correction coding seems to be the right choice.

6 CONCLUSION

In this paper we have investigated the effect of chip synchronization error (chip timing jitter) in MC-CDMA indoor radio systems. In particular we have proposed an accurate model to analytically (or semi-analitycally) evaluate very small error probability values without using long, time consuming computer simulations. In addition, the performance of an ADC has been evaluated assuming a multipath indoor radio channel with slow Rayleigh fading.

Acknowledgements

The authors wish to thank the Telecommunications Research Laboratories (TR*Labs*), and the Natural Sciences and Engineering Research Council of Canada for their support of this work.

APPENDIX: Proof of (6)

The process $x(t)$ (see Fig. 2) is *wide-sense cyclostationary*, i.e. its mean value, $m_x(t)$, and its autocorrelation, $r_x(t,\tau)$, are periodic functions of period T in the variable t. Our objective is the evaluation of the mean value, the average spectral density and the power of $e_n \triangleq e(nT)$, n integer, given by:

$$e_n = x(nT + \xi(nT)) - x(nT) \tag{11}$$

The mean value of e_n is given by:

$$\mathrm{E}[e_n] = \mathrm{E}[\mathrm{E}[x(nT+\xi_n) - x(nT)|\xi_n]] = \mathrm{E}[m_x(nT+\xi_n) - m_x(nT)] = 0 \tag{12}$$

where $m_x(\cdot) = 0$ because we have assumed $\mathrm{E}[c_n] = 0$. To get the average autocorrelation of e_n, first we evaluate the autocorrelation of $e(t)$, $t \in \mathcal{R}$, namely $r_e(t,\tau) \triangleq \mathrm{E}[e(t+\tau)e^*(t)]$. For simplicity of notation, let $\xi_1 \triangleq \xi(t)$ and $\xi_2 \triangleq \xi(t+\tau)$, so that

$$\begin{aligned} r_e(t,\tau) =& \mathrm{E}[x(t+\tau+\xi_2)x^*(t+\xi_1)] - \mathrm{E}[x(t+\tau)x^*(t+\xi_1)] \\ &- \mathrm{E}[x(t+\tau+\xi_2)x^*(t)] + \mathrm{E}[x(t+\tau)x^*(t)] \\ \triangleq& \, r_e^{(1)}(t,\tau) - r_e^{(2)}(t,\tau) - r_e^{(3)}(t,\tau) + r_e^{(4)}(t,\tau) \end{aligned} \tag{13}$$

where the definition of each $r_e^{(\cdot)}$ is apparent. The autocorrelation of e_n, r_{e_n}, does not depend on the reference instant, t, and it is related to the time average of $r_e(t,\tau)$, $\overline{r}_e(\tau)$, by the following relationship:

$$r_{e_n}(\tau) = \overline{r}_e(\tau) = \frac{1}{T}\int_0^T r_e(t,\tau)\,dt \quad , \qquad \tau = nT \tag{14}$$

Like r_e, r_{e_n} is composed of four terms, namely

$$\begin{aligned} r_{e_n}^{(1)}(\tau) =& \overline{\mathrm{E}[\mathrm{E}[x(t+\tau+\xi_2)x^*(t+\xi_1)|\xi_1,\xi_2]]} = \mathrm{E}[\overline{r}_x(\tau+\xi_2-\xi_1)] \\ =& \int_{-\infty}^{+\infty} \overline{R}_x(\lambda)\, e^{j2\pi\lambda\tau}\, \mathrm{E}[e^{j2\pi\lambda(\xi_2-\xi_1)}]\, d\lambda \end{aligned} \tag{15}$$

$$\begin{aligned} r_{e_n}^{(2)}(\tau) =& \overline{\mathrm{E}[\mathrm{E}[x(t+\tau)x^*(t+\xi_1)|\xi_1]]} = \mathrm{E}[\overline{r}_x(\tau-\xi_1)] \\ =& \int_{-\infty}^{+\infty} \overline{R}_x(\lambda)\, e^{j2\pi\lambda\tau}\, \Psi_\xi^*(\lambda)\, d\lambda \end{aligned} \tag{16}$$

and similarly

$$r_{e_n}^{(3)}(\tau) = \int_{-\infty}^{+\infty} \overline{R}_x(\lambda)\, e^{j2\pi\lambda\tau}\, \Psi_\xi(\lambda)\, d\lambda \tag{17}$$

where $\overline{r}_x$ is the time average of the autocorrelation of $x(t)$ and $\overline{R}_x$ its Fourier transform. The average spectral density of e_n, $\overline{R}_{e_n}(f)$, is given by the Fourier transform of r_{e_n}, namely

$$R_{e_n}(f) = \sum_{k=-\infty}^{+\infty} r_{e_n}(kT)e^{-j2\pi f kT} \quad . \tag{18}$$

$R_{e_n}(f)$ is a periodic function with period $F = 1/T$ in the real variable f. With the use of (5) and (15), taking into account the definitions of ξ_1 and ξ_2, $r_{e_n}^{(1)}(\tau)$ is given by

$$r_{e_n}^{(1)}(\tau) = \int_{-\infty}^{+\infty} \overline{R}_x(\lambda)\, e^{j2\pi\lambda\tau} \left\{[1 - \kappa(\tau)]|\Psi_\xi(\lambda)|^2 + \kappa(\tau)\right\} d\lambda \tag{19}$$

where $\kappa(\tau) = 1$ when $\tau = 0$ and 0 elsewhere. From (19), with algebraic manipulations, we get the Fourier transform of $r_{e_n}^{(1)}(\tau)$, namely

$$\begin{aligned} R_{e_n}^{(1)}(f) &= \sum_{k=-\infty}^{+\infty} \int_{-\infty}^{+\infty} \overline{R}_x(\lambda)\, e^{j2\pi(\lambda - f)kT} \left\{[1 - \kappa(k)]|\Psi_\xi(\lambda)|^2 + \kappa(k)\right\} d\lambda \\ &= \sum_{k=-\infty}^{+\infty} F\overline{R}_x(f + kF)|\Psi_\xi(f + kF)|^2 + \int_{-\infty}^{+\infty} \overline{R}_x(\lambda)[1 - |\Psi_\xi(\lambda)|^2]\, d\lambda \, . \end{aligned} \tag{20}$$

In the same way we get

$$R_{e_n}^{(2)}(f) = \sum_{k=-\infty}^{+\infty} F\overline{R}_x(f + kF)\Psi_\xi^*(f + kF) \quad , \tag{21}$$

$$R_{e_n}^{(3)}(f) = \left[R_{e_n}^{(2)}(f)\right]^* \quad , \qquad R_{e_n}^{(4)}(f) = \sum_{k=-\infty}^{+\infty} F\overline{R}_x(f + kF) \quad , \tag{22}$$

where

$$\overline{R}_x(f + kF) = |G(f + kF)|^2\, R_{c_n}(f + kF) = |G(f + kF)|^2\, R_{c_n}(f) \quad . \tag{23}$$

In conclusion, combining the relevant formulas, we get

$$\begin{aligned} R_{e_n}(f) =& R_{c_n}(f) \sum_{k=-\infty}^{+\infty} F\,|G(f + kF)[1 - \Psi_\xi(f + kF)]|^2 \\ &+ \int_{-\infty}^{+\infty} R_{c_n}(\lambda)|G(\lambda)|^2 \left[1 - |\Psi_\xi(\lambda)|^2\right] d\lambda \quad . \end{aligned} \tag{24}$$

REFERENCES

[1] N. Yee, J.P. Linnartz and G. Fettweis, "Multi-carrier CDMA in indoor wireless radio networks," in Proc. *PIMRC '93*, pp. 109–113, Yokohama, Sept. 1993.

[2] L. Vandendorpe, "Multitone spread spectrum multiple access communications system in a multipath Rician fading channel," *IEEE Trans. on Vehicular Technology*, vol. 44, no. 2, pp. 327–337, May 1995.

[3] A. Chini, M.S. El-Tanany, S.A. Mahmoud, "High rate ATM packet transmission over indoor radio channels," in Proc. *VTC '95*, pp. 195–199, Chicago, July 1995.

[4] T. Pollet, M. Van Bladel and M. Moeneclaey, "BER sensitivity of OFDM systems to carrier frequency offset and Wiener phase noise," *IEEE Trans. on Commun.*, vol. 43, no. 2/3/4, pp. 191–193, Feb./Mar./Apr. 1995.

[5] E. Viterbo, K. Fazel, "How to combat long echoes in OFDM transmission schemes: sub-channel equalization or more powerful channel coding," in Proc. *GLOBECOM '95*, pp. 2069–2074, Singapore, Nov. 1995.

[6] J.G. Proakis, *Digital Communications*. 3rd ed., New York, McGraw-Hill 1995.

[7] T. Pollet and M. Moeneclaey, "Synchronizability of OFDM signals," in Proc. *GLOBECOM '95*, pp. 2054–2058, Singapore, Nov. 1995.

[8] G. Wu, A. Jalali, P. Mermelstein, "On channel model parameters for microcellular CDMA systems," *IEEE Trans. on Vehic. Technol.*, vol. 44, no. 3, pp. 706–711, Aug. 1995.

[9] S. Souissi, S. B. Wicker, "A diversity combining DS/CDMA system with convolutional encoding and Viterbi decoding," *IEEE Trans. on Vehicular Technology*, vol. 44, no. 2, pp. 304–312, May 1995.

[10] S. Kaiser, "On the performance of different detection techniques for OFDM-CDMA in fading channels," in Proc. *GLOBECOM '95*, pp. 2059–2063, Singapore, Nov. 1995.

[11] H. Sari, G. Karam, and I. Jeanclaude, "Transmission techniques for digital terrestrial TV broadcasting," *IEEE Commun. Magazine*, vol. 33, no. 2, pp. 100–109, Feb. 1995.

[12] L. Tomba and W.A. Krzymien, "Downlink detection schemes for MC-CDMA systems in indoor environments," submitted to *IEICE Transactions*, Special Issue on Personal Communications.

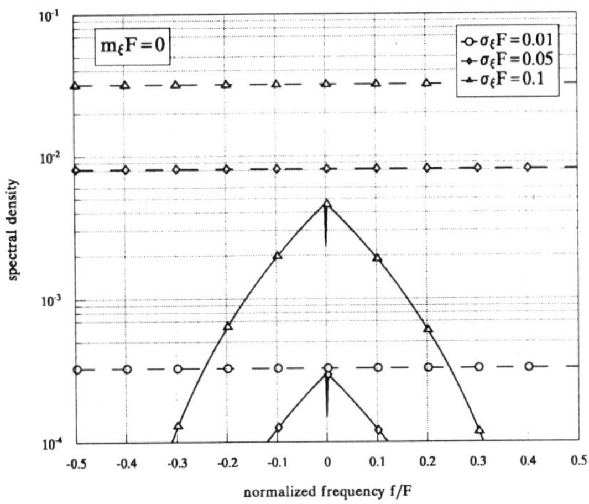

Figure 3 Average spectral density of the chip timing jitter error with Gaussian jitter and $\beta = 0$ (dashed line: white component, solid line: coloured component).

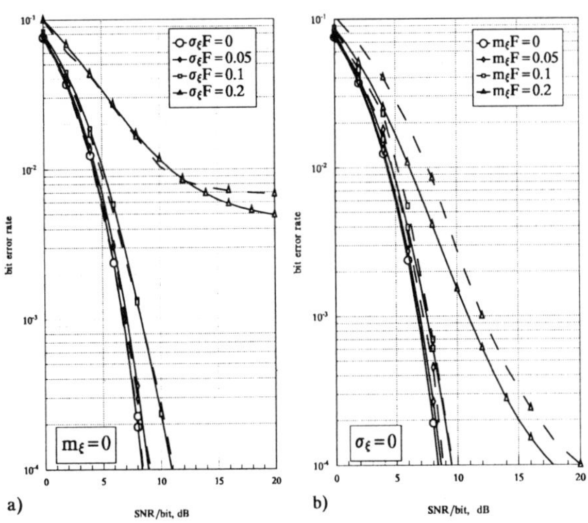

Figure 4 BER vs. SNR/bit for the AWGN channel in the presence of Gaussian jitter: comparison between numerical approximation (solid line) and simulation results (dashed line) ($\beta = 0$, $N = L = N_a = 64$ (worst case)).

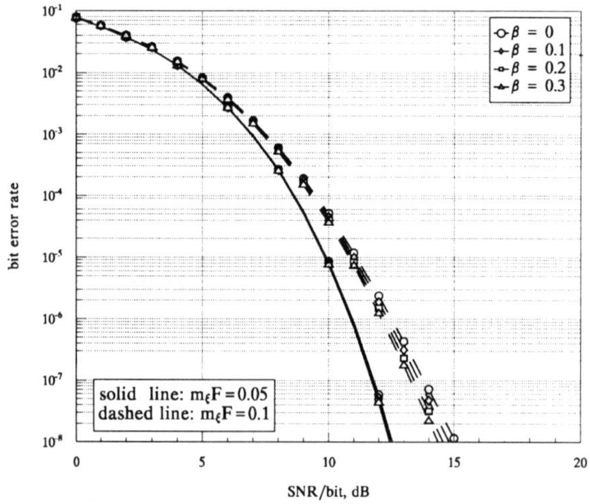

Figure 5 BER vs. SNR/bit for the AWGN channel in the presence of Gaussian jitter for different values of the roll-off factor β and two values of $m_\xi F$ ($N = L = N_a = 64$, $\sigma_\xi = 0$).

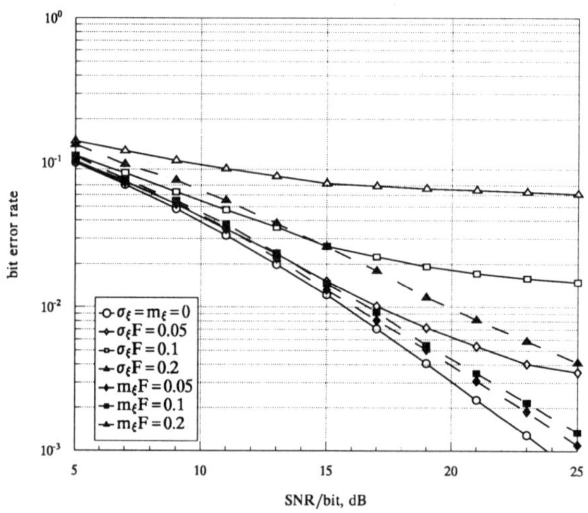

Figure 6 BER vs. SNR/bit for the multipath Rayleigh fading channel in the presence of Gaussian jitter ($\beta = 0$, $N = L = N_a = 64$).

CODED MODULATIONS AND DIVERSITY FOR SATELLITE CELLULAR COMMUNICATIONS

Ezio Biglieri, Giuseppe Caire, Giorgio Taricco*
Javier Ventura-Traveset**

** Politecnico di Torino*
Corso Duca degli Abruzzi 24, I-10129 Torino (Italy)

*** European Space Agency / ESTEC*
P.O. Box 299, 2200 AG Noordwijk (The Netherlands)

ABSTRACT

Two detection schemes are considered for the fading channel with co-channel interference. Both schemes exploit the synergy of diversity and coding with maximum-ratio combining and perfect channel state information. The first one – conventional receiver – reduces the error floor by the use of diversity and coding. The second – multi-user receiver – eliminates the error floor by exploiting the structure of the interfering signal which is no more treated as an unknown source of disturbance [6]. Performance results include bit error probability bounds and simulation results for both schemes envisaged.

Keywords: PCS Systems, Satellite Communications, Coding, Diversity.

1 INTRODUCTION

Most Personal Communication Systems (PCS) include coding and diversity to protect the transmitted signal from the severe performance degradation associated with fading in the mobile radio channel. Their simultaneous use allows the designer to cope with stringent power requirements and co-channel interference (CCI) typical of cellular environments. However, only a few works address the interaction of coding and diversity, which is the main focus of this paper.

This work was supported in part by the Human Capital and Mobility Program of the Commission of the European Union.

Different kinds of diversity have been considered for a long time: space, frequency, direction-of-arrival, polarization, time, and multipath diversity [8]. Coded modulations can be regarded themselves as kind of time diversity [9]. It is then natural to try to exploit coding and diversity together in order to reduce receiver complexity.

The design of coded modulations for the fading channel [3, 4] without diversity is generally more complex than designing coded modulation for the AWGN channel. Codes are selected according to different criteria than the standard minimum Euclidean distance used for the AWGN channel, so that good codes for the AWGN channel may perform poorly on a fading channel and vice versa. However, as the order of diversity increases, a fading channel approaches an AWGN channel [12, 13, 11]. This allows us to use powerful codes designed for the AWGN channel – such as the well known Ungerboeck codes [10] – after introducing a suitable amount of diversity.

Focusing on space diversity, we consider a NORmalized Diversity (NORD) receiver based on an antenna array of M elements, each having equivalent area A_e. Assuming that the transmitted power is P_t, we identify two equivalent cases of NORD: i) $A_e = A/M$, $P_t = P$ (fixed total antenna gain and transmitted power) and ii) $A_e = A$, $P_t = P/M$ (antenna gain increased by M and transmitted power reduced by M). In both cases, the product of each antenna element's gain by the transmitted power is constant. The average signal-to-noise ratio (SNR) per diversity branch is given by Γ/M where $\Gamma = E/N_0$ is the total average SNR at the receiver input, E is the energy per symbol and N_0 is the unilateral power spectral density of the AWGN.

Cellular PCSs are mostly limited by co-channel interference (CCI), which determines the system capacity [5]. Therefore, we will concentrate on system performance vs. CCI. We consider two different receiver schemes based on i) coherent detection with maximum-ratio combining (MRC), hereafter referred to as the *conventional receiver*, suboptimal in the presence of CCI and ii) a novel detection scheme based on the multi-user approach introduced in [6], hereafter referred to as the *multi-user receiver*. In both cases, we present results based on error probability bounds and simulation.

The paper is organized as follows. Section 2 describes the channel model. Section 3 considers the coded system with diversity and the conventional receiver. Section 4 considers the same system with the multi-user receiver. In both cases, error probability performance is studied based on union bounds. Section 5 gives numerical results on the performance of a specific code. Finally, our conclusions are collected in Section 6.

2 SYSTEM MODEL

The basic transmission system under analysis is described in Fig. 1. We have a source generator of binary symbols m_k, mapped by a coded modulator to channel symbols x_k from the q-ary PSK modulation set $\mathcal{X}_q = \{e^{j2n\pi/q}\}_{n=0}^{q-1}$. The encoded symbols are then interleaved and shaped to produce the signal $x(t) = \sqrt{E}\sum_k x'_k p(t-kT)$, where E is the energy per symbol, T is the symbol time, and $p(t)$ is a unit-energy pulse chosen in order that, when passed through a matched filter $p^*(-t)$, the synchronously-sampled output is inter-symbol interference (ISI) -free. The transmitted signal $x(t)$ is sent through M different faded channels (diversity branches). Interleaving allows us to assume the case of stationary slow Rician fading (see, e.g., [4]). The received signal over each branch i is given by

$$y^i(t) \;=\; g^i(t)x(t) + h^i(t)I(t) + n^i(t)$$

where, $g^i(t)$ and $h^i(t)$ are independent complex Gaussian random processes with $\mathrm{E}[g^i(t)] = \mathrm{E}[h^i(t)] = \sqrt{K/(K+1)}$, $\mathrm{E}[|g^i(t)|^2] = \mathrm{E}[|h^i(t)|^2] = 1$, and $\mathrm{var}\,(g^i(t)) = \mathrm{var}\,(h^i(t)) = 1/(2(K+1))$. K is the Rician parameter defined as the line-of-sight to diffuse multipath component power ratio; $n^i(t)$ is an independent white Gaussian complex noise process with two-sided power spectral density $N_0/2$; $I(t)$ is the CCI signal of the same type as $x(t)$ but statistically independent from it and with energy per symbol $E' = E/\beta$ where β is the signal-to-interference ratio (SIR). In our model, we assume a single dominant CCI signal and perfect channel state information (CSI) on both the $g^i(t)$ and the $h^i(t)$. This strong assumption can be well approximated, for instance, by including non-overlapping pilot tones, multiplexed pilot symbols or multibeam receivers in the return link of a satellite mobile communication system.

After demodulation, matched filtering, and synchronous sampling with ideal timing (which provides ISI-free samples), we obtain the following discrete-time channel model:

$$\mathbf{y}_k \;=\; \sqrt{\gamma_0}\,\mathbf{g}_k x_k + \sqrt{\gamma_1}\,\mathbf{h}_k I_k + \mathbf{n}_k \tag{1.1}$$

where

- $\mathbf{g}_k$ and $\mathbf{h}_k$ are independent random M-vectors whose components g_k^i and h_k^j are complex Gaussian random variables (RVs) with unit second moment $\mathrm{E}[|g_k^i|^2] = \mathrm{E}[|h_k^j|^2] = 1$, mean $\mathrm{E}[g_k^i] = \mathrm{E}[h_k^j] = \sqrt{K/(K+1)}$ and covariance $\mathrm{E}[g_k^i(h_\ell^j)^*] - \mathrm{E}[g_k^i]\mathrm{E}[(h_\ell^j)^*] = C_g(k-\ell, i-j)$.
- $\mathbf{n}_k$ is a random M-vector whose components n_k^i are i.i.d. complex Gaussian RVs with zero mean and variance $1/2$.

- $\gamma_0 = \Gamma/M$ and $\gamma_1 = \Gamma_1/M$ are the average SNR per branch of the useful and interfering signal, respectively, and $\Gamma = E/N_0$ and $\Gamma_1 = E'/N_0$ are the corresponding total average SNRs.
- I_k is the CCI given by $I_k = \sum_\ell w_\ell a_{k-\ell}$, where w_ℓ are the samples of the total channel impulse response $w(t) = p(t) * p^*(-t)e^{j\theta}$ sampled with an arbitrary delay with respect to the optimal sampling time and rotated by the phase difference θ between the signal and the interference carriers, and the $a_k \in \mathcal{X}_q$ are symbols from the CCI source.

3 CONVENTIONAL RECEIVER

Here, we assume that CCI is an additive disturbance treated as independent Gaussian noise, thus obtaining a suboptimal detection scheme. From the perfect CSI assumption, $\mathbf{y}_k$ is conditionally Gaussian with mean $\sqrt{\gamma_0}\, x_k$ and covariance matrix $\frac{1}{2}(\gamma_1 \mathrm{E}[|I_k|^2] + 1)\mathbf{I}_M$. Since the modulated symbols x_k have constant magnitude, maximum-likelihood (ML) coherent detection consists of maximizing the additive metric $\sum_k 2\,\mathrm{Re}\,[\mathbf{g}_k^\dagger \mathbf{y}_k\, \widehat{x}_k^*]$ over all the possible coded sequences $\widehat{\mathbf{x}}$ (here the dagger † denotes Hermitian transpose). We can first form the combined diversity-channel output $r_k = \mathbf{g}_k^\dagger \mathbf{y}_k$, and then use it to generate the metric associated to the branch of the code trellis labeled by $\widehat{x}_k$: $m(r_k, \widehat{x}_k) = 2\,\mathrm{Re}\,[r_k \widehat{x}_k^*]$. This is equivalent to perform maximum-ratio combining (MRC) detection before Viterbi decoding. The bit error probability can be obtained by the union bound [1]

$$P_b \leq \frac{1}{mS} \sum_{s=1}^{S} \sum_{\ell=1}^{\infty} 2^{-m\ell} \sum_{(\mathbf{x}_{\ell,s}, \widehat{\mathbf{x}}_{\ell,s})} B(\mathbf{x}_{\ell,s}, \widehat{\mathbf{x}}_{\ell,s}) P(\mathbf{x}_{\ell,s} \to \widehat{\mathbf{x}}_{\ell,s}). \tag{1.2}$$

where m is the number of information bits entering the encoder at each trellis step; S is the number of states; the pair $(\mathbf{x}_{\ell,s}, \widehat{\mathbf{x}}_{\ell,s})$ denotes an error event with span ℓ starting from state s, i.e., a pair of distinct sequences stemming from the same state s and merging after $\ell \geq 1$ discrete-time instants; $B(\mathbf{x}_{\ell,s}, \widehat{\mathbf{x}}_{\ell,s})$ is the Hamming distance between the information bit sequences corresponding to $\mathbf{x}_{\ell,s}$ and $\widehat{\mathbf{x}}_{\ell,s}$; $P(\mathbf{x}_{\ell,s} \to \widehat{\mathbf{x}}_{\ell,s})$ is the pairwise error probability (PEP) of the sequences $\mathbf{x}_{\ell,s}$ and $\widehat{\mathbf{x}}_{\ell,s}$.

Several methods have been proposed to calculate (1.2) in closed form for the AWGN channel [1] but none applies for the fading channel. In this case, we have to resort to numerical approximation and consider a limited number of terms in (1.2) [3].

The PEP $P(\mathbf{x} \to \widehat{\mathbf{x}})$ is calculated as follows. The difference between the metrics of two sequences $\mathbf{x}$ and $\widehat{\mathbf{x}}$ is given by

$$\Delta = \sum_{k \in \mathcal{K}_L} 2\,\mathrm{Re}\,[\mathbf{g}_k^\dagger \mathbf{y}_k d_k^*] \tag{1.3}$$

where we defined $d_k = x_k - \widehat{x}_k$ and $\mathcal{K}_L = \{j_1, \ldots, j_L\} = \{k : x_k \neq \widehat{x}_k\}$. L is the Hamming distance between $\mathbf{x}$ and $\widehat{\mathbf{x}}$. The smallest Hamming distance between two encoded sequences is usually referred to as the "code diversity." We will examine the interaction between antenna and code diversity. If the code is not catastrophic and L is finite, the span of the error event is also finite, and the PEP is given by

$$P(\mathbf{x} \to \widehat{\mathbf{x}}) = P(\Delta \leq 0). \tag{1.4}$$

The right-hand side (RHS) of (1.4) can be computed as follows. First, evaluate the Laplace transform of the probability density function of Δ, $\Phi_\Delta(s) = \mathrm{E}[e^{-s\Delta}]$. Then, we have from [7]

$$P(\Delta \leq 0) = \frac{1}{2\pi j} \int_{c-j\infty}^{c+j\infty} \Phi_\Delta(s) \frac{ds}{s} \tag{1.5}$$

where c is chosen so as to let (1.5) converge. Numerical evaluation of (1.5) can be carried out by means of Gauss-Chebyshev quadrature, as explained in [2].

3.1 Performance of the conventional receiver

Following the approach of [12] with ISI-free CCI ($|I_k|^2 = 1$), we obtain

$$\Phi_\Delta(s) = \prod_{k \in \mathcal{K}_L} \left[1 + \sqrt{\gamma_0}\, |d_k|^2 s - (\gamma_1 + 1)|d_k|^2 s^2\right]^{-M} \tag{1.6}$$

and then we get from (1.5):

$$P(\Delta \leq 0) = \frac{1}{2\pi j} \int_{c-j\infty}^{c+j\infty} \prod_{k \in \mathcal{K}_L} \left[1 + \sqrt{\gamma_0}\, |d_k|^2 s - (\gamma_1 + 1)|d_k|^2 s^2\right]^{-M} \frac{ds}{s} \tag{1.7}$$

Alternatively, applying the Chernoff bound to the RHS of (1.6), we obtain

$$P(\mathbf{x} \to \widehat{\mathbf{x}}) \leq \min_{\lambda \geq 0} \Phi_\Delta(\lambda) = C(\mathbf{x}, \widehat{\mathbf{x}}) = \prod_{k \in \mathcal{K}_L} \left(1 + \frac{1}{4} \frac{\gamma_0}{\gamma_1 + 1} |d_k|^2\right)^{-M} \tag{1.8}$$

Letting $\Gamma \to \infty$, the Chernoff bound (1.8) approaches the limit value

$$\widetilde{C}(\mathbf{x}, \widehat{\mathbf{x}}) = \prod_{k \in \mathcal{K}_L} \left(1 + \frac{1}{4}\beta\,|d_k|^2\right)^{-M} \leq \left(1 + \frac{1}{4}\beta\,|d_{\min}|^2\right)^{-ML} \quad (1.9)$$

This follows from the existence of an irreducible bit error probability, with CCI and fading, as the SNR increases without limits – an error floor – which is exponentially reduced with the product of time and space diversities. The effect is mostly remarkable for weak codes (up to uncoded modulation). In the following, we will see a more powerful approach that is able to *delete* the error floor, which we call multi-user detection.

4 MULTI-USER RECEIVER

With multi-user detection, we recognize that CCI cannot be assimilated to Gaussian noise and deal with it as another modulated signal. We assume that the receiver has perfect knowledge of both $\mathbf{g}_k$ and $\mathbf{h}_k$. The discrete-channel output sample can be written as $\mathbf{y}_k = \mathbf{G}_k \mathbf{b}_k + \mathbf{n}_k$ where $\mathbf{G}_k = [\sqrt{\gamma_0}\,\mathbf{g}_k, \sqrt{\gamma_1}\,\mathbf{h}_k]$ and $\mathbf{b}_k = (x_k, I_k)^T$. Then, the ML decoder branch metric associated to x_k is given by

$$\begin{aligned} m(\mathbf{y}_k, x_k) &\doteq \ln P(\mathbf{y}_k | x_k, \mathbf{g}_k, \mathbf{h}_k) \doteq \ln\{\mathrm{E}_{I_k}[P(\mathbf{y}_k | x_k, \mathbf{g}_k, I_k, \mathbf{h}_k)]\} \\ &\doteq \ln\left\{\mathrm{E}_{I_k}[\exp[-|\mathbf{y}_k - \mathbf{G}_k \mathbf{b}_k|^2]]\right\} \\ &\doteq \ln\left\{\mathrm{E}_{I_k}\left[\exp\left[2\mathrm{Re}\,[\mathbf{y}_k^\dagger \mathbf{G}_k \mathbf{b}_k] - \mathbf{b}_k^\dagger \mathbf{H}_k \mathbf{b}_k\right]\right]\right\} \end{aligned} \quad (1.10)$$

where $\doteq$ stands for metric equivalence ($m_1 \doteq m_2$ if and only if $m_1 = \alpha_1 m_2 + \alpha_2$ for some constant $\alpha_1 > 0$ and α_2) and $\mathbf{H}_k = \mathbf{G}_k^\dagger \mathbf{G}_k$. We further assume that the CCI signal is not affected by ISI, thus wanted and interfering signals are symbol-synchronous, but not frame-synchronous. Therefore the (coded) sequence (I_k) is received as a sequence of independent, uniformly distributed, symbols from $\mathcal{X}_q$ and we have

$$m(\mathbf{y}_k, x_k) \doteq \ln\left[\sum_{I_k \in \mathcal{X}_q} \exp[\Omega(x_k, I_k)]\right] \simeq \Omega_{\max}(x_k) = \max_{I_k \in \mathcal{X}_q} \Omega(x_k, I_k) \quad (1.11)$$

where we defined $\Omega(x_k, I_k) \equiv \Omega(\mathbf{b}_k) = 2\,\mathrm{Re}\,[\mathbf{y}_k^\dagger \mathbf{G}_k \mathbf{b}_k] - \mathbf{b}_k^\dagger \mathbf{H}_k \mathbf{b}_k$ and used the approximation $\ln[\sum_i \exp(z_i)] \simeq \max_i z_i$. The metric (1.11) is formally identical to the multi-user metric derived in [6] for two-user CDMA in AWGN and Gaussian spreading sequences $\mathbf{g}_k$ and $\mathbf{h}_k$. The detection scheme based on (1.11) defines what we call *multi-user receiver*.

4.1 Performance of the multi-user receiver

Our approach to the evaluation of the PEP is based on the union bound. Let us consider the error event $(\mathbf{x}, \widehat{\mathbf{x}})$ with Hamming distance L and let $\mathcal{K}_L = \{j_1, \ldots, j_L\} = \{k : x_k \neq \widehat{x}_k\}$. Let $\mathbf{I}_\ell$ denote the ℓth interfering sequence from $\mathcal{X}_q^L$, for $\ell = 0, 1, \ldots, q^L - 1$. From the branch metric (1.11), we can write the path metric increment as follows[1]:

$$\Omega_{\max}(\mathbf{x}) = \sum_{k \in \mathcal{K}_L} \Omega_{\max}(x_k) = \max_{\mathbf{I}} \Omega(\mathbf{x}, \mathbf{I}) \tag{1.12}$$

where $\mathbf{I} = (I_{j_1}, \ldots, I_{j_L})$ and the last equality derives from the independence of the CCI samples in $\mathbf{I}$. Then, the PEP is given as

$$P(\mathbf{x} \to \widehat{\mathbf{x}}) = P\{\Omega_{\max}(\mathbf{x}) \leq \Omega_{\max}(\widehat{\mathbf{x}})\} \tag{1.13}$$

In order to evaluate this probability, we first condition on the (random) transmitted CCI sequence $\mathbf{I}$ and we call $\mathbf{I}_{\max}$ the CCI sequence attaining the maximum of $\Omega(\mathbf{x}, \mathbf{I})$:

$$\begin{aligned} P(\mathbf{x} \to \widehat{\mathbf{x}} \mid \mathbf{I}) &= P\left\{\Omega(\mathbf{x}, \mathbf{I}_{\max}) \leq \max_{\widehat{\mathbf{I}}} \Omega(\widehat{\mathbf{x}}, \widehat{\mathbf{I}})\right\} && (1.14) \\ &\leq P\left\{\Omega(\mathbf{x}, \mathbf{I}) \leq \max_{\widehat{\mathbf{I}}} \Omega(\widehat{\mathbf{x}}, \widehat{\mathbf{I}})\right\} \\ &= P\left\{\bigcup_{\ell=0}^{q^L-1} \{\Omega(\mathbf{x}, \mathbf{I}) - \Omega(\widehat{\mathbf{x}}, \mathbf{I}_\ell) \leq 0\}\right\} && (1.15) \\ &\leq \sum_{\ell=0}^{q^L-1} P\{\Omega(\mathbf{x}, \mathbf{I}) - \Omega(\widehat{\mathbf{x}}, \mathbf{I}_\ell) \leq 0\} && (1.16) \end{aligned}$$

where (1.15) is greater than (1.14) because $\Omega(\mathbf{x}, \mathbf{I}) \geq \Omega(\mathbf{x}, \mathbf{I}_{\max})$ and (1.16) is obtained from (1.15) by the union bound. The branch metric difference can be written as

$$\Delta_k = \Omega(\mathbf{b}_k) - \Omega(\widehat{\mathbf{b}}_k) = 2\,\mathrm{Re}[\mathbf{z}_k^\dagger \mathbf{d}_k] + \mathbf{d}_k^\dagger \mathbf{H}_k \mathbf{d}_k \tag{1.17}$$

where $\mathbf{d}_k = \mathbf{b}_k - \widehat{\mathbf{b}}_k$ and

$$\mathbf{z}_k = \mathbf{G}_k^\dagger \mathbf{y}_k - \mathbf{H}_k^\dagger \mathbf{b}_k = \mathbf{G}_k^\dagger(\mathbf{G}_k \mathbf{b}_k + \mathbf{n}_k) - \mathbf{H}_k^\dagger \mathbf{b}_k = \mathbf{G}_k^\dagger \mathbf{n}_k$$

since $\mathbf{H}_k = \mathbf{G}_k^\dagger \mathbf{G}_k$. Conditionally on $\mathbf{G}_k$, $\mathbf{z}_k$ is a vector of complex Gaussian RVs with zero mean and covariance matrix $\frac{1}{2}\mathbf{H}_k$. Then, still conditionally on

[1] We do not consider the positions k such that $x_k = \widehat{x}_k$ in the path metric since, in this case, $\Omega_{\max}(x_k) = \Omega_{\max}(\widehat{x}_k)$.

$\mathbf{G}_k$, $\mathbf{z}_k^\dagger \mathbf{d}_k$ is a complex Gaussian RV with zero mean and variance $\frac{1}{2}\mathbf{d}_k^\dagger \mathbf{H}_k \mathbf{d}_k$. It is then clear that the conditional PEP (1.15) does not depend on $\mathbf{I}$ but only on the difference $\mathbf{I}_\ell - \mathbf{I}$, so that, when $\mathbf{I}_\ell$ runs over all possible sequences from $\mathcal{X}_q^L$, we can remove conditioning and choose, arbitrarily, $\mathbf{I} = \mathbf{1}$, the sequence of all ones. Then, we can write

$$P(\mathbf{x} \to \widehat{\mathbf{x}}) \leq \sum_{\ell=0}^{q^L-1} P\left\{\Omega(\mathbf{x}, \mathbf{1}) - \Omega(\widehat{\mathbf{x}}, \mathbf{I}_\ell) \leq 0\right\} \tag{1.18}$$

Now, to calculate the probabilities $P\{\Omega(\mathbf{x}, \mathbf{1}) - \Omega(\widehat{\mathbf{x}}, \mathbf{I}_\ell) \leq 0\}$, we advocate the Laplace transform technique already used in Section 3. First, by iterating expectations, we obtain

$$\begin{aligned} \mathrm{E}[\exp(-s\Delta_k)] &= \mathrm{E}_{\mathbf{G}_k}\left[E_{\mathbf{z}_k}\left[\exp\left(-s\left(2\,\mathrm{Re}\left[\mathbf{d}_k^\dagger \mathbf{z}_k\right] + \mathbf{d}_k^\dagger \mathbf{H}_k \mathbf{d}_k\right)\right) \mid \mathbf{G}_k\right]\right] \\ &= \mathrm{E}_{\mathbf{G}_k}[\exp(-\mathbf{d}_k^\dagger \mathbf{H}_k \mathbf{d}_k (s - s^2))] \\ &= \left[1 + (\gamma_0 |d_k|^2 + \gamma_1 |d_k'|^2)(s - s^2)\right]^{-M} \end{aligned}$$

with $d_k = x_k - \widehat{x}_k$ and $d_k' = 1 - I_k$. Then, from the independence of the Δ_k's, setting $\Delta = \sum_{k \in \mathcal{K}_L} \Delta_k$, we have

$$\Phi_\Delta(s) = \mathrm{E}[\exp(-s\Delta)] = \prod_{k \in \mathcal{K}_L} \left[1 + (\gamma_0 |d_k|^2 + \gamma_1 |d_k'|^2)(s - s^2)\right]^{-M} \tag{1.19}$$

Applying (1.5) to (1.19) and inserting the result in bound (1.18), we get

$$\begin{aligned} P(\mathbf{x} \to \widehat{\mathbf{x}}) &\leq \sum_{\ell=0}^{q^L-1} \frac{1}{2\pi j} \int_{c-j\infty}^{c+j\infty} \prod_{k \in \mathcal{K}_L} \left[1 + (\gamma_0 |d_k|^2 + \gamma_1 |d_k'|^2)(s - s^2)\right]^{-M} \frac{ds}{s} \\ &= \frac{1}{2\pi j} \int_{c-j\infty}^{c+j\infty} \prod_{k \in \mathcal{K}_L} \left\{\sum_{\ell=0}^{q-1} \left[1 + (\gamma_0 |d_k|^2 + \gamma_1 |\delta_\ell|^2)(s - s^2)\right]^{-M}\right\} \frac{ds}{s} \end{aligned} \tag{1.20}$$

where $\delta_\ell = 1 - \exp(j2\pi\ell/q)$. This probability can be evaluated numerically by resorting to Gauss-Chebyshev quadrature, as described in Appendix A. Alternatively, applying the Chernoff bound to each term in (1.18), we get

$$P(\mathbf{x} \to \widehat{\mathbf{x}}) \leq \prod_{k \in \mathcal{K}_L} \left\{\sum_{\ell=0}^{q-1} \left[1 + \frac{1}{4}(\gamma_0 |d_k|^2 + \gamma_1 |\delta_\ell|^2)\right]^{-M}\right\} \tag{1.21}$$

Bound (1.21) shows that the multi-user receiver suppresses the error floor, irrespectively of the value of SIR and for any finite diversity order M. With strong CCI, i.e., for $\beta \to 0$, (1.21) approaches the limit

$$P(\mathbf{x} \to \widehat{\mathbf{x}}) \leq \left[\sum_{\ell=0}^{q-1} \left(1 + \frac{\gamma_1}{4} |\delta_\ell|^2 \right)^{-M} \right]^L$$

with $\delta_\ell = 1 - \exp(j2\pi\ell/q)$. This bound is independent of the code as we assumed that the CCI interferer is not frame-synchronous and therefore appears uncoded. For weak interfering signal, i.e., for $\beta \to \infty$, the performance approaches the CCI-free case, since the effect of CCI becomes negligible.

Bound (1.16) is tight when the interference-to-noise ratio per branch γ_1 is so high that there is a CCI sequence that attains a sharp maximum in $\Omega_{\max}(\widehat{\mathbf{x}})$. This is not true when the SNR is lower than or around the SIR and, in this case, the bound is loose. However, performance cannot be worse than with the conventional receiver, so that we obtain a tighter general bound as follows

$$P(\mathbf{x} \to \widehat{\mathbf{x}}) \leq \min \Big\{ \text{bound (1.7), bound (1.19)} \Big\} \qquad (1.22)$$

5 RESULTS

In this section we collect results obtained by our analysis. We consider an uncorrelated flat Rayleigh fading channel affected by CCI and AWGN.

Performance with conventional coherent detection. In the following, results are relevant to Ungerboeck's rate 2/3 coded 8-PSK scheme with 4 states (U4). Fig. 2 shows the performance of U4 with conventional detection. Results, obtained by analytic bounds and simulation, show excellent agreement below the level BER $= 10^{-2}$. Fixing a target BER $= 10^{-3}$, Fig. 3 shows the minimum E_b/N_0 required to obtain the target performance vs. SIR for different values of the diversity order M. These results are obtained by analytic bounds, which proved to be extremely accurate in this case, as mentioned above. Finally, Fig. 4 shows how diversity progressively eliminates the error floor, following the asymptotic behavior of bound (1.9).

Impact of multi-user detection. An alternative to diversity is multi-user detection. We analyzed it in conjunction with diversity for the same code U4. Figs. 5 to 8 show the BER vs. E_b/N_0 for different values of SIR ($-10, 0, 10$ and 20 dB) and diversity order. First, we notice that the error floor has been deleted for any diversity order. Second, the curves show the same good performance at all values of SIR, as if the negative impact of SIR were totally removed. This can be explained by noting that multi-user detection exploits the structure of the CCI disturbance, provided a reliable overall estimation of the CSI. This also explains why intermediate levels of SIR (around 5 dB) achieve the worst performance, as neither the wanted nor the interfering signal are dominant with this level of CCI. The figures also show the looseness of bound (1.19) at low values of SNR. Simulation results are in good agreement with analytic bounds below the level BER $= 10^{-3}$. The goodness of approximation is better for low diversity order. Fig. 9 shows the minimum E_b/N_0 required to obtain a target BER $= 10^{-3}$ vs. SIR for different values of the diversity order M. As a consequence of the elimination of the error floor we can allways achieve the target performance, irrespectively of the SIR.

As a final observation, the angular points of the bounds for the multi-user receiver derive from the fact that the multi-user bound (1.22) is a minimum of two bounds, the second of which is looser for low SNR. However, from a practical point of view, this is acceptable since it produces a conservative performance analysis.

6 CONCLUSIONS

We have studied diversity receivers for a coded communication system affected by fading and co-channel interference, namely a typical cellular PCS. Two detection schemes have been considered here: coherent detection with maximum-ratio combining (MRC) (conventional receiver) and a novel detection scheme based on the multi-user approach introduced in (multiuser receiver) [6]. The performance of the conventional receiver is limited by the implicit assumption that the co-channel interference is an unknown disturbance, which inevitably introdeces an error-floor in the BER performance. The multi-user receiver overcomes this limitation and provides rather interesting results.

Further study is required to assess how close we can approach the perfect CSI assumption, fundamental to the operation of both receiver, in the multipath environment. This is mostly important for the multi-user receiver which uses the CSI of both useful and interfering signals.

REFERENCES

[1] E. Biglieri, D. Divsalar, P. J. McLane, and M. K. Simon, *Introduction to Trellis-Coded Modulation with Applications.* New York: MacMillan, 1991.

[2] E. Biglieri, G. Caire, G. Taricco, and J. Ventura-Traveset, "Simple method for evaluating error probabilities," *Electronics Letters*, Vol. 32, No. 3, pp. 191-192, Feb. 1996.

[3] P. Ho and D. K. P. Fung, "Error Performance of Interleaved Trellis Coded PSK Modulations in Correlated Rayleigh Fading Channels," *IEEE Trans. Communications,* Vol. 40, No. 12, Dec. 1992.

[4] S. H. Jamali and T. Le-Ngoc, *Coded-Modulation Techniques for Fading Channels.* New York: Kluwer Academic Publishers, 1994.

[5] W. C. Y. Lee, "Spectrum Efficiency in Cellular," *IEEE Trans. Veh. Technol.,* vol. 38, pp. 69-75, May 1989.

[6] H. V. Poor and S. Verdu, "High-speed Digital Signal Processing for Satellite Communications," *Final Report European Space Agency P.O. 134422*, Feb. 1995.

[7] J.Proakis, *Digital Communications.* New York: McGraw-Hill, 1983.

[8] M. Schwartz, W. R. Bennett, and S. Stein, *Communications Systems and Techniques.* New York: McGraw-Hill, 1966.

[9] C.E. W. Sundberg and N. Seshadri, "Coded modulation for fading channels: An overview," *European Transactions on Telecommunications*, Vol. 4, No. 3, pp. 325–334, May-June 1993.

[10] G. Ungerboeck, "Channel Coding with Multilevel/Phase Signals," *IEEE Trans. on Inf. Theory,* Vol. IT-28, Jan. 1982.

[11] J. Ventura-Traveset, G. Caire, E. Biglieri, and G. Taricco, "Impact of diversity reception on fading channels with coded modulation. Part I: Coherent detection," submitted to *IEEE Trans. Communications.*, June 1995.

[12] J. Ventura-Traveset, G. Caire, E. Biglieri, and G. Taricco, "A multi-user approach to combating co-channel interference in narrowband mobile communications," *7th Tyrrhenian International Workshop on Digital Communications*, Viareggio (LU), Italy, 10-14 Sept., 1995.

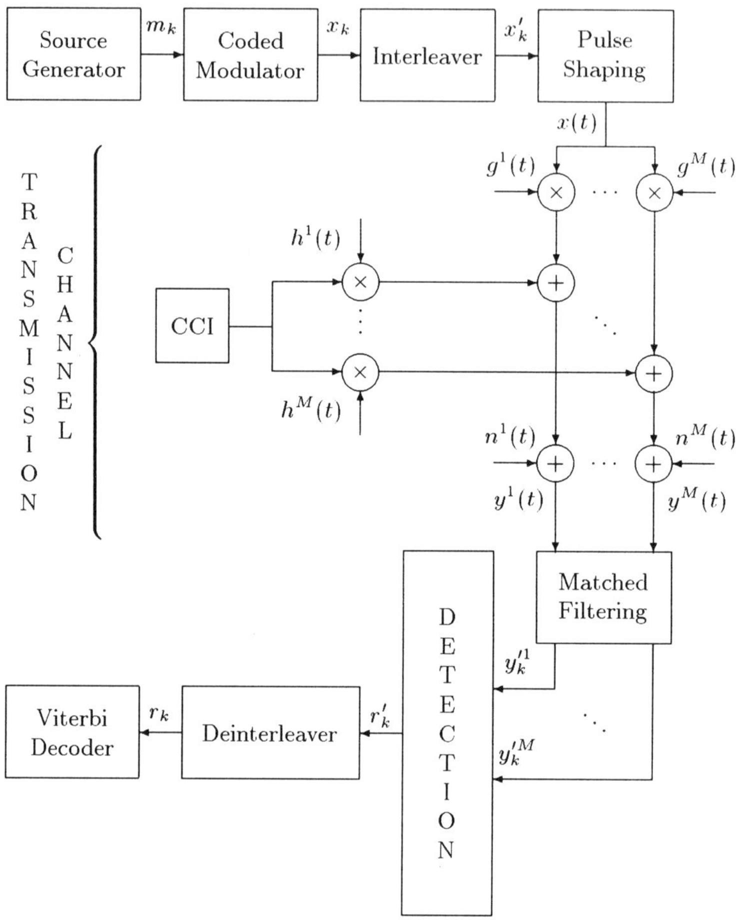

Figure 1 Model of a coded transmission system with fading, diversity, and coherent detection combining, affected by CCI and AWGN.

[13] J. Ventura-Traveset, G. Caire, E. Biglieri, and G. Taricco, "Synergy of diversity reception and coded modulation in mobile communication channels," *European Personal Mobile Communications Conference*, Bologna, Italy, 30 Nov., 1995.

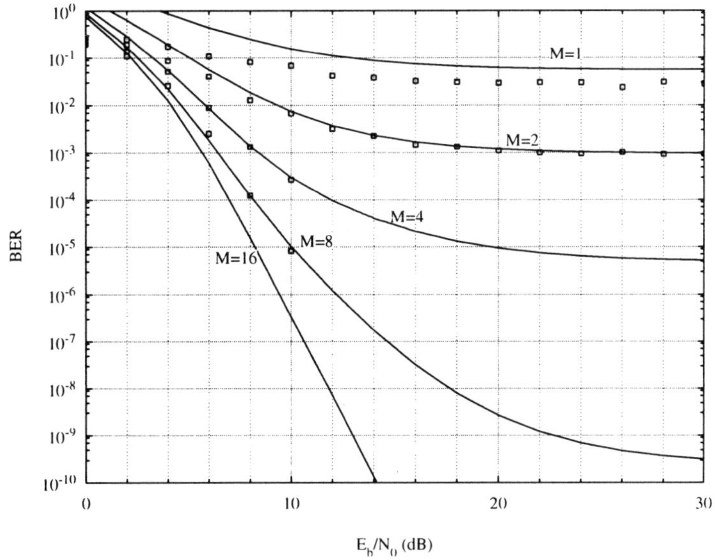

Figure 2 Bit error rate of U4 vs. E_b/N_0 with coherent detection, SIR=10 dB. Solid curves: TUB, dots: simulation results.

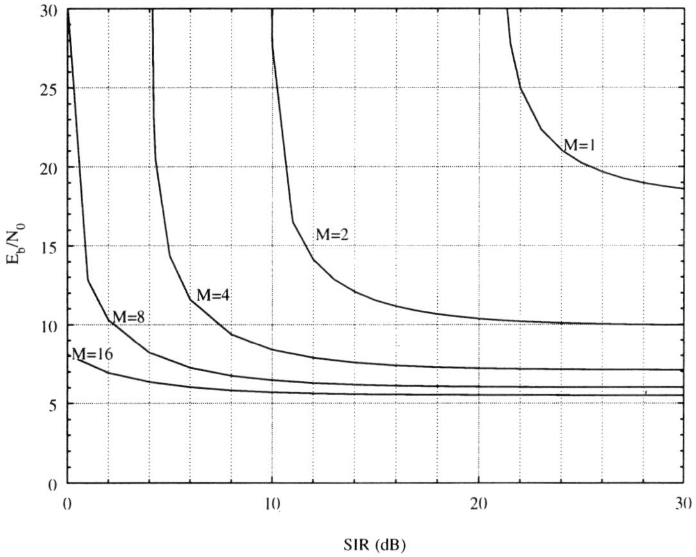

Figure 3 E_b/N_0 vs. SIR at $P_b = 10^{-3}$ of U4 with coherent detection for different values of M (diversity order).

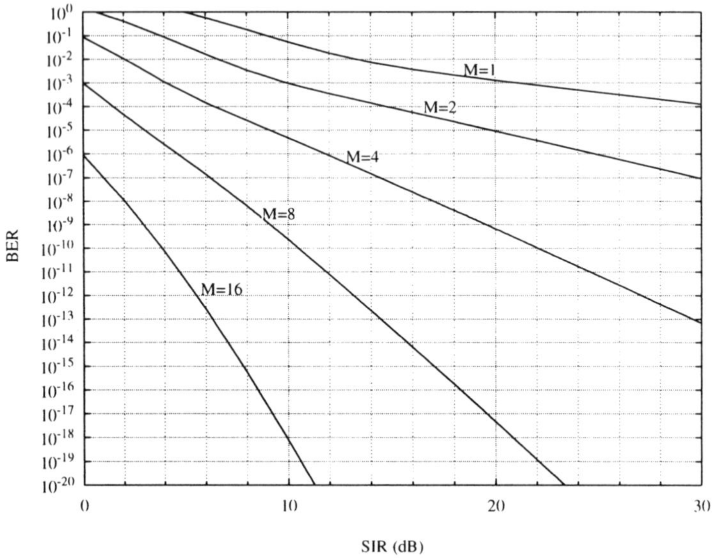

Figure 4 "Error floor" of U4 vs. SIR with coherent detection for diversity order $M = 1, 2, 4, 8, 16$.

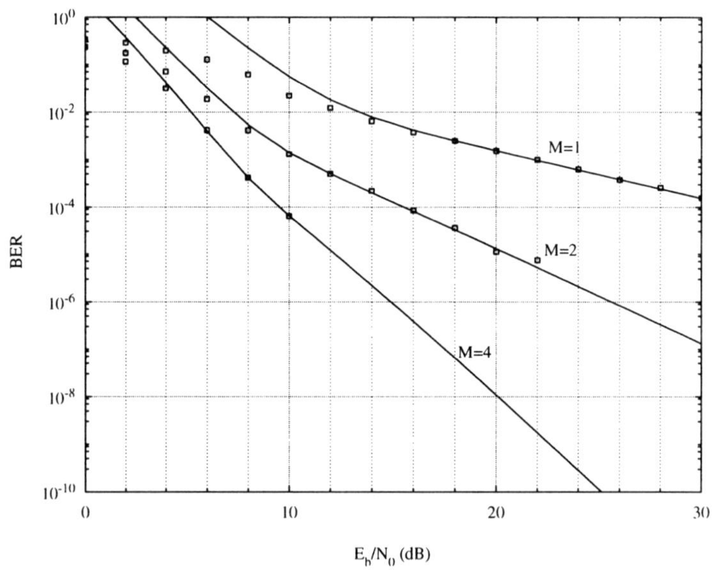

Figure 5 BER vs. E_b/N_0 of U4 for SIR=-10 dB and $M = 1, 2, 4$. Solid curves: minimum upper bounds, dots: simulation results.

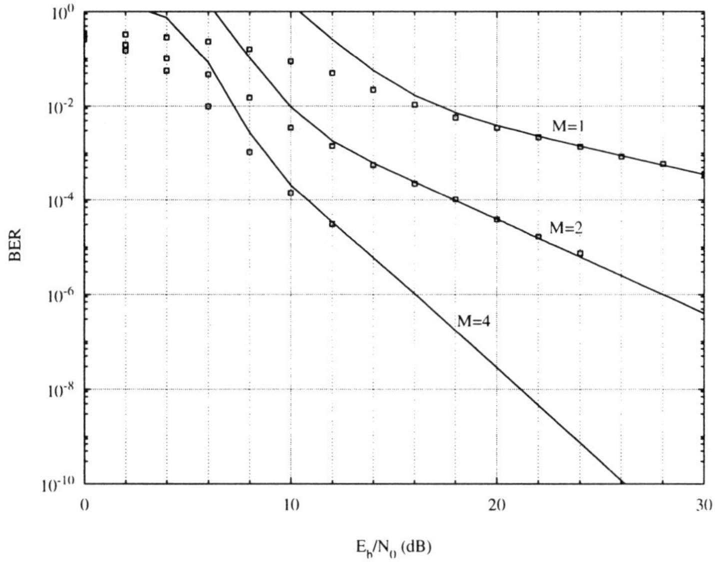

Figure 6 BER vs. E_b/N_0 of U4 for SIR=0 dB and $M = 1, 2, 4$. Solid curves: minimum upper bounds, dots: simulation results.

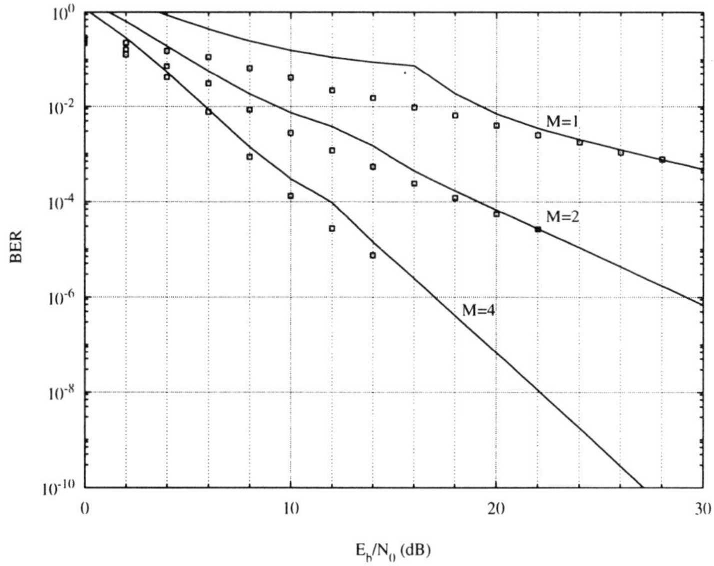

Figure 7 BER vs. E_b/N_0 of U4 for SIR=10 dB and $M = 1, 2, 4$. Solid curves: minimum upper bounds, dots: simulation results.

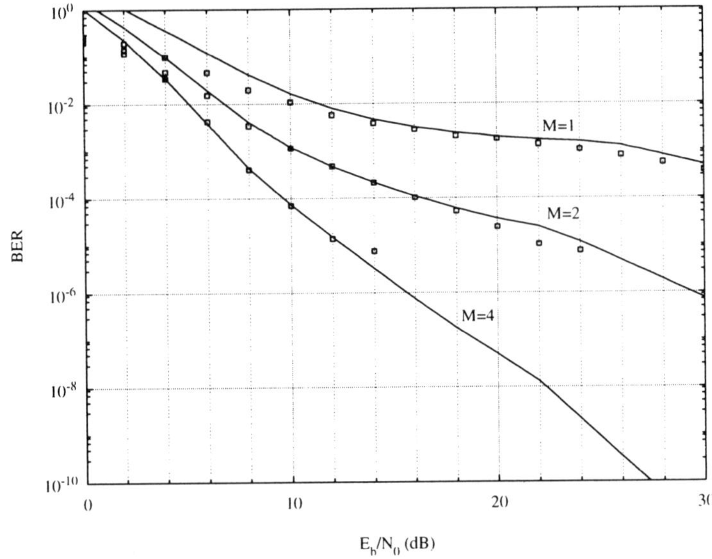

Figure 8 BER vs. E_b/N_0 of U4 for SIR=20 dB and $M = 1, 2, 4$. Solid curves: minimum upper bounds, dots: simulation results.

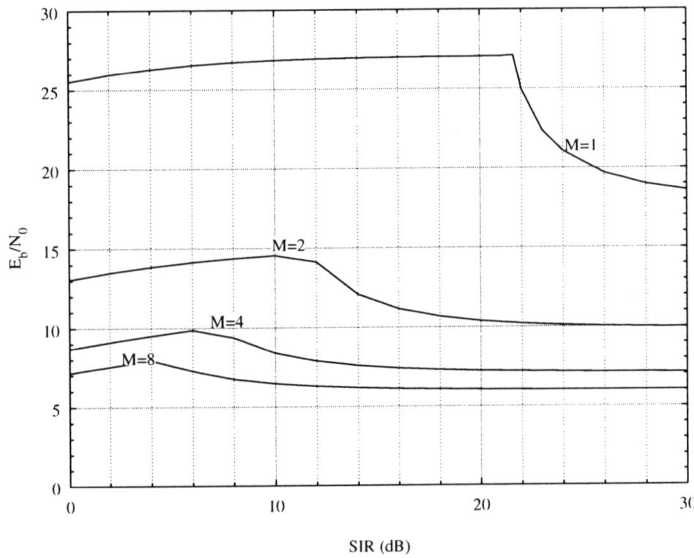

Figure 9 E_b/N_0 vs. SIR at $P_b = 10^{-3}$ of U4 with multiuser detection for different values of M (diversity order).

MULTICODE HIGH-SPEED TRANSMISSION WITH INTERFERENCE CANCELLATION FOR WIRELESS PERSONAL COMMUNICATIONS

Maurice Chan and Justin C-I Chuang[†]

Department of Electrical and Electronic Engineering
The Hong Kong University of Science and Technology
Clear Water Bay, Kowloon
Hong Kong

ABSTRACT

In this paper, we include interference cancellation in a high-speed transmission technique, called "multicode modulation," to improve signal quality over frequency-selective fading channels. In this technique, the high-rate data bits are serial-to-parallel converted into low-rate data streams which are then modulated with Walsh and PN concatenated codes. After using Rake receivers to process the received waveform, interference cancellers are applied to lower the BER floor. Numerical results show that significant performance improvement can be achieved. Using this technique, high transmission rates can be supported over a multipath fading channel having only non-line-of-sight components. Potential applications of this modulation method include those that require high peak transmission rates such as the wireless Ethernet connection for multimedia communications.

1 INTRODUCTION

During the last decade, the telecommunications industry has produced an explosion of wireless technology, which has opened a new dimension to provide universal personal communications [1]. To achieve such an objective, the next generation personal communications networks will need to be capable of supporting high level user traffic with a wide range of high-quality services having varying bit rates. For example, if personal communications were extended for high peak-rate applications such as Eithernet connection using wireless access, a rate of 10 Mb/s or higher would be desirable. In addition, to achieve reasonable mobility, the capability for a personal communicator to properly function under the non-line-of-sight conditions is also very important. However, the ability to achieve high bit rates at low error rates over wireless channels is severely restricted by the propagation characteristics of the wireless environment where signals typically arrive at the receiver via a scattering mechanism resulting in multiple propagation paths with different time delays, attenuation, and phases. This causes time delay spread which imposes a limit on the

[†] Email Address : eechuang@ee.ust.hk

Phone : (+852) 2358 7061, Fax : (+852) 2358 1485

maximum transmission rate. These restrictions manifest themselves as intersymbol interference (ISI) which leads to the introduction of an irreducible error floor [2].

One possible solution proposed is multicarrier modulation [3, 4] in which the transmitted data is divided into several interleaved bit streams which are used to modulate several sub-carriers. However, such an approach requires equalization in frequency domain which can be quite complex for high transmission rates over time-varying wireless channels. An alternative approach proposed is *multicode modulation* [5, 6]. In this approach, the high-rate data stream is divided into a number of parallel low-rate bit streams as in multicarrier modulation. However, each low-rate bit stream is modulated with Walsh codes and Pseudo-noise (PN) codes in order to separate different channels and reduce the interference of multipath, respectively. The processing gain is properly selected so that the total required bandwidth will be of the ***same*** order as that of the original high-rate data stream, thereby gaining the inherent benefit of multipath rejection without expanding the bandwidth of the original high-rate stream.

In this paper, an Interference Cancellation scheme [7] is employed in the multicode system to lower the irreducible bit error rate (BER). This technique was originally proposed and employed in direct-sequence spread-spectrum multi-access (DS/SSMA) receiver over multipath fading channels. The receiver first coherently demodulates the received signal to produce correlator outputs and initial data estimates. Based on these estimates, the cancellation scheme essentially creates replicas of the contributions of the interference embedded in the correlator outputs and removes them for a successively improved data decisions. Unlike the DS/SSMA system, this interference cancellation scheme is much simpler to implement because all bit streams are transmitted from the same source and thus encounter the same channel impulse response. We proposed a method which employs a ***single*** pilot to estimate channel parameters for ***all*** bit streams to perform interference cancellation. In comparison, multiple pilots are often inserted [4] in multicarrier systems to obtain channel response at each sub-carrier.

Section 2 illustrates the system model, including channel model, multicode system and interference cancellation scheme. Section 3 presents some numerical results, followed by conclusions in Section 4.

2 SYSTEM MODEL

In this section, the models used in this paper including frequency-selective fading channel, multicode system and interference cancellation are presented.

2.1 Channel Model

The impulse response of the channel can be described as [8]

$$h(t) = \sum_{m=1}^{L} \beta_m e^{i\phi_m} \delta(t - \tau_m) \tag{1}$$

where L is the number of paths, $i = \sqrt{-1}$, and β_m, τ_m and ϕ_m are the mth path amplitude, delay and phase respectively. A commonly accepted model suggests that β_m be a random variable with a Rayleigh distribution when line-of-sight paths are not present and ϕ_m be a random variable uniformly distributed from 0 to 2π.

A function called *power delay profile* characterizes a multipath channel and it is

$$p(t) = \left\langle |h(t)|^2 \right\rangle \tag{2}$$

where < > denotes the ensemble average. At a specific time delay τ_m the channel impulse response, denoted by $h(\tau_m)$, is a zero-mean complex Gaussian random variable with variance $p(\tau_m)$. A measure of the width of $p(t)$ is the root-mean-square (rms) delay spread τ, defined as the square root of the second central moment [9]. That is,

$$\tau = \left[\frac{\int (t - D)^2 p(t) dt}{\int p(t) dt} \right]^{1/2} \tag{3}$$

where the average delay D, i.e., the centroid of $p(t)$, is

$$D = \frac{\int t p(t) dt}{\int p(t) dt} . \tag{4}$$

It was found that the irreducible BER for transmission through a multipath channel depends strongly on the normalized rms delay spread which is defined as [2]

$$d = \frac{\tau}{T} \tag{5}$$

where T is the symbol period.

For the numerical results, an equal-amplitude two-ray profile

$$p(t) = \frac{1}{2} \left[\delta(t - \tau) + \delta(t + \tau) \right] \tag{6}$$

is considered. In this case, $L = 2$ and $d = \tau / T$. This means we will assume that the channel consists of two independently faded paths separated by 2τ. In addition, the channel is assumed to be quasi-static so that the channel random parameters do not change significantly over many consecutive symbol intervals. The overall performance is evaluated by considering the ensemble of channel impulse response $h(t)$.

2.2 Multicode System

In the multicode system, the incoming data bits with bit duration T_b are serial-to-parallel converted into K parallel bit streams with symbol duration $T = KT_b$,, similar to multicarrier modulation. A baseband model of the multicode system is shown in Fig. 1. After the serial-to-parallel conversion, the symbols on each low-rate branch, called *Traffic Channel*, are modulated using a *Concatenated Sequences* $a_k(t)$, where $k \in [1, K]$, which is a combination of ***individual*** orthogonal sequences (Walsh codes) and a ***common*** PN sequence. After combining all traffic signals together, the resultant signal is passed through a multipath channel, described in the above section. At the receiver, a Rake receiver with the corresponding concatenated sequence is employed to obtain the desired signaling bits for each traffic channel. For practical implementations, the combined signal will be used to modulate an RF carrier prior to transmission over the air.

It is worth noting that the orthogonal and PN sequences are employed for channel separation and for multipath rejection over fading channels, respectively. This concept of concatenated sequences has been successfully applied in the downlink transmission of the Qualcomm Code Division Multiple Access (CDMA) system [10].

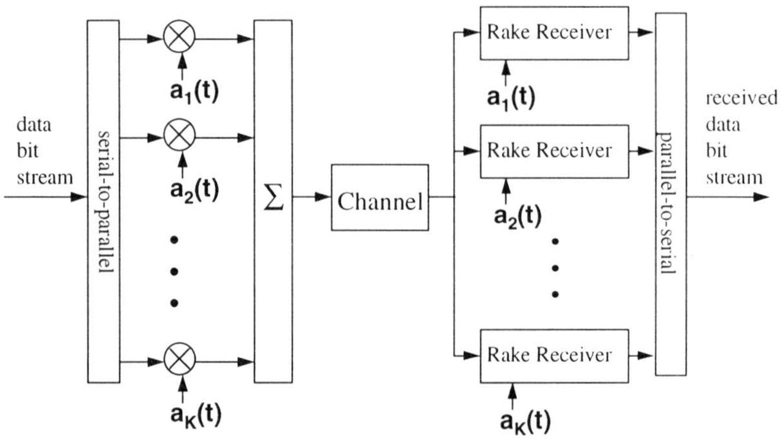

Figure 1 : Multicode System Model with *K* parallel Traffic Channels

Considering the baseband signal representation, let $s_k(t)$ be the transmitted signal for the kth traffic channel and

$$s_k(t) = \sqrt{P} b_k(t) a_k(t) \tag{7}$$

where P is the signal power of the kth traffic channel. Correspondingly,

$$b_k(t) = \sum_{n=-\infty}^{\infty} b_n^{(k)} p_t(nT, (n+1)T) \tag{8}$$

is the binary data signal, and

$$a_k(t) = \sum_{n=-\infty}^{\infty} a_n^{(k)} p_t(nT_c, (n+1)T_c) \tag{9}$$

is the concatenated sequence signal with $p_t(t_1,t_2)$ being a unit rectangular pulse on $[t_1,t_2)$. $b_n^{(k)} \in \{-1,1\}$ where $\Pr(b_n^{(k)} = -1) = \Pr(b_n^{(k)} = 1) = 1/2$, and the kth stream's code sequence $a_n^{(k)} \in \{-1,1\}$ with $a_n^{(k)} = a_{n+N}^{(k)}$ for all n and k and for some integer N. The integer N is the minimum period of the spreading sequence. The chip period T_c is given by $T_c = T / N = T_b$. Therefore, one concatenated sequence per data symbol is $a_k = \left(a_0^{(k)}, a_1^{(k)}, \ldots, a_{N-1}^{(k)}\right)$.

The received signal can be written as

$$r(t) = \sum_{k=1}^{K} \sum_{m=1}^{L} \sqrt{P}\beta_m e^{i\phi_m} b_k(t-\tau_m) a_k(t-\tau_m) + n(t) \tag{10}$$

where $n(t)$ is the channel noise which is assumed to be a white Gaussian noise process with double sided power spectral density $N_0/2$. In this study we have restricted our consideration to the high signal-to-noise ratio (SNR) or the irreducible BER performance. In other words, our main interest is the "irreducible' bit errors. With increasing SNR, an irreducible "floor" of BER is approached because intersymbol interference (ISI) and interchannel interference (ICI) increase in proportion to the signal level. Therefore, the effects of $n(t)$ can be ignored.

After having completed acquisitions of the signal, K Rake receivers [11] coherently demodulates the signal. The Rake receiver correlating nth bit data with L paths of the kth traffic channel is shown in Figure 2.

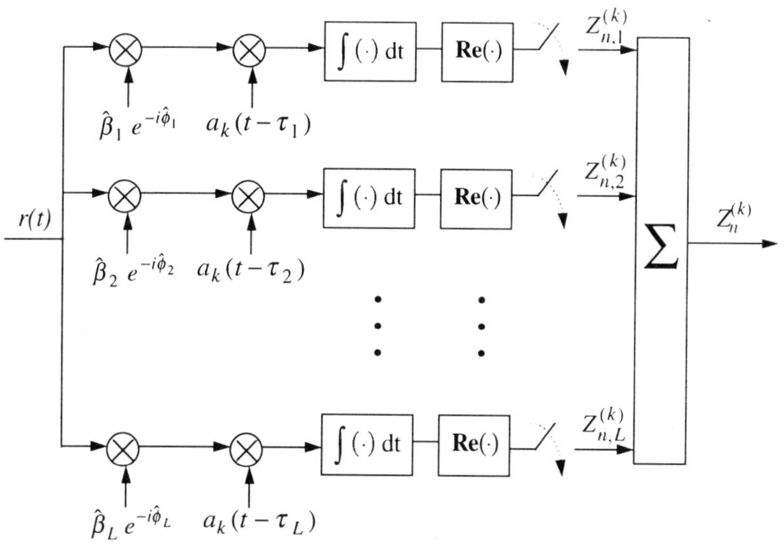

Figure 2 : Rake Receiver correlating the nth bit data of the kth Traffic Channel

The nth bit output of the kth traffic channel for the nth bit can be written as

$$Z_n^{(k)} = \sum_{m=1}^{L} Z_{n,m}^{(k)} = \sum_{m=1}^{L} \mathbf{Re}\left(\int_{(n-1)T+\tau_m}^{nT+\tau_m} r(t) a_k(t-\tau_m) \hat{\beta}_m e^{-i\hat{\phi}_m} dt \right) \tag{11}$$

where $\hat{\beta}_m$ and $\hat{\phi}_m$ represent the amplitude and phase estimates of β_m and ϕ_m respectively and **Re**(x) denotes taking the real part of x. $Z_{n,m}^{(k)}$ is the correlator output aligned with the signal arriving on the mth path and it is equal to

$$Z_{n,m}^{(k)} = \int_{(n-1)T+\tau_m}^{nT+\tau_m} \sum_{k'=1}^{K} \sum_{m'=1}^{L} \sqrt{P}\beta_{m'}\hat{\beta}_m \cos(\phi_{m'} - \hat{\phi}_m) b_{k'}(t-\tau_{m'}) a_{k'}(t-\tau_{m'}) a_k(t-\tau_m) dt$$

$$= T\sqrt{P}\beta_m\hat{\beta}_m \cos(\phi_m - \hat{\phi}_m) b_n^{(k)} + T\sqrt{P} \cdot \sum_{\substack{m'=1 \\ m'\neq m}}^{L} \beta_{m'}\hat{\beta}_m \cos(\phi_{m'} - \hat{\phi}_m) X_{n,m'}^{(k,k)}$$

$$+ T\sqrt{P} \cdot \sum_{\substack{k'=1 \\ k'\neq k}}^{K} \sum_{m'=1}^{L} \beta_{m'}\hat{\beta}_m \cos(\phi_{m'} - \hat{\phi}_m) X_{n,m'}^{(k,k')}$$

$$= D_{n,m}^{(k)} + F_{n,m}^{(k)} + I_{n,m}^{(k)} \tag{12}$$

where $D_{n,m}^{(k)}$ represents the correlator's output of the kth traffic channel of the mth path signal, $F_{n,m}^{(k)}$ represents the ISI terms and $I_{n,m}^{(k)}$ is the ICI term. Moreover,

$$X_{n,m'}^{(k,k')} = \frac{1}{T}\left[b_{n-1+j}^{(k)} R^{(k,k')}(\Delta + jT) + b_{n+j}^{(k)} \hat{R}^{(k,k')}(\Delta + jT)\right] \tag{13}$$

for k, k' = 1, 2, ..., K and m = 1, 2, ..., L. $\Delta = \tau_{m'} - \tau_m$ and j = 0 or 1 depending on whether $\Delta \geq 0$ or $\Delta < 0$, respectively. Likewise,

$$R^{(k,k')}(\tau) = \int_0^{\tau} a_{k'}(t-\tau) a_k(t) dt \tag{14a}$$

$$\hat{R}^{(k,k')}(\tau) = \int_{\tau}^{T} a_{k'}(t-\tau) a_k(t) dt \tag{14b}$$

for $0 \leq \tau \leq T$ and k, k' = 1, 2, ..., K. Therefore, the output of the kth traffic channel's Rake receiver is

$$Z_n^{(k)} = \sum_{m=1}^{L} \left(D_{n,m}^{(k)} + F_{n,m}^{(k)} + I_{n,m}^{(k)}\right)$$

$$= \sum_{m=1}^{L} D_{n,m}^{(k)} + \sum_{m=1}^{L} F_{n,m}^{(k)} + \sum_{m=1}^{L} I_{n,m}^{(k)} \tag{15}$$

$$= D_n^{(k)} + F_n^{(k)} + I_n^{(k)}$$

where $D_n^{(k)}$ represents the Rake receiver's output of the kth traffic channel, $F_n^{(k)}$ represents the ISI terms and $I_n^{(k)}$ is the ICI term. The initial estimate of $b_n^{(k)}$, nth bit data of kth traffic channel, is

$$\hat{b}_n^{(k)}(0) = \mathrm{sgn}\left[Z_n^{(k)}\right]. \tag{16}$$

2.3 Model of Interference Cancellation

The basic idea of the interference canceller [7] is to create replicas of the interference terms in each Rake receiver output based on the data estimates and wireless channel-parameter estimates. The structure of the interference cancellers is shown in Figure 3.

$\hat{F}_n^{(k)}(v)$ and $\hat{I}_n^{(k)}(v)$ represent the estimated ISI terms and ICI terms of the nth data bit of the kth traffic channel at the vth stage interference canceller respectively. They can be expressed as

$$\hat{F}_n^{(k)}(v) = \sum_{m=1}^{L} \hat{F}_{n,m}^{(k)}(v) = \sum_{m=1}^{L} T\sqrt{P} \sum_{\substack{m'=1 \\ m' \neq m}}^{L} \beta_{m'} \hat{\beta}_m \cos(\phi_{m'} - \hat{\phi}_m) \hat{X}_{n,m'}^{(k,k)}(v) \tag{17}$$

and

$$\hat{I}_n^{(k)}(v) = \sum_{m=1}^{L} \hat{I}_{n,m}^{(k)}(v) = \sum_{m=1}^{L} T\sqrt{P} \sum_{\substack{k'=1 \\ k' \neq k}}^{K} \sum_{m'=1}^{L} \beta_{m'} \hat{\beta}_m \cos(\phi_{m'} - \hat{\phi}_m) \hat{X}_{n,m'}^{(k,k')}(v) \tag{18}$$

where

$$\hat{X}_{n,m'}^{(k,k')}(v) = \frac{1}{T}\left[\hat{b}_{n-1+j}^{(k)}(v-1) \cdot R^{(k,k')}(\Delta + jT) + \hat{b}_{n+j}^{(k)}(v-1) \cdot \hat{R}^{(k,k')}(\Delta + jT)\right]. \tag{19}$$

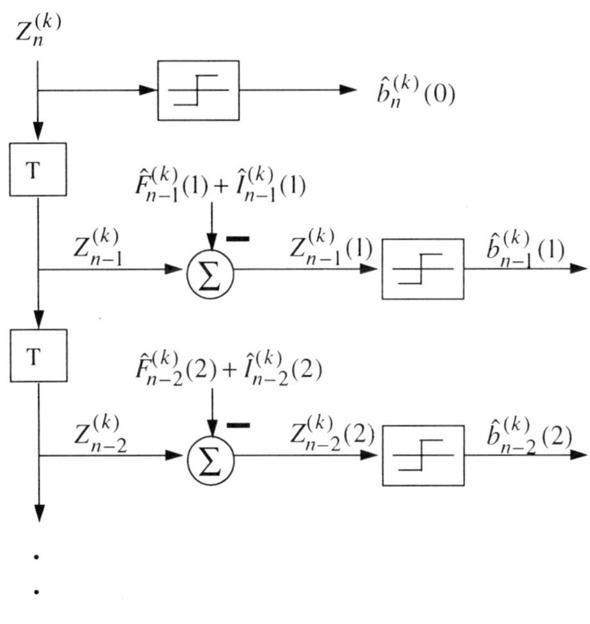

Figure 3 : Cascade of Interference Cancellers for the kth Traffic Channel

Since the Rake receivers are capable of estimating channel parameters, i.e., amplitudes, phases and delays, the receiver only needs to compute the cross-correlation functions, ISI and ICI terms to perform interference cancellation. Moreover, only ***one set*** of channel parameters is used for ***all*** channels. Therefore, the complexity is much lower than that required for a typical CDMA system [7].

2.4 Idealized and Practical Cases

Similar to the Qualcomm CDMA System [10], the multicode system needs one traffic channel to act as a Pilot Channel which is used to estimate channel parameters. In the following, we consider two cases in the computer simulations:

1. Idealized Case

Except for the first channel, all the other K - 1 channels are used as Traffic Channels and we assume that the channel parameters: amplitudes, time delays, and phases, are estimated without error.

2. Practical Case

The first channel is used as the Pilot Channel for the receiver to estimate channel parameters, and all the other K - 1 are Traffic Channels and they are used for data transmissions. The mutual interference between pilot and traffic channels is considered in the simulations.

In order to estimate the channel parameters accurately, the pilot signal level should be higher than the other K - 1 channels. The *power ratio* of the pilot channel to that of a traffic channel is defined as

$$PR = 10 \cdot \log_{10}\left(\frac{P_1}{P}\right) \quad \text{(in dB)} \tag{20}$$

where P_l is the power of the pilot channel. As PR increases, better channel estimates are obtained; however, higher ICI will be introduced to the traffic channels.

The model of the Pilot Channel receiver is shown in Figure 4. **ABS**(x) and **PH**(x) denotes taking the amplitude and phase of the complex number x. $\rho_{n,m}$, which is a complex number, represents the estimated value of the mth path of the channel impulse response when nth bit data is received. During the estimation of channel parameters, the amplitude and phase estimate of a particular path will be affected by ISI and ICI. Hence, a function called *Averaging Window* **AW**$(\cdot)$ is used to find out the average value of $\rho_{n,m}$, i.e.,

$$\overline{\rho}_{n,m} = \mathbf{AW}(\rho_{n,m}) = \frac{\sum_{j=0}^{W_s-1} \rho_{n-j,m}}{W_s} \tag{21}$$

where W_s is called *Averaging Window Size*. This process can reduce the effects of interference during a particular symbol period. If the system has higher averaging window size, the channel parameters can be more accurately estimated, provided that the channel is quasi-static.

In this paper, we assume parameters τ_1, τ_2, ..., τ_L are accurately estimated by the correlators in the Rake receiver.

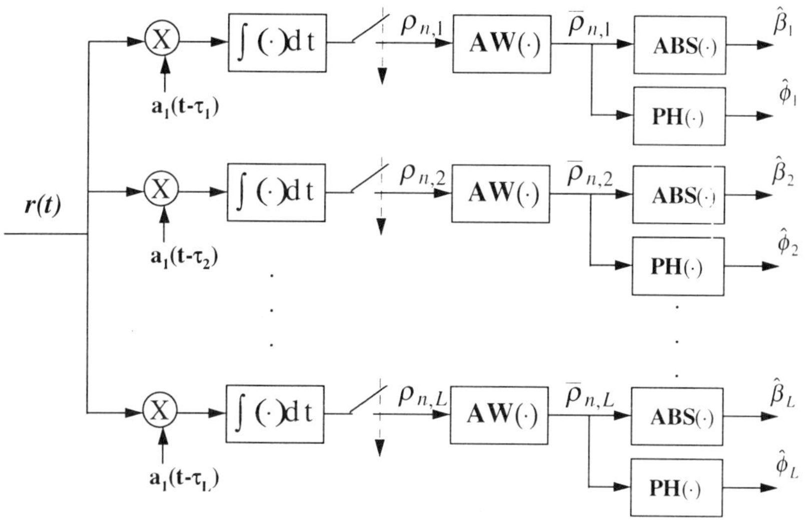

Figure 4 : Using Pilot Channel to estimate Channel Parameters

3 NUMERICAL RESULTS

In this section, some numerical results are presented. Throughout, $(K\text{-}1) = 63$ traffic channels are considered. $K = 64$ Walsh sequences are multiplied or concatenated with 64 bits of a m-sequence, which is generated by appending one extra bit "0" to a $2^6 - 1 = 63$-bit PN sequence. For example, if the bit rate of the multicode system is $1/T_b = 25$Mbps. Then the transmission rate of each traffic channel is $1/T = 25/(K\text{-}1) = 397$ksps. For a typical rms delay spread value $\tau = 200$nsec which is approximately the upper bound of the delay spread in the personal communications environment at 1.5GHz [12], the normalized delay spread will be $d = \tau/T_b = 5$. Clearly, it is impossible to send such a high bit rate with a single traffic channel (K=1) [2] unless one uses an alternative modulation scheme such as the one proposed here.

3.1 Idealized Case

Figure 5 shows a comparison of the average irreducible BER of different interference cancellation stages employed as a function of resolution r, which is defined as

$$r = \frac{2\tau}{T_c} \tag{22}$$

Resolution r represents the separation of two paths with respect to chip duration. Clearly, for $r \geq 1$, the irreducible BERs reach the ceiling when r takes on integer values, i.e. $r = 1, 2, 3, \ldots$ On the other hand, the irreducible BERs drop to the floor when $r = 1.5, 2.5, 3.5, \ldots$ This is caused by the non-zero cross-correlation functions of the concatenated sequences which reach the maximum and minimum values when the separation between two sequences is integer and integer-and-half multiple of chip duration, respectively. Furthermore, the irreducible BER is almost the same when r takes on integer values; while the same situation occurs when r takes on integer-and-half values. We thus define the irreducible BER at r taking on integer and integer-and-half values of r to be *Upper Bound* and *Lower Bound* respectively.

It is also clear that the interference cancellation scheme can improve the multicode system with much lower irreducible BER for $r \geq 1$. Conversely, the interference cancellation scheme cannot improve the system performance for $r < 1$ because of the high cross-correlation function values when the separation of two path is less than one T_C.. In practice, a Rake receiver is also unable to separate two paths for interference cancellation. Note that the Lower Bounds for 2- and 3-stage interference cancellation schemes are not shown because the results are much less than 10^{-5}. Only the asymptotic irreducible BER for $r \geq 1$ will be discussed in the remaining part of this paper.

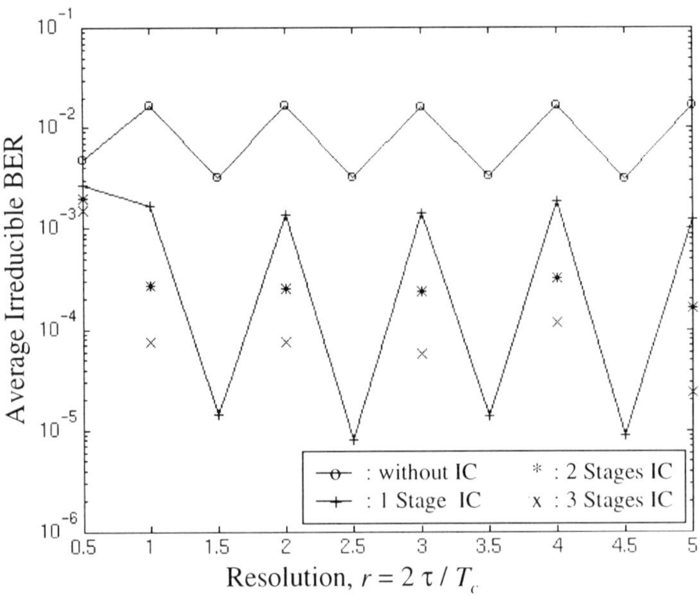

Figure 5 : Comparison of the average irreducible BER performance of multicode system as a function of resolution for different number of Interference Cancellation (IC) Stages for the idealized Case

Figure 6 shows the upper and lower bound of irreducible BER performance for different numbers of interference cancellation stages employed. It indicates that the irreducible BER can be reduced with increasing of interference cancellation stages. The upper bound irreducible BER can be significantly decreased from 2×10^{-2} to 7×10^{-5} if 3 stages of interference cancellation are applied.

3.2 Practical Case

Figure 7 illustrates the use of averaging window which can reduce the irreducible BER with increasing window size. Clearly, the improvement diminishes for $W_S \geq 5$. It is also clear that the irreducible BER is higher than that of the idealized case due to the ICI between pilot and traffic channels.

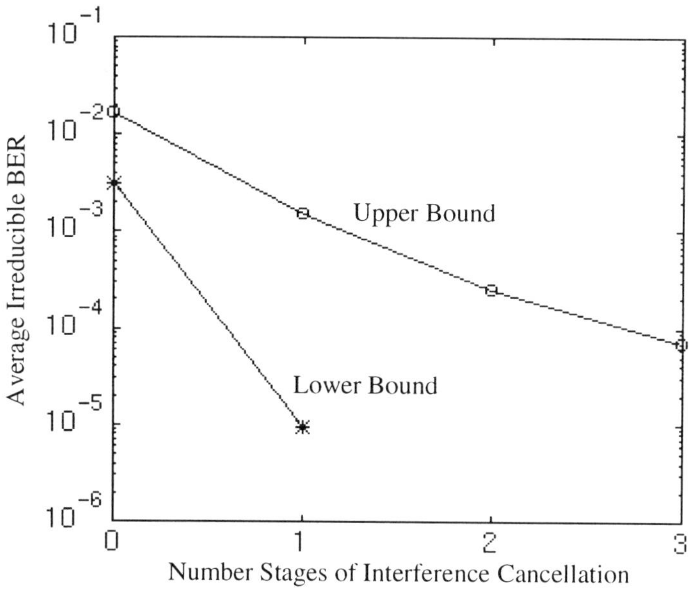

Figure 6 : The irreducible BER performance as a function of the number of Interference Cancellation Stages in the Idealized Case

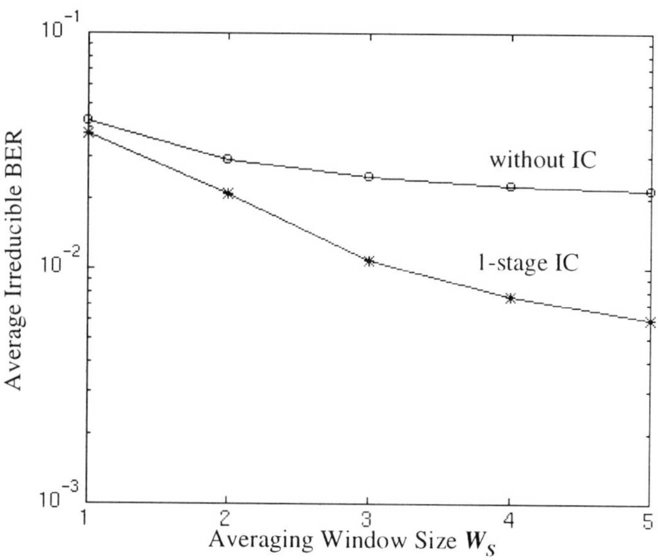

Figure 7 : The Upper Bound irreducible BER performance as a function of Averaging Window Size W_S for the Practical Case (PR = 6.0 dB)

Figure 8 shows the upper bound irreducible BER performance as a function of power ratio. We can see that the system without interference cancellation performs better when PR is between 5 to 10dB but the irreducible BER increases for $PR > 10$dB. The reason is that the ICI from pilot channel increases if the pilot channel has higher power. Nevertheless, the systems having higher *PR* perform better if more interference cancellation stages are employed. The reason is that if the pilot channel has greater signal level, it can estimate the channel parameters more accurately. It has been proved [7] that accurate estimation of channel parameters is required for the interference cancellation scheme. However, increasing the power ratio also means reducing the available power for the traffic channels. This will further reduce the signal to interference-plus-noise ratio, $S/(I+N)$, for each traffic channel. Therefore, a suitable *PR* value should be chosen such that the system can be operated efficiently.

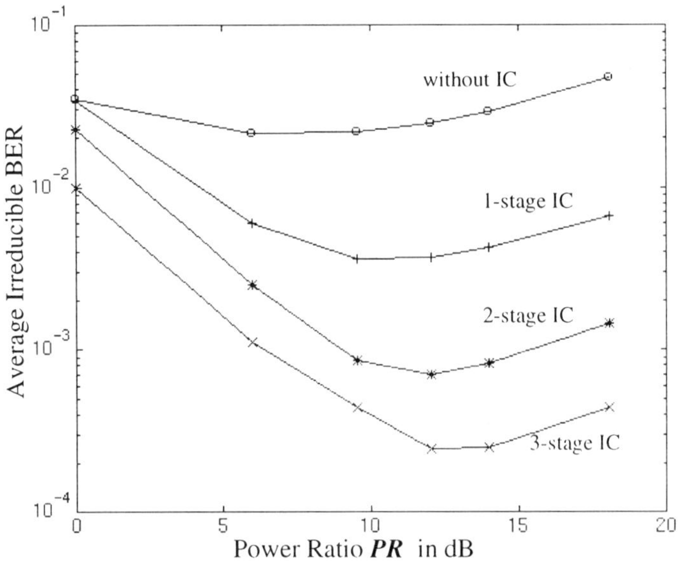

Figure 8 : The Upper Bound irreducible BER performance as a function of Power Ratio ***PR*** for the Practical Case ($W_S = 5$)

A comparison of the idealized and practical cases is shown in Figure 9. For the practical case, two systems are illustrated, one using $PR = 6.0$dB and the other using 9.5dB. The same averaging window size $W_S = 5$ is used. Performance under different *PR* values is similar for cases without interference cancellation while the system using higher *PR* performs better if interference cancellation is applied. Moreover, performance difference increases with increasing number of interference cancellation stages due to a higher sensitivity to the accuracy of channel parameter estimation. Considering cases with $PR = 9.5$dB, the upper bound irreducible BER is reduced from 3×10^{-2} to 4×10^{-4} when 3-stage interference cancellation is used while the lower bound irreducible BER becomes practically negligible after two stages of interference cancellation.

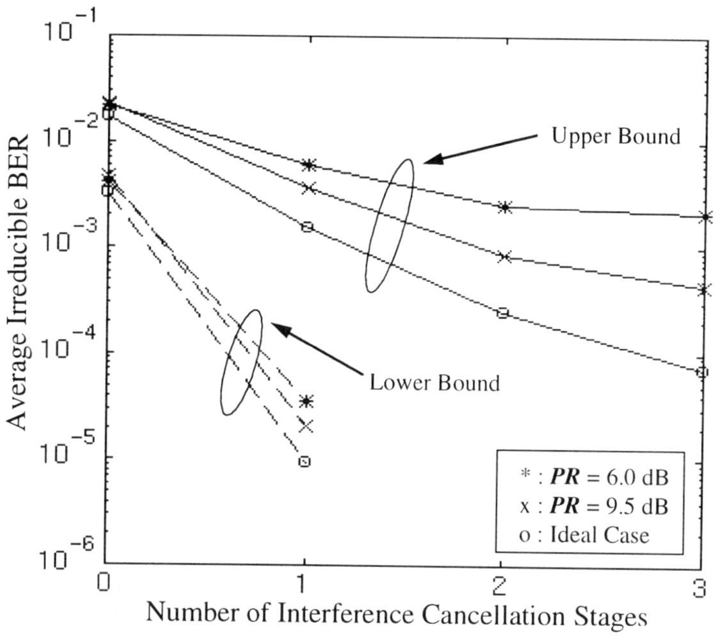

Figure 9 : Comparison of the average irreducible BER performance as a function of Interference Cancellation Stages for Idealized and Practical Cases ($W_S = 5$)

4 CONCLUSIONS

We have studied a multicode system using interference cancellation techniques to improve performance in non-line-of-sight frequency-selective fading channels. Possible applications include wideband wireless access such as Eithernet connections where 10 Mb/s or higher rates and full mobility are desirable. Implementation aspects of this technique have also been addressed. The high-rate data bits are serial-to-parallel converted into low-rate data streams which are then modulated with Walsh and PN concatenated sequences. An unmodulated pilot channel is employed to estimate relative delays, amplitudes and phases of all radio paths, which are then used to compute interference components for successive cancellation. Numerical results illustrate that significant performance improvement is gained by interference cancellation and transmission rates at several times the coherence bandwidth is achievable. Since only a ***single*** pilot signal is required for ***all*** traffic channels, the system overhead is very low. These preliminary results indicate that multicode modulation employing interference cancellation appears to be a good candidate for high-rate air interface for future wireless multimedia communication systems.

ACKNOWLEDGMENT

The authors would like to thank K. Ben Letaief, Curtis C. Ling and Roger Cheng for their useful and valuable suggestions and comments.

REFERENCES

[1] D. Cox, "Wireless Network Access for Personal Communications," Special Issue on Personal Communications, *IEEE Comm. Mag.*, December 1992.

[2] J. C-I Chuang, "The Effects of Time Delay Spread on Portable Radio Communications Channels with Digital Modulation," *IEEE Journ. on Selected Areas in Communications*, vol. SAC-5, No. 5, pp. 879-889, June 1987.

[3] R. W. Chang and R. A. Gibbey, "A Theoretical Study of Performance of an Orthogonal Multiplexing Data Transmission Scheme," *IEEE Trans. Commun. Technol.*, vol. COM-16, pp. 529-540, Aug. 1968.

[4] L. J. Cimini, Jr., "Analysis and Simulation of a Digital Mobile-channel using Orthogonal Frequency Division Multiplexing," *IEEE Trans. Commun.*, vol. COM-33, No. 7, pp. 665-675, July 1985.

[5] K. Ben Letaief, J. C-I Chuang, and R. D. Murch, "Multicode High-Speed Transmission for Wireless Personal Communications," *IEEE 1995 Global Telecommunications Conference*, Singapore, November 1995.

[6] H. Zhang, D. Rutkowski, "Orthogonal Sequence Division Modulation -- A Novel Method for Future Broadband Radio Services," *1995 IEEE 45th Vehicular Technology Conference*, pp. 810-814.

[7] Y. C. Yoon, R. Kohno and H. Imai, "A Spread-Spectrum Multi-access System with Cochannel Interference Cancellation for Multipath Fading Channels," *IEEE Journ. Selected Areas in Commun.*, vol. SAC-11, No. 7, Spet. 1993.

[8] J. G. Proakis, *Digital Communications*. New York: McGraw-Hill, 1983.

[9] P. A. Bello, "Characterization of randomly time-variant linear channels," *IEEE Trans. Commun. Syst.*, vol. CS-11, pp. 360-393, Dec. 1963.

[10] R. Steele (Ed), *Mobile Radio Communications*, IEEE Press, N.Y., 1992, pp. 70-74

[11] G. L. Turin, "The effects of multipath and fading on the performance of direct-sequence CDMA systems," *IEEE Trans. Vech. Technol.*, vol. VT-33, pp. 213-219, Aug. 1984.

[12] A. A. M. Saleh and R. A. Valenzuela, "A Statistical Model for Indoor Multipath Propagation," *IEEE Journ. on Selected Areas of Commun.*, vol. SAC-5, No. 2, Feb. 1987.

PACKET ACCESS FOR CELLULAR SYSTEMS: THE CAPACITY OF S-ALOHA AND CDPA WITH SLOW SHADOWING

Michele Zorzi*†, Flaminio Borgonovo* and Luigi Fratta*

**DEI, Politecnico di Milano, Piazza Leonardo da Vinci 32, 20133 Milano, ITALY*
fax: +39 2 2399 3413; e-mail: `last_name@elet.polimi.it`
†Center for Wireless Communications, UC San Diego, La Jolla, CA 92093-0407
fax: +1 619 534 2486; e-mail: `zorzi@ece.ucsd.edu`

ABSTRACT

The paper discusses the features and compares the performance of two packet oriented channel-access schemes that appear particularly suitable for the PCN cellular environment. These schemes exploit the packet capture and can deal with inter-cell interference without any a priori bandwidth-consuming techniques such as those used in FDMA and CDMA. The first scheme is the well known S-ALOHA, whereas the second one is the Capture Division Packetized Access, recently proposed. The analytical model accounts for several features of real-world systems, e.g., the exact traffic pattern that must be offered in order to achieve throughput fairness, the exact interference caused by the traffic in the same cell and in other cells, the fading, the slow shadowing attenuation and some means to counteract it, such as best base station selection and power control. We also discuss the effects of outage, i.e., a mechanism which prevents a user from attempting more than a predetermined number of transmissions of the same packet.

1 INTRODUCTION

Personal Communications Networks (PCNs) are expected to provide extensive and integrated access to the wired network, a formidable task given the limited available bandwidth and the lack of flexibility of existing wireless systems. To avoid the capacity bottleneck at the wireless interface, access must be obtained by adopting an architecture based on microcells or even picocells, which, in turn, raises several problems. Microcellular systems must be easy to deploy, flexible enough to adapt to traffic changes, and able to support frequent and seamless handovers and perhaps the site diversity capability intrinsic in the wireless access. Existing proposals meet only in part these requirements. CDMA appears to be flexible enough since its inherent robustness with respect to cochannel interference avoids the need of cell and frequency planning. On the other hand, TDMA access is more flexible in integrating different kinds of traffic since slots can be assigned to meet bandwidth requirements. In all cases, the access technology proposed is inherently circuit oriented, i.e., it is required to provide to the requesting user an access channel of suitable (possibly flexible) bandwidth with adequately low bit error rate for an indefinite amount of time. Since

*This work was partially supported by MURST 40% 1995, Italy.

†The work of M. Zorzi was supported by the Center for Wireless Communications, UC San Diego.

the wireless environment is highly hostile, the task of obtaining such channels is a very hard one indeed, and leads to complicate and somewhat inefficient mechanisms.

Two crucial issues to be addressed in the design of PCNs, and of cellular access systems in general, are the channel reuse technique and the multiple access technique. The *channel reuse*, or cell multiplexing, is the technique by which the same channel can be used in different cells, provided that these cells are sufficiently apart [1, 2], thus achieving transmission parallelism. Transmission parallelism occurs when the transmission in one cell can be successfully detected because the power of transmissions in other cells, called interference, is sufficiently small. The ability of the receiver to detect a signal in the presence of interference noise is known as *capture effect* [1, 3], and the signal-to-noise threshold needed to allow capture is denoted as *capture ratio*. With the usual values of the capture ratio for narrowband systems, concurrent transmissions in adjacent cells can not be tolerated. The solution to this problem can be obtained either by decreasing the interference or by lowering the capture ratio. The first solution is adopted in narrowband systems, e.g., GSM [4], where the system bandwidth is divided into a number of channels, and the same channel is reused according to a periodic regular pattern chosen in such a way that the distance between co-channel cells, i.e., cells using the same channel, is sufficiently large. The second solution, i.e., lowering the threshold, has been adopted in systems such as CDMA, where a coding (Spread-Spectrum in this case) is employed in order to allow capture even with a small Signal-to-Interference Ratio (SIR) at the receiver [5, 6]. The price to be paid is a bandwidth subdivision in the first case, and the need for a bandwidth spreading in the second.

The second factor that affects the system efficiency is the *multiple-access technique* used within a cell to share the assigned channel. The systems proposed today, which are based on circuit-oriented access, use either FDMA and TDMA, as in GSM, or code division, as in CDMA, where spread spectrum is used to achieve both channel reuse and multiple access.

In a cellular system the bandwidth collectively required in a cell is rapidly changing in time, not only because of the bursty nature of the integrated traffic, but also because of the limited number of users that are expected to be served in a cell. Furthermore, the cochannel interference is a highly variable process since propagation conditions are also affected by phenomena rapidly changing in time. This strongly suggests that circuit-oriented reuse techniques, that must be designed to face the worst interference condition, can be very inefficient.

In principle, a flexible TDMA technique could be envisaged that assigns slots to multiple cells in such a way that transmissions within the same time slot do not disturb each other and are always captured by the proper receiver. In practice, perfect coordination among different cells is impossible to achieve because instantaneous requirements and transmitter-to-receiver propagation conditions can not be made available instantaneously at all transmitters involved.

A classic solution to the lack of coordination among transmitters has been the ALOHA technique, which fights transmission conflicts by retransmitting collided packets. In the wireless environment this technique is particularly attractive because it meets all the requirements we have cited above. In particular, it does not require any planning of transmission resources, allows a flexible growth of the cell number according to needs, and, being a packet technique, it offers integration flexibility to all kinds of traffic. The inefficiency of ALOHA is mitigated by the capture effect and by the fact that the available bandwidth is not reduced by "a priori" interference-fighting techniques such as frequency splitting or spread spectrum [7].

Because of retransmissions and the impossibility of guaranteeing a tight control over the transmission delay, the ALOHA technique is usually considered unsuited to real time traffic. Furthermore, other issues, such as stability and fairness, prevent the ALOHA technique, in its simplest form, to be successfully used.

The above observations, together with the fact that a cellular system is inherently a centralized system where the BS is destined to perform a plethora of centralized functions, have lead to the proposal of a centralized access scheme, called Capture-Division Packetized Access (CDPA) [8, 9, 10, 11], that takes advantage of the ALOHA with capture characteristics in cochannel access coordination, while providing delay controlled and collision free access among users of the same cell. To exploit the multipacket nature of most kind of traffic, CDPA uses a hybrid reservation-polling (HRP) mechanism that lets mobiles transmit upon a command issued by the BS, as in polling systems. Unlike polling systems, however, only active stations, i.e., stations that have packets to transmit, are polled. An integrated random access mechanism is used to make "reservations". To this purpose, some slots are declared for access-requests transmission, i.e., the transmissions of packets that carry to the BS the mobile service requirements. Based on this information, that includes polling rate, maximum delay, etc., the BS, slot by slot, optimizes the packet transmissions from mobiles.

The master/slave HRP mechanism of CDPA has been chosen because it is much simpler and more flexible than a reservation mechanism, such as PRMA [12], that grants access slots to MTs by reserving a fixed slot in a constant-length frame. Since the BS expects immediate response to the command, the CDPA HRP scheme, if coupled with a smart scheduling protocol, e.g., the C-PRMA protocol [13], allows a prompt retransmission of erroneous packets, so that the total packet delay can be kept within the requested values. This feature makes CDPA an attractive system even for delay-sensitive traffic, such as voice. Moreover, it eliminates contention in the intracell traffic and the corresponding loss in throughput, thus increasing the overall efficiency with respect to the ALOHA protocol.

The basic idea of the CDPA architecture, together with a mathematical model to evaluate the throughput of the uplink and downlink channels, was presented in [8, 9]. In [14] a simulation study of an 81-cell system using CDPA shows that voice traffic can be transmitted, within a maximum access delay of 32 ms and without loss, up to a value very close to the channel capacity. A further throughput analysis of CDPA, including partial frequency reuse and power control methods, is given in [10, 11].

In this paper we compare the performance of S-ALOHA and CDPA using a new analytical approach that accounts for a complex cellular environment which comprises lognormal shadowing and some ways to counteract it, namely best Base Station assignment and power control. Lognormal shadowing constitutes a severe impairment for cellular systems since its additional attenuation, accounting for obstacles that obscure the line of sight to the receiving antenna, can be as large as tens of dB. However, as in a cellular system more than one BS are usually candidate to support mobile communications, means can be easily provided to use the most suitable BS, e.g., the one presenting the minimum long term attenuation.

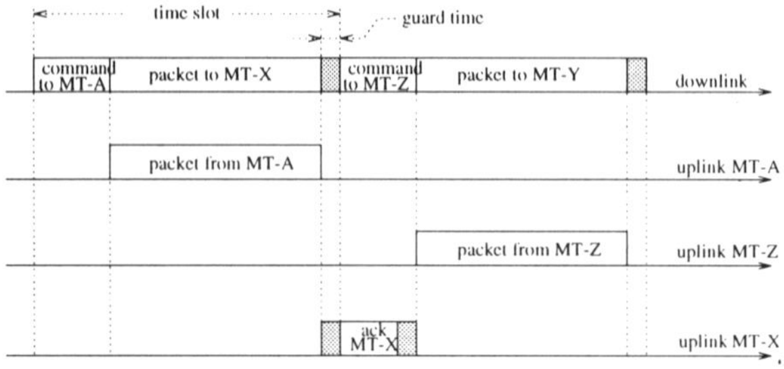

Figure 1 Uplink and downlink-channel time relationship.

Another new issue investigated in this paper is "outage". Outage is well known in circuit-oriented wireless systems and occurs when a connection experiences so severe propagation conditions that it can not be sustained. In the packet systems here considered, where fairness imposes that disadvantaged users retransmit more often, outage must be declared when the number of unsuccessful transmission attempts reaches a predefined value $N_{\max}$. Otherwise, users with strongly adverse propagation conditions could seize most or even all of the transmission resources, causing the throughput to collapse.

The paper is organized as follows. In Section 2 we shortly recall the basic features of CDPA and discuss the main differences with S-ALOHA. In Section 3 we present the model of the cellular environment we used for the analysis of both S-ALOHA and CDPA. The performance analysis is presented in Section 4, and in Section 5 numerical results are discussed. Conclusions are given in Section 6.

2 CDPA BASICS

CDPA is a novel access architecture, especially devised for cellular systems [10, 11]. While ALOHA and its derivations are usually indicated for single packet traffic, CDPA is conceived for multipacket sources. These sources, when ready to transmit packets, signal their need to the BS via a signaling channel, e.g., using slots explicitly reserved for this purpose, by means of some random access protocol, e.g., ALOHA or stack algorithms (see, for example, [15]). The BS coordinates the transmissions of reserved MTs within its cell, whereas no attempt is made to coordinate the transmissions of MTs belonging to different cells [10, 11]. In other words, perfect coordination is achieved among transmissions within a cell, whereas an ALOHA-like mechanism regulates conflicts among packets transmitted in different cells.

In CDPA the uplink transmission is logically driven by the BS, which at each slot releases, on the downlink channel, a *command* that specifies the use of that slot. This technique can be implemented in different ways. One is exemplified in Figure 1, where all signals are observed by three MTs, assumed at the same distance from the BS. Commands are used like in polling systems and trigger the immediate response of the polled station (processing times are ignored). An unsuccessful MT transmission is promptly recognized by the BS, which reschedules a new attempt by issuing the appropriate command in a later slot. (More details, omitted here due to space constraints, are given in [10].)

Note that, since the whole system may operate under strong co-channel interference, also the commands may be incorrectly received if no means to protect them from the co-channel interference is assumed [10, 11]. However, to be consistent with the S-ALOHA model, in which acknowledgments are assumed error-free, here too we assume that commands are always correctly received. Commands, in fact, carry out the same function as acknowledgments in S-ALOHA: they confirm the correct reception of previously transmitted packets and solicit the transmission of the next one. The only difference is that a command specifies the address of the next transmission instead of the address of the previous one.

The basic CDPA mechanism provides considerably more throughput than S-ALOHA, as it will be shown in the following sections, and in addition it presents some operational advantages that are shortly discussed next.

Stability. Because of collisions that are possible in both systems, S-ALOHA and CDPA are prone to congestion and therefore need some control to operate in the best conditions. In practice, the control procedures are very different. S-ALOHA is a completely distributed system, and a control scheme coherent to its philosophy would call for distributed procedures able to estimate the channel traffic or the backlog [15].

In CDPA the traffic control procedure is easily implemented because of its centralized architecture. In fact, the channel traffic is completely under control of the BS, which can easily measure the channel capture performance and manage the traffic accordingly. As an example, a probabilistic throttle which allows the command release with probability β, being β the load at which maximum throughput is achieved, has been successfully used in the CDPA simulation system described in [14].

Fairness. The S-ALOHA scheme is inherently unfair if the transmission probability is not appropriately related to the success probability experienced by the MT (as discussed in Section 4). Power control could be a solution since all users would experience the same success probability. However, power control is hard to implement if the MT has a very low activity because measurements become difficult.

In CDPA, fairness and any priority scheme can be easily enforced by a suitable scheduling mechanism (e.g., the one proposed in [13]). Power control enhances the performance but is not necessary. This fact helps in keeping the system simple.

Signaling. In CDPA, where the command/response mechanism must be explicitly established by signaling before the channel itself can be used, an additional signaling channel is needed. This can be integrated by having the BS to declare "signaling slots" to be accessed according to some random multiple access techniques, possibly the S-ALOHA technique itself [9, 10].

The signaling channel constitutes an overhead whose impact depends on the traffic characteristics. It can be negligible if transmission sessions are composed of long bursts of packets, as happens when considering circuit switching service or voice service, even with silence suppression. On the other side, the S-ALOHA slots used for signaling could also be used, still with S-ALOHA, for single packet service.

Transmission delay. When dealing with real-time traffic, throughput performance must refer to the maximum throughput allowed under a given delay constraint. This requirement in practice prevents S-ALOHA from providing any throughput at all.

On the contrary, in CDPA a proper scheduling algorithm can provide prompt retransmission of packets which are not correctly received and then guarantee with high probability the correct transmission within a given delay [13].

The suitability of CDPA for delay constrained traffic has been discussed in [14], where the results of a detailed simulation of CDPA in a multi-cellular system with users that generate constant rate traffic have been analyzed. It has been shown that, with continuous sources, the system can operate close to the maximum throughput with no loss and no packets exceeding a transmission delay greater than twice their generation period.

3 PROTOCOL AND ENVIRONMENT ASSUMPTIONS

In the cellular environment taken into account, BSs are evenly spaced on the plane, at the center of ideal hexagonal cells, and operate with omni-directional antennas. This choice is made in order to have a reference model which is widely accepted in the literature. However, the following analysis is by no means critically dependent on this assumption, and can be modified to incorporate other features such as, for example, lineal cells or different topologies.

Each MT is assumed to communicate with a base station according to the protocol under consideration, either S-ALOHA or CDPA.

In this paper we focus only on the throughput of uplink data channel, i.e., the channel used for data transmissions from MT to BS. We ignore the overhead that both protocols need to operate in real systems, such as the bandwidth used by the reservation channel in CDPA, mainly because this overhead largely depends on the particular implementation and traffic characteristics. In this paper, we focus on an analytical approach to the performance evaluation, common to both protocols, rather than on a detailed comparison. Thus, we assume that each packet transmitted and correctly received on this channel is acknowledged by some perfect signaling on the downlink channel.

We assume that the same frequency is used in all cells and that the time axis is subdivided into time slots, which are aligned in all cells. Packet transmission is aligned to the slots so that all transmissions in a slot, both within and outside the cell, overlap completely. This involves little overhead due to the presence of adequate guard times, as discussed in [10].

The propagation model takes into account fading due to multipath, log-normal shadowing, due to terrain irregularities and obstructions, and an η-th power loss law. The propagation loss exponent, η, typically takes values close to 4 [16]. The power, W_R, received from a transmitter located at distance r, is therefore given by

$$W_R = \alpha^2 10^{0.1\xi} K r^{-\eta} W_T, \tag{1}$$

where in the general case of Rician fading α^2 is a noncentral chi-square random variable with unit mean; in the important (and analytically tractable) case of Rayleigh fading, α^2 is exponentially (central chi-square) distributed. ξ is a Gaussian random variable with zero mean and variance σ^2, $Kr^{-\eta}$ accounts for the power-loss law, and W_T is the transmitted power, which may not be the same for all MTs if power control is used. Since ξ is the lognormal attenuation in decibels, the shadowing parameter, σ, is usually given in dB. Note that, with this notation, $\sigma = 0$ dB means that the lognormal attenuation in dB is Gaussian with zero standard deviation, i.e., it is a constant.

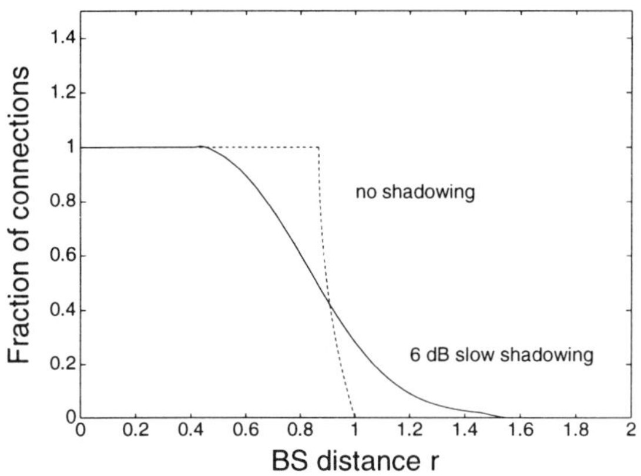

Figure 2 Fraction of users at distance r from the BS attached to the BS itself under the best BS selection and assuming $r = 1$ the radius of the hexagonal cell.

The traffic model assumes that each active MT has only one packet to transmit and retransmits the same packet until capture occurs. Alternatively, an outage mechanism could be implemented, which denies service to a user after $N_{\max}$ unsuccessful transmission attempts.

The capture model assumes that a receiver can correctly detect a packet whose received power is W_{R0} if

$$\frac{W_{R0}}{\sum_i W_{Ri}} > b \tag{2}$$

where W_{Ri} is the power received at the intended BS from the intended user (index $i = 0$) or from an interfering transmission (index $i > 0$), either from the same cell or from different cells, and b is the *capture ratio*. This capture model is rather crude, since in practice capture is influenced also by other parameters associated to the interfering signals. Nevertheless, this model is useful in attaining preliminary evaluation figures, as it leads to tractable computation complexity. Values of b between 6 and 10 dB are usually considered. Recent investigations [17] about the capture in CDPA showed that a suitable value for the capture threshold value is close to 6 dB.

As for the time-related characteristics of the introduced attenuations, fading is rapidly changing with time and we assume that the fading attenuation is drawn independently in each slot. MT's position and lognormal attenuation are slowly varying in time and we assume that they are actually constant all the time the MT is trying to successfully transmit its packet.

Active MTs are located uniformly all over the plane and select their BS according to one of the two following criteria. The first criterion selects the "closest" BS, whereas the second selects the BS that presents the lowest average attenuation ("best" BS). If no shadowing is present ($\sigma = 0$) the two procedures coincide and the "logical" cell, i.e., the area to which BS's potential users belong, coincides with the geographical cell. If shadowing is present, the BS that suffers the lowest average attenuation is not always the nearest one, i.e., the BS at the center of the geographical cell where the MT is located. This means that the probability $\delta(r, \theta)$ that a user in (r, θ), with respect to the BS, is attached to the BS itself is still a decreasing function of r, but is not negligible even at distances from the BS well beyond the geographic cell. Figure 2 shows the average $E[\delta(r, \theta)]$ with respect to θ as function of r in the case of $\sigma = 6$ dB.

The geographical cell border is between $\sqrt{3}/2$ and 1 times the cell radius, depending on the angular position. If no shadowing is assumed, a user within a geographical cell is always attached to the BS in the center of the cell and the probability $\delta(r, \theta)$ is 1 within the cell and zero outside (see Figure 2).

By using (1), Eq. (2) can be expressed as

$$\frac{\alpha_0^2 10^{0.1\xi_0} K r_0^{-\eta} W_{T0}}{\sum_i \alpha_i^2 10^{0.1\xi_i} K r_i^{-\eta} W_{Ti}} > b \tag{3}$$

The introduction of the slowly varying shadowing model, and especially of the best BS selection procedure, makes the analysis much more realistic than those available in the literature [7, 10, 11, 18, 19, 20, 21, 22], but also much harder. In fact, even if the user's location and the propagation conditions experienced are independent, as here assumed, the choice of the best BS introduces a statistical dependence between them, which is to be taken into account.

Finally, in the following analysis we assume that each MT is granted equal throughput. This renders the throughput density on the plane $s(\wp)$ uniform ($s(\wp) = s$) and the throughput of each cell S the same in all cells. Note, however, that the S-ALOHA scheme is inherently unfair if the transmission probability is not appropriately related to the distance r from the BS and possibly to the shadowing attenuation experienced.

4 ANALYSIS

In systems like those under consideration, where the success of a transmission is not granted a priori, for each packet correctly received many more might need to be transmitted, and the average offered traffic in a cell G (pkt/cell), which represents the transmission attempt intensity, can be expressed as

$$G = NS \tag{4}$$

where N represents the average number of transmission attempts per successful transmission. Alternatively, (4) can be rewritten in the more usual form

$$S = P_s G \tag{5}$$

where $P_s = 1/N$ is interpreted as average probability that capture occurs.

In general, however, the average number of transmission attempts depends on several long term random variables, such as the position and shadowing of users and interferers. Let us now represent these variables collectively by X and introduce the traffic and throughput densities $g(x)$, with $G = \int_0^\infty g(x)dx$, and $s(x)$, with $S = \int_0^\infty s(x)dx$. More specifically, $g(x)dx$ and $s(x)dx$ are the offered traffic and throughput demand per cell, respectively, of the users for which $x \le X \le x + dx$. We have

$$g(x) = n(x)s(x) \tag{6}$$

where $n(x)$ represents the conditional average number of transmission attempts per successful transmission given $X = x$. Because of the assumed throughput fairness, the throughput density given $X = x$ must be constant for all x, and

$$s(x) = S\lambda(x) \tag{7}$$

where $\lambda(x)$ is the p.d.f. of the long term power, X. In these conditions, (6) gives by integration

$$N = E[n(X)] = \int_0^\infty n(x)\lambda(x)dx = \frac{1}{S}\int_0^\infty g(x)dx, \tag{8}$$

which, given S and $\lambda(x)$, provides the required relationship between S and G. Thus, the performance is completely described by $n(x)$, which is a function of the *conditional* capture probability, $p_s(x)$. (Note that the event that a packet is captured, conditioned on x, is independent from slot to slot, since all long term variables are summarized in x.) In the absence of any outage mechanism, a user will keep transmitting until success, and $n(x) = 1/p_s(x)$. In outage is implemented, the relationship between $n(x)$ and $p_s(x)$ is changed, as discussed in subsection 4.2 below.

Turning to our environment, let us consider the sequence of slots in which a user is attempting to successfully transmit a packet. Its position and shadowing conditions remain the same by assumption during all the transmission attempts. Thus, the long term power level that the intended user causes at the intended BS (cell 0), given by

$$X = 10^{0.1\xi_{00}} r_{00}^{-\eta} W_{T00},$$

is a long term variable, and the only one relevant to our problem. In fact, during the transmission attempts the capture condition (3) can be rewritten as

$$\frac{X\alpha_{00}^2}{\sum_{j=0}^{C}\sum_{i=1}^{k_j} I_{ji}\alpha_{ji}^2} > b \tag{9}$$

where

$$I_{ji} = 10^{0.1\xi_{ji}} r_{ji}^{-\eta} W_{Tji}$$

represents the long term power level that, in any slot, the i-th transmission in the j-th cell causes at the intended BS (cell 0). In (9) k_j is the number of transmissions in cell j, and is a random variable, and C is the total number of interfering cells considered.

According to the propagation model, the α_{ji}'s in (9) are i.i.d. random variables, independent from slot to slot. Furthermore, the variables I_{ji}'s can also be considered independent from slot to slot. Since this fact can not be formally proved, we state it as an explicit assumption:

Basic Assumption: *The r.v.'s k_j and I_{ji}, $j = 0, \ldots, C, i = 1, \ldots, k_j$, are drawn independently in each slot.*

Note, however, that this assumption is quite reasonable because, even though for a given interferer, say interferer h in cell j, the long term power level at the intended BS, Y_{jh}, is itself a long term variable, the term I_{ji} in (9) corresponds to different users in different slots. Also, the cumulative interference is composed of many contributions that constitute a limited fraction of all potential transmissions since, for both ALOHA and CDPA, each user transmits in a small fraction of slots. Moreover, transmissions in different cells occur independently so that the correlation among the interferers and between the intended user and the interferers can be neglected. On the other hand, the Basic Assumption is essential in making the problem analytically tractable.

The accuracy of the Basic Assumption for the CDPA protocol can be checked with the results presented in [14], which give the throughput of the CDPA protocol obtained by simulation where each user generates a constant rate traffic and maintains a fixed position throughout the simulation run. These simulation results, obtained in the case without shadowing, are in very good agreement with the analytical ones, as discussed in [23].

From the above discussion it is clear that the performance is obtained by evaluating the conditional capture probability p_s, given by

$$p_s(x) = P\left[\frac{X\alpha_{00}^2}{\sum_{j=0}^{C}\sum_{i=1}^{k_j} I_{ji}\alpha_{ji}^2} > b \middle| X = x\right]. \tag{10}$$

The probability in (10) is averaged over the distribution of the number k_j of interferers in cell j, which is determined according to the protocol adopted. In CDPA we have $k_0 = 0$, and $k_j, j = 1, \ldots, C$, are i.i.d. binary random variables, taking the values 1 and 0 with probability G and $1 - G$, respectively. For Slotted ALOHA the $k_j, j = 0, \ldots, C$, are i.i.d. Poisson random variables with mean G. The distribution of each I_{ji} does not depend on x, and is related to the distribution of Y_{jh} in a rather involved way, because of the retransmission mechanism and the correlation among the levels received at different stations.

In the following we evaluate distribution of the r.v. I_{ji}. However, since the distributions of all the variables involved do not depend on the index denoting the user, we will only retain the index j denoting the cell to which the user is attached.

Let now denote by Z_j the long term powers received in cell j from a randomly picked user in cell j, and Y_j the corresponding level received at the intended BS. Define

$$\Lambda_j(z, y) = \frac{\partial^2 P[Z_j \leq z, Y_j \leq y]}{\partial z \partial y} \tag{11}$$

$$\lambda_j(z) = \int_0^\infty \Lambda_j(z, y) dy, \tag{12}$$

where $\lambda_j(z) = \lambda(z)$ for all j, due to the symmetry of the environment.

Let $n_j(z, y)$ be the conditional average number of transmissions per successful packet of a user in cell j given that it experiences a power level $Z_j = z$ at its own BS and a power level $Y_j = y$ at the intended BS. This number depends on the interference that overlaps the level z at BS$_j$, which in turn depends on the activity going on in the surrounding cells, other than the intended cell. Thus, it is reasonable to assume that $n_j(z, y)$ is independent of y and j, so that $n_j(z, y) = n(z)$.

Therefore, similarly to (7) and (6), we can define the densities of throughput and offered traffic in cell j as

$$s_j(z, y) = S\Lambda_j(z, y), \qquad g_j(z, y) = n(z)S\Lambda_j(z, y). \tag{13}$$

Then, the rate at which I_j, $y < I_j \leq y + dy$, is generated is

$$\gamma_j(y) = \int_0^\infty g_j(z, y)dz = S\int_0^\infty n(z)\Lambda_j(z, y)dz, \tag{14}$$

and the probability density of I_j is $\gamma_j(y)/G$.

In particular, since $\Lambda_0(z,y) = \lambda(z)\delta(z-y)$ (with $\delta(z)$ the Dirac's δ-function), we have

$$\gamma_0(y) = Sn(y)\lambda(y) = g(y). \tag{15}$$

From the statistics of the I_{ji}'s given by Equations (14) and (15), and using the appropriate relationship between $n(x)$ and $p_s(x)$, the conditional capture probability (10) can be expressed as

$$p_s(x) = f(x, S, p_s(.), \Lambda_j(.), j = 0, \ldots, C, \mathcal{P}) \tag{16}$$

where $\mathcal{P}$ represents the protocol used. Note that the functions $\Lambda_j(.), j = 0, \ldots, C$, are only determined by long term transmission and propagation variables (in particular, they are independent of the protocol), and can be determined by geometrical evaluations. (More details about their derivation, which is beyond the scope of the present paper, can be found in [24]. Also, details for the simpler case without shadowing have been presented in [23].) After that, (16) can be solved iteratively for each S, providing $p_s(x)$ and finally G.

In the systems under analysis, however, the solution does not exist for any value of S. Furthermore, when a solution exists, it does not correspond to a unique choice of G and $p_s(x)$, the solution depending on the starting values assumed in the iteration procedure [23]. Therefore, it is preferable to derive S having fixed a predetermined value of G. This can be done by replacing S in (15) and (14) by G/N at each step, where $N = \int_0^\infty \lambda(x)n(x)dx$. Our computations show that the above procedure converges to the solution for any value of G, starting from a uniform $p_s(x)$, even though, at present, we can not prove it formally.

If we consider Rayleigh fading, the analytical expression of $p_s(x)$ can be found as (see [24] for details)

$$p_s(x) = \prod_{j=0}^{C} e^{-B(x)} \quad \text{(ALOHA)}, \qquad p_s(x) = \prod_{j=1}^{C} (1 - B(x)) \quad \text{(CDPA)}, \tag{17}$$

where

$$B(x) = S \int_0^\infty n(z)dz \int_0^\infty \frac{\Lambda_j(z,y)dy}{(1 + x/by)}. \tag{18}$$

4.1 Analysis with power control

If the power transmitted by each MT is constant, it turns out that the curve $p_s(x)$ is an increasing function of x, so that users that experience worse long term conditions suffer higher collision probability and must attempt transmission more often. Since we require uniform throughput, performance is dominated by the worst case. If we provide some means to improve the performance of these users, we expect an overall performance increase.

Power control is a feature that allows each MT to transmit a power level such that the long term power received at the proper BS is kept at the same constant level for all transmitting MTs. In this case, the transmitted power, W_{Ti}, is inversely proportional to $10^{0.1\xi_i} r_i^{-\eta}$, and the received power varies only because of fading. This is a way to eliminate the performance unbalancing cited above. On the other hand, eliminating the differences in the received power is expected to degrade the capture effect, and it is not clear which of the two conflicting effects will prevail. Furthermore, the interference distribution is also changed.

The extension of the analysis to perfect power control is simple, because the value of x (and z) experienced by all users is the same. Thus all the analysis becomes independent of x and z and equation (16) reduces to

$$P_s = P_s(G, \Gamma_j(.), j = 0, \ldots, C, \mathcal{P}) \tag{19}$$

where the functions $\Gamma_j(.)$ are the analogous of $\Lambda_j(.)$ in the presence of power control, and are functions of y only. The simplification obtained is remarkable since now no recursive solution is required.

4.2 Analysis with outage

According to the basic protocol, MSs keep transmitting until a packet is successfully received. Thus, users that experience severe propagation conditions (or interference) have a great impact on the amount of interference generated and, if not inhibited, they can reduce the system throughput to very low values. To avoid this situation, an "outage" condition is declared when the number of failed transmissions of the same packet reaches a preselected value, $N_{\max}$. when such a condition occurs, the user is denied further service.

In this case, the conditional average number of transmissions, $n(x)$, can be written as

$$n(x) = \sum_{k=1}^{N_{\max}} k(1-p_s)^{k-1}p_s + \sum_{k=N_{\max}+1}^{\infty} N_{\max}(1-p_s)^{k-1}p_s = \frac{1}{p_s}(1 - P_o(x)) \tag{20}$$

where p_s is the same as in the previous section and

$$P_o(x) = \sum_{N_{\max}+1}^{\infty} (1-p_s)^{k-1}p_s = (1-p_s)^{N_{\max}} \tag{21}$$

is the conditional outage probability. The effective throughput density is

$$s'(x) = s(x)(1 - P_o(x)), \tag{22}$$

and the effective cell throughput as

$$S' = \int_0^{\infty} s'(x)dx = S\left(1 - \overline{P}_o\right), \tag{23}$$

where the unconditional outage probability, $\overline{P}_o$, is given by

$$\overline{P}_o = \int_0^{\infty} \lambda(x)P_o(x)dx. \tag{24}$$

Note that, as shown by (22), the actual throughput density is no longer proportional to $\lambda(x)$, and the outage mechanism introduces some unfairness. This was to be expected, since avoiding absolute fairness is precisely the reason why outage was introduced.

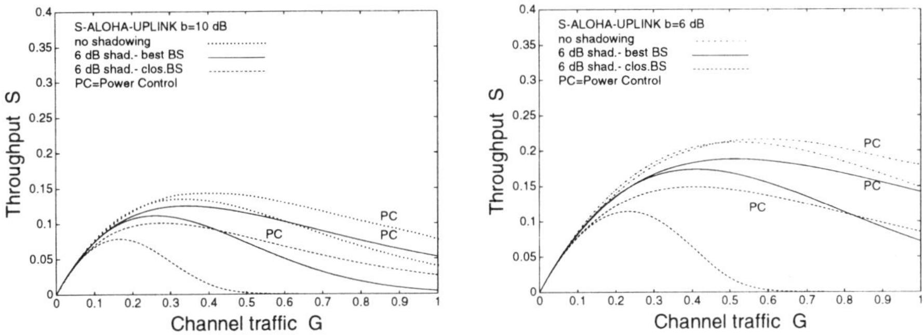

Figure 3 Throughput versus the offered channel traffic of the S-ALOHA protocol for capture threshold $b = 10$ and 6 dB and different anti-shadowing mechanisms.

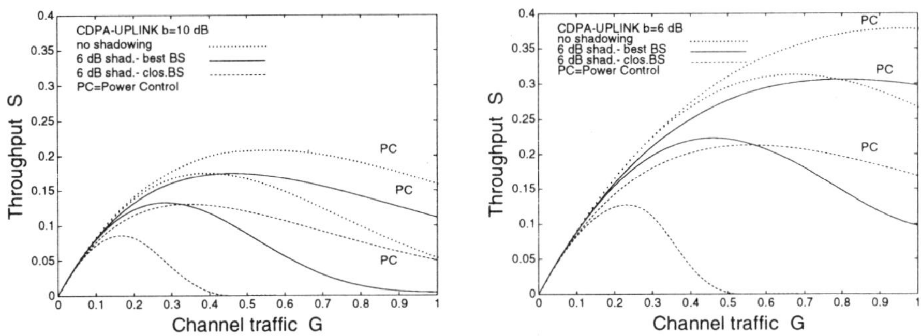

Figure 4 Throughput versus the offered channel traffic of the CDPA protocol for capture threshold $b = 10$ and 6 dB and different anti-shadowing mechanisms.

5 NUMERICAL RESULTS

In the following, we present the throughput curves for the two protocols obtained following the analysis presented in the previous section. Because of the rather involved nature of the analytical expressions (which does not allow closed-form solutions), evaluations have been performed using accurate numerical quadrature methods, accounting for the interference coming from up to the third layer of cells, summing to a total of $C = 37$ cells, including the intended one. More details on the computation can be found in [24].

In Figure 3 we report the throughput of the S-ALOHA protocol as function of the global channel load G for the capture threshold $b = 10$ dB and $b = 6$ dB. Figure 4 reports the same curves for the CDPA technique.

In each figure we compare six throughput curves, i.e., the three curves representing the cases with no shadowing, shadowing with best BS and shadowing with closest BS, respectively, and the corresponding curves when power control is used. Note that for the S-ALOHA protocol the counteracting effects cited above balance each other and no practical gain in the maximum throughput is provided by the power control. On the contrary, in CDPA the power control is more effective because of the absence of intracell interference.

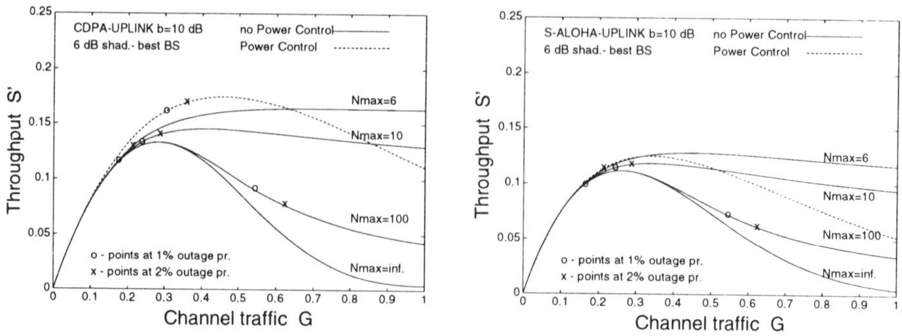

Figure 5 Throughput versus the offered channel traffic of the S-ALOHA and CDPA protocols with 6 dB shadowing and the outage mechanism.

As a general behavior, observed in all four graphs, we note that the slow shadowing severely impairs the performance if the closest BS is chosen. This impairment is limited if either power control or the best BS criterion is adopted. The combined use of power control and best BS selection is the most effective strategy in fighting the effect of shadowing. In all cases, the power control helps stabilizing the throughput curves. In fact, without power control, the unbalancing in the capture probability causes a severe throughput bottleneck for the users that experience higher global attenuation (e.g., those at the cell border). These users need to retransmit more often than others and, as G increases, the difference in the number of retransmissions becomes very large, causing the throughput to collapse.

As an example of the numerical results obtained in the presence of outage we show in Figure 5 the throughput performance of CDPA for $b = 10$ dB and for different values of $N_{\max}$. The value $N_{\max} = \infty$ represents the case of no outage and the corresponding performance is the same as in Figure 4. For a given outage condition ($N_{\max}$), the outage probability $\overline{P}_o$ clearly depends on the offered traffic intensity G and increases as G increases. As a reference for this probability we have marked, on each of these curves, the points that correspond to an outage probability of 1% and 2%. If the system operation is delay constrained in such a way that, for instance, no more than 6 transmissions are tolerated, and if we require an outage probability no greater than 2%, then the offered traffic must be smaller than $G = 0.21$, and the maximum achievable throughput is $S = 0.129$. If the delay constraint is relaxed, and $N_{\max}$ can be set equal to 10, then the same outage probability is guaranteed up to a maximum throughput $S = 0.140$. The curve for $N_{\max} = 100$ is reported to show the convergence as $N_{\max} \rightarrow \infty$.

In Figure 5 we have also reported the performance of CDPA with power control. The throughput for this case can be simply derived from the previous analysis, recalling that, p_s, n, and P_o do not depend on x. As a result, the throughput curve $S'(G)$ does not depend on $N_{\max}$. This result can be easily understood by observing that outage is effective only when users with different long term performance are present. In this case, in fact, the occurrence of the outage condition is negatively correlated to the success probability $p_s(x)$ experienced by the user and the shutdown of the users that operate in adverse conditions avoids a number of retransmissions greater than the average. The two points marked on this curve merely represent the operation points at which the probability that a packet is transmitted no more than 6 times is equal to 1% and 2%. Similar curves for S-ALOHA are also shown in Figure 5.

6 CONCLUSIONS

In this paper we have discussed the use of the packet technology together with the retransmission of corrupted packets in PCNs, and have shown that it provides an effective cell multiplexing technique alternative to FDMA, TDMA and CDMA. In particular the advantages of the CDPA technique have been discussed and compared to S-ALOHA.

To investigate the performance of the packet techniques we have introduced a unified analytical approach applicable to all systems that adopt packet transmission with collisions and capture. The analytical model accounts for several features of real-world systems, e.g., the exact traffic pattern that must be offered in order to achieve throughput fairness, the exact interference caused by the traffic in the same cell and in other cells, the fading, and the slow shadowing attenuation. Means to counteract shadowing, such as the best base station selection and the power control, are also taken into account. We have also analyzed the performance in the presence of outage.

The results here obtained show that the throughput of CDPA is superior to the throughput of ALOHA in all the conditions considered, the gain ranging from 30% to 40%. This gain, together with other attractive features of CDPA, such as the stable operation and the effective ability to deliver uniform throughput and to guarantee a bounded delay, make this system suitable for a large range of service classes as expected in wireless ATM. Moreover, CDPA can integrate S-ALOHA to provide signaling and single packet transmission.

We have not attempted a comparison of CDPA with more classical circuit-oriented techniques, such as TDMA and CDMA, mainly because a non-biased comparison should take into account a real traffic scenario and all the overheads due to implementation details, e.g., the coding of commands for CDPA. We maintain however that the CDPA technique is inherently very simple, and does not need any planning nor power control. Thus it is appealing for PCNs systems in which simplicity may be a key factor.

REFERENCES

[1] J. M. G. Linnartz, *Narrowband Land-Mobile Radio Networks*, Artech House, 1993.

[2] V. H. McDonald, "The cellular concept," *Bell System Technical Journal*, vol. 58, no. 1, pp. 15–41, Jan. 1979.

[3] C.T. Lau, C. Leung, "Capture models for mobile packet radio networks", *IEEE Trans. Comm.*, vol. COM-40, pp. 917-925, May 1992.

[4] J. Walker, editor, *Mobile Information Systems*, Artech House, 1990.

[5] K. S. Gilhousen, I. M. Jacobs, R. Padovani, A. J. Viterbi, L. A. Weaver, Jr, C. E. Wheatley III, "On the capacity of a cellular CDMA system", *IEEE Trans. on Vehic. Tech.*, vol. VT-40, pp. 303-312, May 1991.

[6] R. L. Pickholtz, L. B. Milstein, D. L. Schilling, "Spread spectrum for mobile communications," *IEEE Trans. on Vehic. Tech.*, vol. 40, no. 2, pp. 313-322, May 1991.

[7] M. Zorzi, S. Pupolin, "Slotted ALOHA for high capacity voice cellular communications", *IEEE Trans. Veh. Tech.*, vol. 43, pp. 1011-1021, Nov. 1994.

[8] F. Borgonovo, L. Fratta, M. Zorzi, "Capture-Division Packetized Access (CDPA) for cellular systems", in *Proc. WCN'94*, pp. 810-815, The Hague, NL, Sep. 1994.

[9] F. Borgonovo, L.Fratta, M. Zorzi, "Performance analysis of capture-division packetized access (CDPA) for cellular systems", in *Proc. Fifth WINLAB Workshop on 3rd Generation Wireless Information Networks*, pp. 121-138, Apr. 1995.

[10] F. Borgonovo, M. Zorzi, L. Fratta, V. Trecordi, G. Bianchi, "The Capture-Division Packetized Access (CDPA) for Wireless Personal Communications", *IEEE J. Selected Areas Comm.*, Apr.-May 1996.

[11] F. Borgonovo, L.Fratta, M.Zorzi, "Performance of Capture-Division Packetized Access (CDPA) with Partial Frequency Reuse and Power Control", *IEEE GLOBECOM'95*, Singapore, pp. 1141-46, Nov. 1995.

[12] D.J. Goodman, "Cellular packet communications", *IEEE Trans. Commun.*, vol. COM-38, Aug. 1990, pp.1272-1280.

[13] G. Bianchi, F. Borgonovo, L. Fratta, L. Musumeci, M. Zorzi, "C-PRMA: the Centralized Packet Reservation Multiple Access for local wireless communications", in *Proc. IEEE GLOBECOM'94*, San Francisco, CA, pp. 1340-1345, Nov. 1994.

[14] G. Bianchi, F. Borgonovo, M. De Marco, V. Trecordi, "Capture-Division Packetized Access (CDPA) for cellular systems: an assessment by simulation" in *Proc. IEEE ICC'95*, Seattle, WA, June 1995.

[15] D. Bertsekas, R. Gallager *Data Networks*, Prentice-Hall International, 1992.

[16] W.C. Jakes, Jr., ed., *Microwave mobile communications*, New York: John Wiley & sons, 1974.

[17] F. Borgonovo, A. Capone, M. Zorzi, *Receiver capture in CDPA*, Politecnico di Milano, internal report, Jan. 1996.

[18] C. Namislo, "Analysis of mobile radio slotted ALOHA networks", *IEEE Journal on Select. Areas in Comm.*, vol. SAC-2, pp. 199-204, Aug. 1984.

[19] J.C. Arnbak, W. van Blitterswijk, "Capacity of slotted ALOHA in Rayleigh-fading channels", *IEEE Journal on Select. Areas in Comm.*, vol. SAC-5, pp. 261-269, Feb. 1987.

[20] A.U.H. Sheikh, Y.D. Yao, X.P. Wu, "The ALOHA systems in shadowed mobile radio channels with slow or fast fading", *IEEE Trans. on Veh. Tech.*, vol. VT-39, pp. 289-298, Nov. 1990.

[21] C. Vanderplas, J.P.M. Linnartz, "Stability of mobile slotted ALOHA network with Rayleigh fading, shadowing, and near-far effect", *IEEE Trans. Veh. Tech.*, vol. VT-39, pp. 359-366, Nov. 1990.

[22] M. Zorzi, R.R. Rao, "Capture and retransmission control in mobile radio", *IEEE J. Selected Areas Comm.*, vol. SAC-12, pp. 1289-1298, Oct. 1994.

[23] F. Borgonovo, M. Zorzi, "Slotted ALOHA and CDPA: a comparison of channel access performance in cellular systems", in *Proc. IEEE INFOCOM'96*, San Francisco, CA, Mar. 1996.

[24] M. Zorzi, F. Borgonovo, *An analytical framework for the performance evaluation of packet access in wireless systems*, Politecnico di Milano, internal report, Dec. 1995.

Dynamic Slot Allocation for TDMA-Systems with Packet Access

Thorsten Benkner
University of Siegen, Dep. for Electrical Engineering & Informatics
Hoelderlinstr. 3, D-57068 Siegen, Germany
e-mail: benkner@nue.et-inf.uni-siegen.de

Abstract — **To fully exploit the inherent flexibility of packet access protocols like PRMA++ for cellular telecommunications systems also highly flexible slot assignment strategies are necessary. In this paper two dynamic interference adaptive slot assignment algorithms are investigated. The schemes fulfil the requirements of packet access protocols. They are decentralized and need no frequency preplanning, which is a major concern in microcell environments.**

I. INTRODUCTION

In mobile cellular telecommunications systems with packet access protocols in the MAC (Medium Access Control) layer and a microcellular structure the requirements upon slot assignment algorithms are very high. Packet access schemes are very flexible and can also offer activity controlled transmission. This can be used for statistical multiplexing in order to get a gain in subscriber capacity. In typical speech connections the voice coder delivers data packets in only about 45 % of the total connection time. This can result in a statistical multiplexing gain factor of 2.2.

Due to the activity controlled mode of operation, i.e. the short discontinuous transmission intervalls, the slot assignment algorithm has to react very quickly. In order not to worsen speech quality because of excessive clipping the slot assignment process has to be finished after some 10 ms. Centralized DCA (dynamic channel assignment) algorithms are therefore not suitable for applications with fast slot assignment. The necessary signaling and control efforts would be too high. So only decentralized slot assignment schemes are useful.

There is no load averaging in microcellular networks resulting in strong and fast load fluctuations. Within a few minutes temporary load inhomogenities can occur that static slot allocation schemes are not able to cope with. In these situations the critical access delay of packet access protocols can become very high. The slot assignment algorithm has to be very flexible and dynamic to react upon fast load changes.

The interference situation in microcellular networks can be very complicated and unpredictable. It can also change over the time of operation. There are three main reasons for this. Microcell base station antennas are typically mounted rather low (street lamp elevation) so that the electromagnetic wave propagation is influenced very much by medium term alterations in the near environment. This can make a frequency preplanning process as it is common in macrocellular networks hard or even impossible. Furthermore sudden interference changes can be caused by a relatively high user mobility. Packet access schemes reallocate time slots on an activity basis, so interference changes much more rapidly than it is the case in second generation TDMA systems which allocate a slot for the whole duration of a call. There are also medium term alterations like new buildings or changing weather conditions that lead to changes of the propagation conditions in an unpredictable way after the network planning process is over.

The requirements stated above lead to fast slot assignment schemes that are adaptive to both interference and traffic, that are decentralized and need not frequency preplanning.

The paper is organized in the following way. After a brief introduction into packet access schemes, especially the PRMA++ protocol [2], in section II, the analyzed system will be presented in section III. This includes a description of the slot allocation algorithms DINF and CS+ [6], [7] for PRMA++. In section IV performance results will be shown and discussed. Section V concludes the paper.

II. PRMA++

PRMA++ is a very flexible medium access control protocol intended for operation on the air interface of third generation TDMA systems like the RACE ATDMA (Advanced Time Division Multiple Access) system [1]. It meets the requirements of UMTS with respect to variable bitrates, different delay constraints and traffic source statistics. The proposed dynamic slot assignment strategies in this paper are investigated in a PRMA++ environment but could also be adapted to other packet access protocols.

There are two major differences between PRMA [3] and the PRMA++ protocol used in ATDMA. At first resource request and usage periods are separated by using different types of slots. So it is possible to use different modulation and coding schemes for reservation access and payload transmission. Secondly PRMA++ has the ability to do statistical multiplexing also on the downlink by using "fast paging".

The payload in a PRMA++ system is carried in information slots (I-slots). Fig. 1 shows the structure of an uplink I-slot (normal burst). Each slot consists of two times 33 payload bits, 2·5 inband signalling bits, 4 header and 4 tail bits, a 29 bit long training sequence and a guard period which corresponds to 12 bit lengths. The carrier symbol rate can be up to 1.8 Mbit/s.

H	DATA	Si.	Training Sequence	Si.	DATA	T	Guard
4	33	5	29	5	33	4	12

Fig. 1: PRMA++ normal burst

A mobile that wants to gain access to one or more I-slots transmits a reservation packet (ACI-burst) in an uplink R-slot. R-slot access is performed according to the Slotted-ALOHA protocol and is therefore subject to packet collisions. After a collision a mobile enters a contention resolution mode and tries a new reservation on the next available R-slot with a certain probability (retransmission probability).

If the reservation attempt was successful and there is enough capacity available at the base station the mobile's request will be acknowledged on a downlink acknowledgement slot (A-slot). If resource is not available requests will be queued and in the case of speech transmission voice packets will be successively dropped after each 10 ms. The mobile monitors the A-slots until it receives an acknowledgement from the base station. The BS has centralized control over the I-slot queue. This is particularly important in the case of services with different priorities and also for dynamic slot allocation as will be explained later.

The statistical multiplexing on the downlink is achieved by special FP-slots (fast paging slots) which all mobiles listen to. Here the BS informs MSs on which downlink slots it can find data. FP-messages will be confirmed on an uplink fak-slot (fast acknowledgement slot).

In ATDMA four cell types have been defined ranging from macrocells to picocells. The PRMA++ protocol differs in these cell types. In particular the number of slots per 5 ms frame varies between 15 and 72. Details can be found in [1].

III. Simulated System

A. Environment

The investigations presented here were performed in a planar cellular system of 30 hexagonal microcells with 1 km radius each. Fig. 2 shows the geography of the scenario. In the case of a static carrier assignment (FCA) a cluster size of $N_C = 4$

was chosen. Simulation results were only taken from the four central cells in order to diminish border effects.

User mobility has a large impact on the performance of slot allocation schemes as these investigations showed. So a mobility model was included. Mobiles choose a random speed $v_0 \in [5\ km/h; 60\ km/h]$ and a random direction $\alpha_0 \in [0;2\pi)$. These initialization values can be updated randomly in the range $\Delta v/\Delta t \in [-10\ km/h/s; +10\ km/h/s]$ and $\Delta\alpha/\Delta t \in [-\pi/4/s; +\pi/4/s]$. Mobility implicitly requires handover possibilities. Therefore an optimized intercell handover at the cell border was implemented.

A major problem in cellular systems with packet access is speech clipping when the traffic load becomes too large. There is no blocking as it occurs in conventional systems when all time slots or carriers respectively are busy. As these investigations are primarily focused on the I-slot access performance the number of lines between BS and MSC (Mobile Switching Center) was dimensioned very large (100). A blocking because of a lack of lines is therefore unlikely.

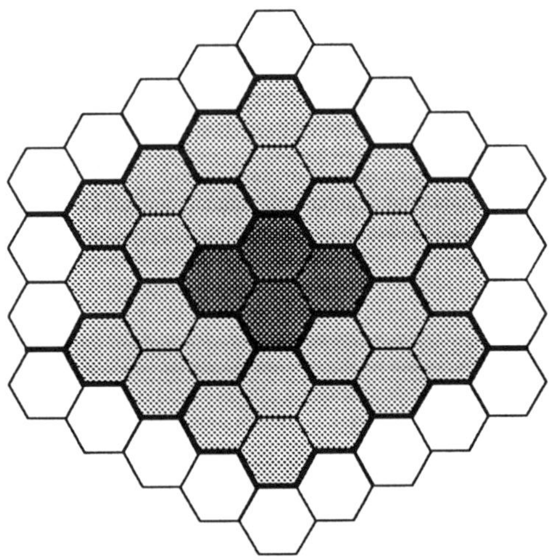

Fig. 2: Simulation environment

B. Propagation

The received signal strength was assumed to be proportional to $r^{-\gamma}$ (r is the distance between BS and MS) with a pathloss slope of $\gamma = 4$. This corresponds to a simple two-ray propagation model with ideal wave reflection on the ground.

Lognormal distributed shadow fading s(σ) with a standard deviation of $\sigma = 4$ dB was included, too. Fast fading was not explicitly modelled because appropriate measures like diversity and interleaving were assumed. Furthermore the coherence bandwidth in microcell environments is typically much lower than the transmission bandwidth (≈ 2 MHz) (frequency selective channel), so that the effects of multipath propagation are not very strong.

With these assumptions the received carrier level C can be modelled as

$$C = A r^{-\gamma} s(\sigma) \tag{1}$$

where A is a constant depending on transmission power and antenna gain. The minimum required cochannel interference ratio CIR_{min} for good quality speech transmissions was assumed 8 dB.

C. Load Model

The traffic load model consists of poisson distributed call arrivals, negative exponential distributed call durations of 120 s mean and also includes slow DTX (Discontinuous Transmissison).

Speech is modelled as a two state Markoff chain with negative exponential distributed transition rates, see fig. 3. The mean talkspurt activity duration T_{act} is 1.41 s, the gap between two talkspurts T_{sil} is 1.74 s resulting in a speech activity factor of 45 %. The bandwidth requirement of a speech signal is therefore 0.45 slots/frame on the average.

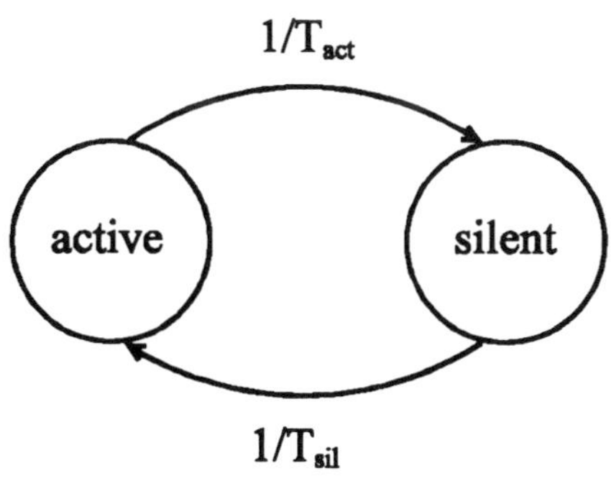

Fig. 3: Speech model

The speech coder rate is 13 kbit/s. The coder generates speech packets of 130 bits after all 10 ms. These packets are split up into four sub-packets of 2.5 ms length which are then interleaved with a depth of 4 giving new speech blocks of again 10 ms duration.

D. MAC-Protocol

The implemented medium access control protocol is PRMA++ with a frame length of 5 ms and 18 slots per frame [2]. The maximum packet holding time is 10 ms corresponding to one voice coder block.

Each uplink frame consists of 16 I-slots, 1 R-slot and 1 fak-slot. Downlink A-slots are delayed by three time slot durations against their corresponding uplink R-slot. The investigations are restricted to the uplink as this is the more critical link in PRMA++, only here a contention process occurs.

Nevertheless as investigations in [2] and also our simulation results showed, not the contention delay is usually the dominant factor in PRMA++ but the allocation delay due to a lack of free I-slots. For simplicity reasons perfect capture of reservation packets was therefore assumed.

There are four carriers available in the systems. In each cell one carrier is available in the case of FCA. When the dynamic slot allocation schemes are investigated, all base stations principally can have access to all carrier/slot combinations provided that the current traffic and interference situation allows that.

E. Slot Allocation with DINF

DINF [6] is a dynamic interference adaptive slot allocation scheme in which assignment decisions are made according to C/I estimates of the considered slots. Four different allocation strategies were investigated and will be explained in the following.

<u>BEST</u>: The BEST strategy assigns the carrier/slot combination I(c,s) ($c \in [1,n_C]$, $s \in [1,n_I]$) with the highest C/I of all available combinations wich are currently not in use at the corresponding base station. n_C and n_I are the total numbers of carriers and I-slots respectively. The C/I has to be at least as high as the setup threshold CIR_T:

$$CIR_T = CIR_{min} + CIR_S \tag{2}$$

where CIR_S is a safety fade margin which shall be high enough to achieve a low interference rate. Investigations in [6] showed that CIR_S has to be at least 14 dB in order to obtain an interference rate comparable to that of an equivalent static slot assignment. The BEST strategy is similar to the DCS algorithm of the DECT system for cordless telephones.

WORST: This scheme is exactly opposite to the BEST scheme. Not the I(c,s) ($c \in [1,n_C]$, $s \in [1,n_I]$) with the highest estimated C/I is chosen but the one with the lowest C/I above the threshold CIR_T (2).

RAND: A combination I(c,s) ($c \in [1,n_C]$, $s \in [1,n_I]$) is randomly chosen according to a uniform distribution among all carrier/slot combinations with a C/I which is at least as high as CIR_T.

ORDER: All base stations scan all combinations of I(c,s) in the same order. The first slot which is currently not in use in the own cell and that has a C/I which is above CIR_T will be assigned.

All of the four schemes have in common that a talkspurt will be queued when no I(c,s) can be found that fulfils the threshold requirement of (2). In this case voice blocks will be successively dropped after each 10 ms.

F. Slot Allocation with CS+

The CS+ strategy [7] is a novel decentralized DCA scheme which can effectively reduce the high safety margins CIR_S of DINF by an adaptive learning process of slot allocation patterns. CS+ is based on a channel segregation approach (CSEG) [4], [5] but was improved for an operation in highly changing load and interference situations as they will occur in microcellular systems applying packet access schemes.

The basic idea of channel segregation (CSEG) is that each channel has a priority P. During the channel allocation process the free slot with the highest P will be assigned. P will be incremented if the last transmission was successful, i.e. if it was not interfered. If there was interference P will be decremented. The steps in which P can change become smaller proportional to 1/n, where n is the number of channel usage attempts. After a certain convergence period a reuse pattern will be obtained automatically.

The original channel segregation strategy [4] only adapts to medium-term interference/traffic alterations. Channel reusability will not be significantly improved as CSEG is in principle a kind of 'adaptive FCA'. Improvements of CSEG can be obtained if not the establishment of a quasi static reuse pattern will be the premier target of the algorithm but a higher flexibility. CS+ establishes a 'weaker' reuse pattern and uses priority dependent C/I values to decide if a slot is free or not.

The novel CS+ strategy uses interference results of previous talkspurt transmissions to determine the threshold CIR_T. Considering the history of previous transmissions in the slot assignment process leads to significant lower safety margins CIR_S compared to DINF. By using a logarithmic function of P to determine CIR_T it is possible to learn from the past in a sense that the CIR_S of low priority slots, which usually have a higher interference probability, can be increased. On the other hand high priority slots can be assigned with a lower 'safety margin'. Therefore a new expression for CIR_T is introduced

$$CIR_T = CIR_{min} + CIR_S - log(P) \qquad (3)$$

By this measure the relatively high CIR_S of DINF can be reduced to 4...6 dB without increasing the interference rate.

A drawback of CSEG is that it is not able to adapt to varying interference situations after a certain convergence intervall has elapsed. The 'learning process' of CSEG stops after a while and a more or less stable reuse pattern is obtained.

CS+ uses a different approach as it has the ability to slowly 'forget' the interference results of the past. This is achieved by an IIR-filter (infinite impulse resonse filter) in the priority update function. The filter transfer function H(z) is:

$$H(z) = \frac{1-a}{1-az^{-1}} \qquad (4)$$

The structure of this filter is shown in fig. 4.

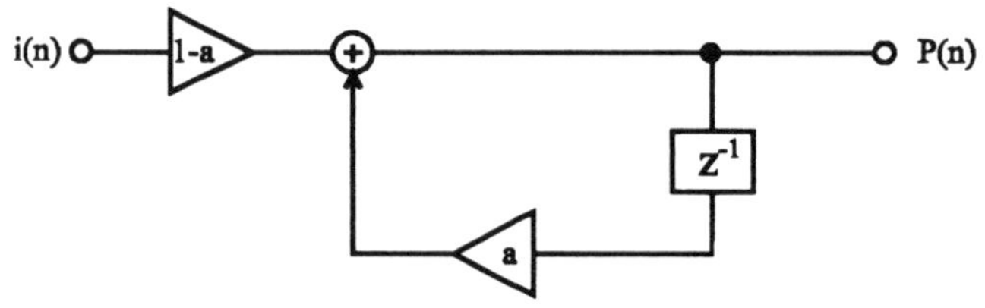

Fig. 4: CS+ priority filter

The input signal $i(n) \in [0,1]$ represents if the last talkspurt transmission on the slot I(c,s) was interfered (i(n) = 0) or not (i(n) = 1). Fig. 5 shows an example of the working principle of the priority update function in CS+.

The speed of CS+ responding to changing interference situations can be adjusted with the location of the pole a in (4) ($|a| < 1$). The value of a depends on the environment where CS+ is intended to operate. The faster interference can change the smaller a has to be. The best value in the environments investigated here was found to be a = 0.9. Fig. 6 shows the working principle of CS+ in form of a flow-chart.

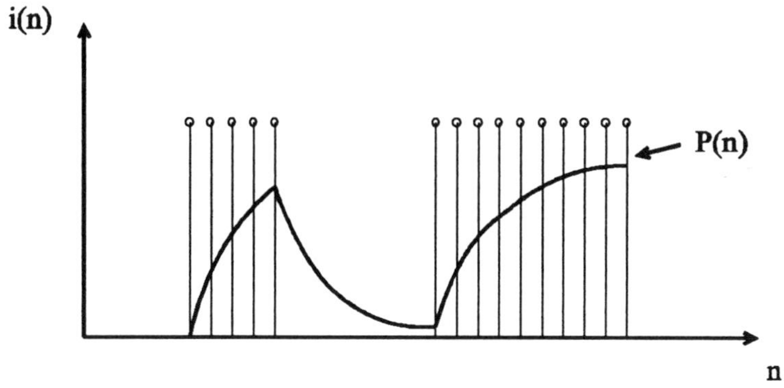

Fig. 5: priority update of CS+

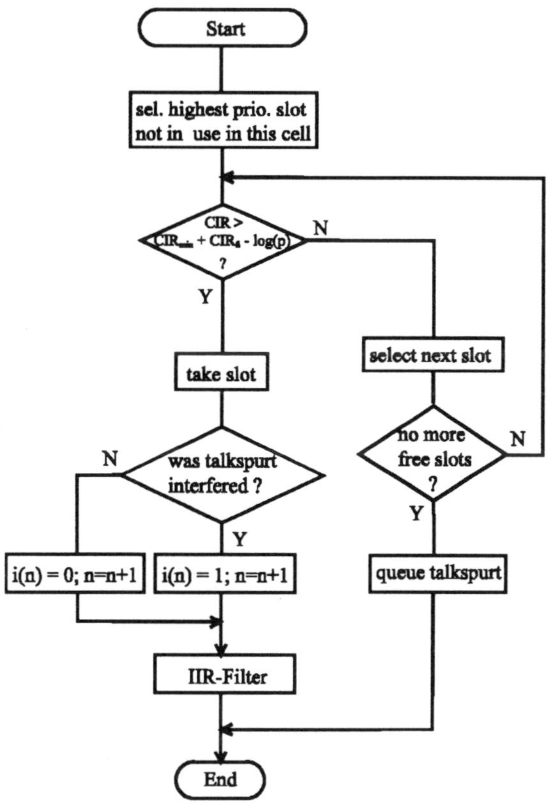

Fig. 6: Flowchart of CS+

IV. Results

A. Performance of DINF

At first the I-slot access performance of the different DINF schemes shall be studied. Fig. 7 shows the I-slot access delay (contention delay + allocation delay) as a function of the offered traffic per cell. Independently of the traffic the BEST strategy leads to the highest I-slot access delay of all investigated DINF schemes. If a delay limit of 20 ms for a speech transmission in good quality (low front-end clipping) is assumed only about 15.4 Erl traffic can be carried.

This result is easy to understand as with the selection of the slots I(c,s) that have the highest C/I the average reuse distance in the system becomes the largest, too. In a FCA system this would be equal to increasing the cluster size N_C resulting in a smaller subscriber capacity of the network when the total available spectrum is kept constant.

The opposite regarding the slot selection process is performed with the WORST strategy. By choosing the slot with the lowest C/I above CIR_T the average reuse distance will be smaller compared to BEST. The 20 ms limit is reached at 22.3 Erl, that is nearly 45 % more than it was with BEST. As will be shown later this advantage has to be 'paid' with a worse transmission quality.

RAND has a delay performance which is in between that of BEST and WORST. With equal probability there are I(c,s) selected that have a rather high C/I and also combinations with a low quality. The average reuse distance is therefore in the middle between BEST and WORST.

All decentralized interference adaptive DCA strategies have in common that the channel (slot) assignment is more or less uncoordinated. This is because channel demands occur at random time distributed across the cell system. The DINF ORDER scheme tries to order the assigned slots according to their interference level. Slots higher in the selection order are chosen more frequently for an assignment. So the interference level on these slots is higher than that of slots at the end of the selection order. Only mobiles with a high carrier level have a chance to overcome the C/I setup threshold CIR_T (2). These mobiles are usually close to their base stations. This results in a frequent reuse of slots at the beginning of the search order.

In the more 'complicated' cases in which mobiles are far away from their base stations and therefore offer only small carrier levels C the slot assignment algorithm has to search longer for a suitable slot. Primarily the slots at the end of the order which have a low interference level will be assigned to these connections. This results in larger reuse distances for far away mobiles on an average.

The ORDER strategy can exploit the carrier signal variations due to different distances between mobiles and their bases in a much better way than the other investigated DINF schemes. ORDER is therefore a kind of combination between signal- and interference adaptive DCA schemes. The I-slot access delay of ORDER is always below that of the other strategies. The 20 ms limit is reached at 25.8 Erl, more than 67 % above the traffic with BEST.

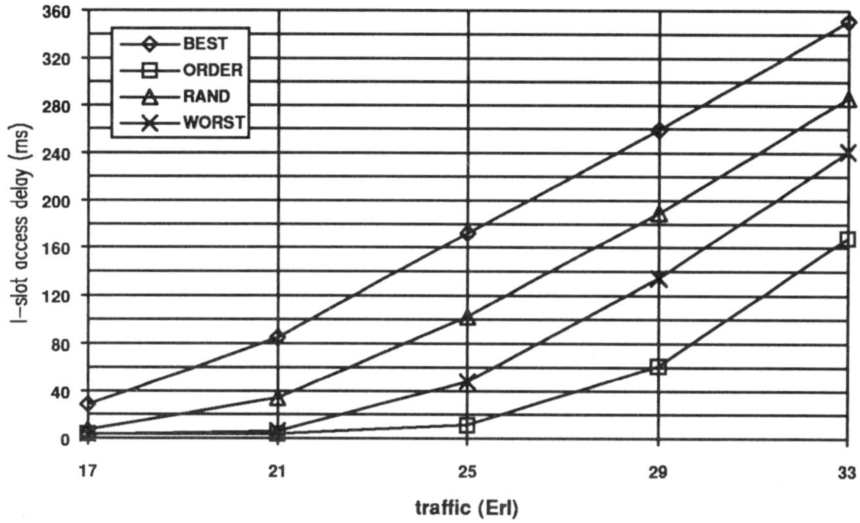

Fig. 7: I-slot access delay of different DINF schemes

Next the interference behaviour of the four different DINF strategies shall be investigated. Fig. 8 shows the average interfered talkspurt transmission time in relation to the total transmission time as a function of the offered traffic.

It can be seen that, as expected, the BEST strategy has the lowest interference rate of all investigated schemes. It is always lower than 2 %. On the other hand the WORST strategy, which had a much better delay performance than BEST, has the highest interference rate (about 5 %).

Here and also with RAND a typical effect of packet access schemes that drop speech packets can be observed. In very high load ranges where the I-slot access delay is high the interference rates become lower with increasing traffic. The reason for this at first glance surprising effect is that at very high loads the packet dropping rate becomes very high so that a large amount of the talkspurt is dropped while it is waiting in the I-queue of PRMA++. Excessive dropping of speech packets results in an effective load decrease on the air interface, so that the interference rate decreases, too.

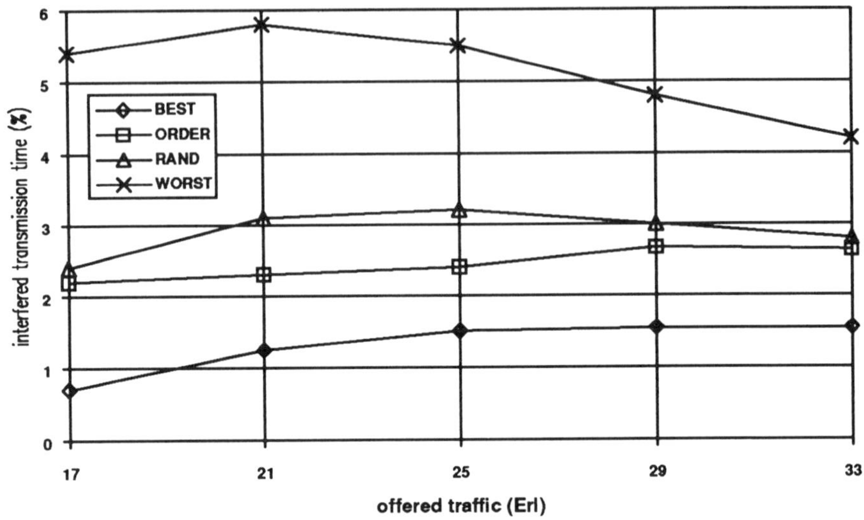

Fig. 8: Interference rate of different DINF schemes

The interference rate of ORDER is in between those of BEST and RAND. It is always lower than 3 %. So the good delay performance of ORDER does not result in a high interference rate as it is in the case of WORST.

B. Comparison with CS+ and FCA

The performance of FCA, DINF(ORDER, CIR_S = 14 dB) and CS+ (CIR_S = 6 dB) was compared in a homogeneous load situation of up to 37 Erl per PRMA++ cell. Fig. 9 shows the I-slot access delay of the slot allocation schemes as a function of the offered traffic per cell.

At low traffic loads DINF has lower access delays than FCA but at a certain point (about 27 Erl) it becomes worse. This effect is well known as it occurs in many DCA algorithms. The reason is that the channel reuse pattern with FCA is very dense. At a certain high load this becomes dominant in contrast to the missing flexibility of FCA compared to DCA. Nevertheless the 20 ms limit is reached at a higher load with DINF. Whereas FCA can carry 23.8 Erl DINF reaches 25.8 Erl (8.4 % more).

Better results can be obtained with CS+. The I-slot access delay is always below that of FCA and DINF in the investigated load range. The 20 ms limit is crossed at 29.1 Erl, corresponding to a load increase of 22.2 % compared to FCA.

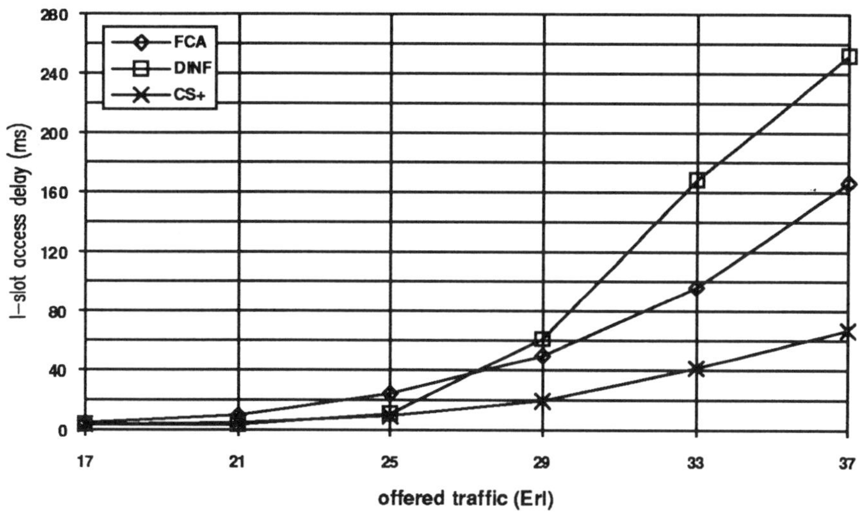

Fig. 9: I-slot access of FCA, DINF and CS+

A cellular system with PRMA++ is not directly comparable with an Erlang-B system, but if a tolerable blocking probability of 2 % is assumed in the case of Erlang-B, the system can carry about 11.5 Erl per cell. The combination of PRMA++ and CS+ can be loaded with 29.1 Erl in the same cell. This corresponds to an improvement of 153 % in subscriber capacity.

V. CONCLUSIONS

Two interference adaptive dynamic slot allocation strategies that fulfil the special requirements of packet access protocols for third generation TDMA systems were investigated in a microcell scenario.

DINF is a self-adaptive scheme in which slot assignment decisions are primarily based on C/I estimates. It was found out that the order in which slot selection occurs within the DINF algorithm has a large impact on the performance both of the I-slot access delay and the interference rate. The best results of all DINF schemes were obtained with the ORDER strategy in which slots are ordered according to their interference level. But even with this scheme only small capacity advantages over FCA are achievable in a homogeneous load situation.

Better results could be obtained with the CS+ strategy. CS+ is a decentralized algorithm based on adaptive 'learning' of allocation patterns. The high setup thresholds of DINF are drastically reduced with CS+ so that higher capacity gains are possible. It was found out that the combination of the MAC protocol PRMA++ and CS+ can offer a cpacity advantage of about 150 % in a homogeneous load scenario.

ACKNOWLEDGEMENTS

The author wishes to thank Dr. K. David, DeTeMobil GmbH, Germany, for many valuable and fruitful discussions. Acknowledgement is also made to Prof. Dr. C. Ruland, University of Siegen, for furthering these investigations.

REFERENCES

[1] Urie A., Streeton M., Mourot C.: "An Advanced TDMA Mobile Access System for UMTS", IEEE Personal Communications, February 1995

[2] Dunlop J., Irvine J., Robertson D., Cosimini P.: "Performance of Statistically Multiplexed Access Mechanism for a TDMA Radio Interface", IEEE Personal Communications, June 1995

[3] Goodman D.J., Valenzuela R.A., Gayliard K.T., Ramamurthi B.: "Packet Reservation Multiple Access for Local Wireless Communications", IEEE Vehicular Technology Conference, VTC'88, 1988

[4] Furuya Y., Akaiwa Y.: "Channel Segregation, A Distributed Adaptive Channel Allocation Scheme for Mobile Communications Systems", DMR II, Stockholm, 1987

[5] Frullone M., Riva G., Grazioso P., Carciofi C.: "Self-Adaptive Channel allocation strategies in Cellular Environments with PRMA", 44th IEEE Vehicular Technology Conference, VTC'94, Stockholm, Sweden, June 8 - 10, 1994

[6] Benkner T., David K.: "Autonomous Slot Assignment Schemes for PRMA++ Third Generation TDMA Systems", IEEE Third Symposium on Communications and Vehicular Technology in the Benelux, Eindhoven, The Netherlands, Oct. 1995

[7] Benkner T., David K.: "CS+: A Novel Self-Adaptive Slot Assignment Scheme for ATDMA", 46th IEEE Vehicular Technology Conference, VTC'96, Atlanta, USA, April 28 - May 1, 1996

DESIGN AND EVALUATION OF AN RRA SCHEME FOR VOICE-DATA CHANNEL ACCESS IN OUTDOOR MICROCELLULAR WIRELESS ENVIRONMENTS

Allan C. Cleary
CIS Dept.
Univ. of Delaware
Newark, DE, USA
accleary@cbdcom.apgea.army.mil

Michael Paterakis
Electronics & Computer Engrg. Dept.
Technical Univ. of Crete
731-00 Chania, Crete, Greece
pateraki@telecom.tuc.gr

1. INTRODUCTION

Personal communication services (PCS) embodies the notion that access to land based public networks should reside with the individual rather than with a fixed location (e.g., a wall phone jack), thereby allowing any user to initiate or accept calls from anywhere. An overview of PCS is provided in [1] and a survey of the research and development in PCS can be found in [2].

Reservation random access (RRA) protocols provide a way to exploit the on/off characteristics of packetized speech traffic; and, various forms have been proposed and studied [3-8] for use in PCS networks. In this work, we assume that voice and data traffic is generated by mobile pedestrians who access the network with small, lightweight and low-power devices (i.e., low tier PCS [1]). In Section 2, we propose an RRA scheme along the general lines of [7] and describe several random access algorithms for transmitting the request packets. In Section 3, we formulate a Markovian model and derive expressions for the steady state voice packet dropping probability, access delay and throughput. The system parameters are included in Section 4; along with the method for calculating the steady state distribution and a brief discussion of the simulations. Representative analytic and simulation results are presented in Section 5 and the paper is concluded in Section 6.

2. RRA SCHEME

Within the microcell, spatially dispersed source terminals share a radio channel that connects them to a fixed base station. The base station allocates channel resources, delivers feedback information and serves as an interface to the mobile switching center. Since base to mobile transmissions can be supported by a TDM broadcast channel, we focus on the mobile to base channel transmissions. Speech alternates between periods of talk and silence; and, voice packet transmissions are subject to a real time constraint. We assume that each voice terminal uses voice activity detection [3] to suppress its silence transmissions. Thus, voice terminals only require channel access during talkspurt. Data traffic, on the other hand, is often bursty and more tolerant of delays (e.g., delays up to 200 ms are often acceptable); but 100% delivery of correct packets is often required.

The bandwidth is organized into periodic time frames of fixed duration. The frame duration is selected such that a voice terminal in talkspurt generates exactly one packet per frame. As illustrated in Figure 1, each frame consists of one voice request interval, one data request interval and an information interval. Within the information interval, each slot accommodates exactly one, fixed length, packet that contains voice (or data) information and a header. Both of the request intervals are subdivided into mini-slots; and each mini-slot accommodates exactly one, fixed length, request packet. For both voice and data traffic, the request must include a source identifier. For data traffic, the request might also include message length and quality of service parameters such as priority and required slots/frame. We simplify our presentation by assuming that both request intervals contain an equal number of mini-slots; and, that data terminals need only one information slot per frame.

Voice (data) terminals with packets, and no reservation, contend for channel resources using a random access protocol to transmit their request packets only during the voice (data) request interval. The base station broadcasts a short binary feedback packet at the end of each mini-slot that indicates only the presence or absence of a collision within the slot (collision (C) versus no collision (NC)). Since the feedback packet is small (several bits) and the transmission delay within a microcell is negligible, we assume that the feedback information is available to the terminals before the next mini-slot. Upon successfully transmitting a request packet the terminal waits until the end of the frame to learn of its reservation slot. If unsuccessful within the request interval, the terminal attempts again in the request interval of the next frame (and, in the case of voice traffic, drops one packet). A terminal with a reservation transmits freely during its reserved slot.

To allocate channel resources, the base station maintains a dynamic table of the active terminals within the microcell. For example, information within the table might include the terminal identifier, the virtual circuit identifier, the channel resources allocated, and quality of service parameters. Upon successful receipt of a request packet, the base station provides an acknowledgment and queues the request. The base station allocates channel resources at the end of the frame, if available. If the resources needed to satisfy the request are unavailable, the request remains queued. Voice terminals with queued requests and data terminals with packets must continuously monitor the base-to-mobile channel. Upon call completion, or when an active terminal exits the microcell (hand-off), the base station will delete the table entry after some prescribed period of time.

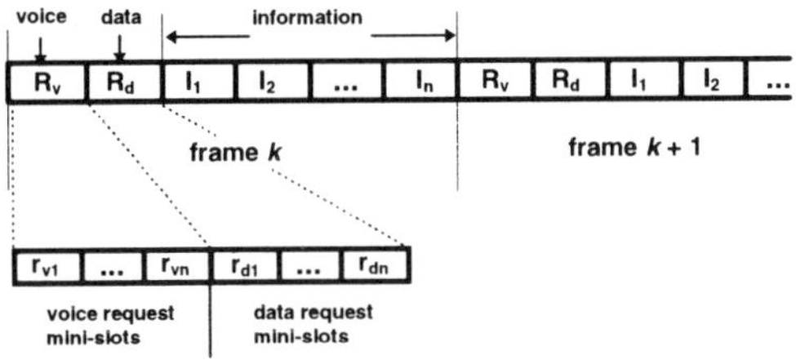

Figure 1. The voice, data, and information intervals within a frame.

We focus on steady state channel access and assume the following about the allocation and queuing policies. When servicing a request, the base station always allocates the earliest empty information slot within the frame. Next, we assume that voice has a higher priority than data traffic (this is reasonable for the near future). Thus, the base station will service all of the outstanding voice requests before servicing any data requests. Within each priority class, the queuing discipline is assumed to be First Come First Served. Finally, if any new voice requests are received and the frame is full: the base station attempts to service the voice requests by canceling the appropriate number of reservations belonging to data terminals (if any). When data reservations are canceled, the base station notifies the effected data terminal and places an appropriate request at the front of the data request queue.

The algorithms below were selected to investigate performance aspects of the proposed RRA scheme, rather than optimization for a set of system parameters.

2.1 Ideal

All request packets present at the start of the voice request interval are correctly received by the base station. This protocol provides an upper bound for the voice system capacity and a lower bound for the voice access delay.

2.2 Slotted aloha, [8]

A contending terminal transmits its request packet only if it has permission to transmit. Permission is issued with probability, $p = 1/3$, in each request slot. This value was selected to ensure stable operation of the algorithm. Our voice traffic simulations with $p = 0.5$ indicate that with N larger than about 75 voice terminals the system tends to oscillate between two equilibrium points: one with high throughput; and, the other with low throughput caused by collisions during contention. When the system is characterized by this bistability, the iterative technique together with the initial distribution discussed below computes the steady state distribution corresponding to the high throughput equilibrium point.

2.3 Two-Cell Stack, [9]

Each terminal uses a counter, **r**, as follows.

1. At the start of the request interval the contending terminals initialize their counter, **r**, to 0 or 1 with equal probability.
2. contenders with **r** = 0 transmit into the request slot.
3. Let x be the feedback for that transmission. Then
   a. if x = no collision:
      if **r** = 0, the request packet was transmitted successfully.
      if **r** = 1, then **r** = 0.
   b. if x = collision:
      if **r** = 0, then reinitialize **r** (i.e., **r** = 0 or 1 with equal probability).
      if **r** = 1, then **r** = 1.
4. Repeat steps 2 and 3, until either two consecutive no collisions occur or the request interval ends.

2.4 Two Cell Stack for Data Terminals, [9]

The term message is used to distinguish between information and request packets and to imply that a data arrival event may consist of several information packets. Over consecutive data request intervals, the contending data terminals perform the following algorithm [9]. A collision resolution period (CRP) is defined as the interval of time that begins with an initial collision (if any) and ends with the successful transmission of all data request packets involved in that collision (or, if no collision occurred, ends with that mini-slot). Terminals involved in a collision follow rules 2-4 in part 2.3. Terminals with new message arrivals may not transmit during a CRP. In the first mini-slot after a CRP, all terminals whose message arrived within the allocation interval (of maximum length Δ) transmit with probability one. We set Δ equal to the reciprocal of the message arrival rate, so that the expected number of message arrivals within the allocation interval is equal to one.

3. VOICE TRAFFIC ANALYSIS

The primary model assumptions are as follows.

(1) Voice terminals use a slow voice activity detector [3] that only responds to principal talkspurts (e.g., > 20 ms) and gaps due to listening and pausing (e.g., > 200 ms). Thus, speech activity is modeled with a two-state discrete time Markov chain as follows. The talkspurt/silence and silence/talkspurt transition probabilities are p_{TS} and p_{ST}, respectively. The talkspurt and silence periods are geometrically distributed with mean $1/p_{TS}$ and $1/p_{ST}$, respectively. Therefore, at steady state, the probability that a terminal is in talkspurt (speech activity), p_T, or silence, p_S, is obtained from, $p_T = p_{ST}/(p_{ST} + p_{TS})$ or $p_S = 1 - p_T$, respectively. Additionally, we assume that the number of active voice terminals within the system is constant and that all of the voice transitions occur at the frame boundaries.

(2) The channel is error-free and without capture, and the base station correctly broadcasts the pending resource allocations before the start of the next frame. Thus, collisions within a request slot are the only system errors.

(3) A terminal with a reservation signals the base station upon the completion of a talkspurt (or data message).

An active voice terminal will be in one of four states: silent, contender, queued, or reserved. A silent terminal does not require channel resources. Upon transition into talkspurt, the terminal enters the contender state and remains there until it successfully transmits a request packet. Since requests are queued at the base station, the terminal enters the queued state and remains there until the base station provides it with a reservation (as a minimum, the end of the frame in which the request was received). After receiving a reservation, the terminal enters the reserved state and transmits one voice packet per frame into its allotted slot until it exits talkspurt (i.e., returns to the silent state).

We define the pairs of random variables T^-_k, T_k; C^-_k, C_k; Q^-_k, Q_k; *and* R^-_k, R_k; as the number of terminals in the talkspurt; contender; queued; and reserved state just before and just after the start of frame k, $k > 0$, respectively. The superscript denotes the former. Given N active voice terminals, the number of terminals in the silent state is equal to $N - T^-_k$ (just before the start of frame k or, alternatively, at the end of frame k - 1); and, $T^-_k = C^-_k + Q^-_k + R^-_k$. Thus, the state of the system is completely described by three random variables (e.g., T^-_k, C^-_k, and R^-_k). Unfortunately, the size of the state space grows very large for any interesting system.

Because the base station allocates reservations at the end of each frame, the Q^-_k term will often be subsumed by the R^-_k term. We make our model tractable by assuming that the base station allocates resources according to the transitions that occur at the frame boundaries (i.e., if a reserved terminal transitions to silence, the base station may allocate its slot). We let $X^-_k = Q^-_k + R^-_k$, and $X_k = Q_k + R_k$. Thus, since $T^-_k - C^-_k = X^-_k$, the state of the system can be described by T^-_k and C^-_k; and, we can use aspects of the methodology in [4] to formulate the Markovian process $Z^- = \{Z^-_k = (T^-_k, C^-_k), k > 0\}$. With our allocation policy, a voice terminal in the queued state receives a reservation for the next frame unless every information slot is already allocated to a voice terminal. Therefore, given I information slots per frame and the value of X^-_k, the values of Q^-_k and R^-_k are obtained as follows: if $X^-_k \leq I$, then $R^-_k = X^-_k$ and $Q^-_k = 0$; and, if $X^-_k > I$, then $R^-_k = I$ and $Q^-_k = X^-_k - I$.

By assuming that the voice transitions occur only at frame boundaries, we can separate them from the random access process and view Z^-, as evolving in two steps [4]. The following equations: $T_k = T^-_k + A_k - S^C_k - S^X_k$ and $C_k = C^-_k + A_k - S^C_k$, account for the voice transitions. Where, A_k is the number of terminals entering talkspurt at the start of frame k; and, S^C_k and S^X_k are the number of contending and queued and/or reserved terminals exiting talkspurt over frame k. Let $E[W_k]$ be the expected number of successfully transmitted request packets within the voice request interval of frame k, then $\bar{C}_{k+1} = C_k - E[W_k]$ accounts for the outcome of the random access process (since $\bar{T}_{k+1} = T_k$). The evaluation of $E[W_k]$ is discussed below.

3.1 Ideal random access

According to this protocol, each contending voice terminal in frame k successfully transmits its request within the voice request interval of frame k. Thus, for all $k > 0$, $C_k = A_k = E[W_k]$ and $\bar{C}_{k+1} = 0$. Since T_k includes A_k, we need only consider the process $\bar{Y} = \{\bar{Y}_k = (\bar{T}_k),\ k > 0\}$. The limiting distribution of $\bar{Y}$ corresponds to the probability that m out of N terminals are in talkspurt, P_m. With our speech activity model, the steady state distribution of $\bar{Y}$ is as follows:

$$\Pr\{Y^{-} = m\} = P_m = \mathrm{B}(N, m, p_T) \tag{1}$$

where, $\mathrm{B}(n,m,p) = \binom{n}{m} p^m (1-p)^{n-m}$.

3.2 Slotted aloha and two-cell stack random access

In both the slotted aloha and the two-cell stack algorithms, a contending voice terminal may fail to successfully transmit a request packet within the voice request interval. In that case, a voice terminal remaining in talkspurt attempts to transmit requests in succeeding frames, until it either succeeds or exits talkspurt. Thus, for all $k > 0$, $\bar{C}_{k+1} = C_k - E[W_k] \geq 0$ and we must consider the process $\bar{Z}$. The evaluation of $E[W_k]$ requires the following probability distribution: $F_{r,c}(w)$ = Pr{exactly w out of c contenders succeed within the voice interval (r mini-slots) of the frame}. $F_{r,c}(w)$ depends on the random access protocol and the methodology used to calculate $F_{r,c}(w)$, for both of these protocols is in [4].

With N active voice terminals in the system and r mini-slots in the voice request interval, the one step transition probability, $p_{mn,ij}$, is expressed as follows.

$$\Pr\{T_{k+1}^{-} = i, C_{k+1}^{-} = j \mid T_k^{-} = m, C_k^{-} = n\} =$$

$$\sum_{h=j}^{\min(N-m+n,\ r+j)} \Pr\{T_k = i, C_k = h \mid T_k^{-} = m, C_k^{-} = n\} \Pr\{T_{k+1}^{-} = i, C_{k+1}^{-} = j \mid T_k = i, C_k = h\} \tag{2}$$

The first term on the right hand side of equation (2) accounts for the voice transitions and it can be calculated by conditioning on A_k as follows.

$$\sum_{a=\max(h-n,0)}^{\min(N-m,h)} \Pr\{T_k = i, C_k = h | T_k^- = m, C_k^- = n, A_k = a\} \Pr\{A_k = a | T_k^- = m, C_k^- = n\} =$$

$$\sum_{a=\max(h-n,0)}^{\min(N-m,h)} \mathrm{B}(n, a+n-h, p_{TS})\, \mathrm{B}(m-n, m-n-i+h, p_{TS})\, \mathrm{B}(N-m, a, p_{ST}) \tag{3}$$

where, $a+n-h$ and $m-n-i+h$ are equal to the number of contender and queued/reserved terminals that exit talkspurt, respectively. The second term on the right hand side of equation (2) accounts for the contention and it is equal to $\mathrm{F}_{r,c}(h-j)$.

3.3 Voice Performance Measures

Our discussion is limited to the process of $\bar{Z}$, since the application to $\bar{Y}$ is straightforward. To formulate expressions for the steady state voice packet dropping probability, access delay and throughput, we let E[$\bar{T}$], E[$\bar{C}$], E[$\bar{Q}$] and E[$\bar{R}$] be the steady state mean number of voice terminals in the talkspurt, the contender, the queued and the reserved state per frame, respectively. E[$\bar{T}$] = Np_T, since it depends only on the speech activity. Given the limiting distribution of $\bar{Z}$, $\pi(i,j)$:

$$\mathrm{E}\left[C^-\right] = \sum_{i=0}^{N} \sum_{j=0}^{i} j \cdot \pi\left(T_k^- = i, C_k^- = j\right) \tag{4}$$

$$\mathrm{E}\left[Q^-\right] = \sum_{i=0}^{N} \sum_{j=0}^{\max(i-j-I,0)} \left(i-j-I\right) \cdot \pi\left(T_k^- = i, C_k^- = j\right) \tag{5}$$

where, I represents the information slots/frame. And, E[$\bar{R}$] = E[$\bar{T}$] - E[$\bar{C}$] - E[$\bar{Q}$].

The steady state voice packet dropping probability is the ratio of the mean number of voice packets dropped per frame to the mean number of voice packets generated per frame. By our assumptions, any terminal in the contender or queued state at the end of the frame drops one voice packet. Thus, the steady state voice packet dropping probability percent, P_{drop}, is: P_{drop} = 100(E[$\bar{C}$] + E[$\bar{Q}$])/E[$\bar{T}$].

Let the voice throughput, η, be the proportion of time slots that successfully carry voice packets from terminals to the base station. Then η is equal to the mean number of successfully transmitted voice packets per frame. Since a contending terminal in frame k is not eligible to receive a reservation until frame k+1, terminals that are in the contender and queued states do not transmit a voice packet during the frame. Because we assume an error-free channel, it follows that η = E[$\bar{R}$].

The access delay for a voice terminal is the time between the start of a talkspurt and the end of the first voice packet transmission into a reserved slot. For the proposed RRA scheme, the mean access delay, D, can be expressed as follows: $D = D_c + D_q + D_r$. Where, D_c and D_q are the mean random access and queuing delays, respectively; and, D_r is the mean time between the start of the frame in which the reservation is granted and the end of the transmission into the reserved slot. We simplify the following by expressing time in information slots and by defining the minimum value of D_c to be the entire voice interval. For example, a terminal that successfully transmits a request during the voice request interval and receives a reservation for the next frame experiences: D_c = 1 slot; D_q = F-1 slots, where, F is the number of slots per frame; and, D_r such that $2 < D_r \leq F$ slots (i.e., 2 slot delay for the request intervals plus a delay equal to some number of information slots).

Since a terminal must be in one of the four states, the input and output of each state will be equal at steady state. Thus, the mean delay due to terminals that fail to successfully transmit a request during the frame is the ratio of the average number of terminals in the contender state to the average rate of terminals transitioning into talkspurt. Therefore, $D_c \approx 1 + F(\mathrm{E}[C^-]/Np_S p_{ST})$.

When all of the slots within the frame are reserved for voice terminals, newly arriving voice requests are queued until resources become available. This delay, on average, is equal to the ratio of the number of terminals in the queued state to the number of terminals entering the queued state (Little's result, [8]). Therefore, $D_q \approx F - 1 + F(\mathrm{E}[\mathrm{Q}^-]/Np_S p_{ST})$.

Recall our assumption that the base station will always allocate the earliest empty information slot within the frame. Due to independent talk/silence transitions, terminals with reservations will exit talkspurt randomly. Thus, by assuming that the reserved slot will be uniformly distributed over the interval between the first information slot and the number of reserved slots in the frame, D_r is approximated as follows: $D_r \approx 2 + \sum_{i=0}^{N} \sum_{j=0}^{i} (0.5\,\mathrm{g}(i,j)+1) \cdot \pi\left(T_k^- = i, C_k^- = j\right)$.

Where, $\mathrm{g}(i,j) = \begin{cases} i-j, & \text{if } \; i-j \leq I \\ I, & \text{otherwise} \end{cases}$.

4. PERFORMANCE EVALUATION

The parameters in Table 1 were used for the analysis and simulations. The channel rate is from [5,6]. The speech rate assumes the use of 32 kb/s adaptive differential pulse code modulation (ADPCM) in the microcellular environment [1]. The packet size was selected for compatibility with ATM networks. The values for the mean talkspurt/silence duration produce a speech activity of approximately 44 percent [5]. The frame duration of 12 ms accommodates approximately 51 information slots and the voice limit delay is equal to the duration of two frames. To account for guard time and synchronization, we assume that 50 slots are available per frame. The first and second slots are dedicated to the voice and data request intervals, respectively. We chose six mini-slots per request interval (accommodates the transmission of approximately 70 bits) to allow for guard time and synchronization overheads and for the transmission of a generic request packet that contains the source identifier, along with some data (e.g., priority, slots required, etc.). The remaining 48 slots comprise the information interval.

Parameter	Value	Parameter	Value
Channel rate	1.8 Mb/s	Voice/Data request interval	1 slot/frame
Speech rate	32 kb/s	Information interval	48 slots/frame
Information packet:	424 b	Mini-slots per request interval	6
Frame duration	12 ms	Mean talkspurt duration	1.41 s
Voice delay limit	24 ms	Mean silence duration	1.78 s

Table 1. Experimental system parameters.

The process $\bar{Z}$ is ergodic, because it is an irreducible, aperiodic Markov chain, with finite state space. Therefore, with P being the one step transition probability matrix, the limiting distribution, π satisfies the equations $\pi = \pi P$ and $\sum_{i=0}^{S} \pi_i = 1$; where, S is the size of the state space. An iterative solution is possible, since, with $\pi(0)$ being any arbitrary distribution, the limiting distribution of a Markov chain may be expressed as $\pi = \lim_{n\to\infty} \pi(0)\, P^n$, [10].

For $\bar{Z}$, given N active voice terminals, $S = (N+1)(N+2)/2$. With the parameters in Table 1, the system accommodates N up to about 100 (S is over 5000 states). The limiting distribution can be obtained via iteration with a desktop workstation; and, a considered choice of $\pi(0)$ makes convergence very rapid (approximately 5 iterations with $\varepsilon = 10^{-6}$). For the process $\bar{Z}$, a good random access algorithm minimizes $\bar{C}_k$. Thus, for $0 \leq m \leq N$, we construct $\pi(0)$ as follows. For all of the states with $\bar{T}_k = m$ and $\bar{C}_k = 0$ set $\pi(0) = P_m$, as calculated in (1); and, for all of the states with $\bar{T}_k = m$ and $\bar{C}_k > 0$ set $\pi(0) = 0$.

All simulations consist of ten independent runs of 305,000 frames each. To reduce start up effects, the first 5000 frames serve as the warm up period. During each run: the number of terminals within the system is held constant; the terminals are initially silent; and, the results for specific performance measures are calculated with cumulative data.

5. RESULTS AND DISCUSSION

For the system under investigation, the steady state voice packet dropping probability, access delay and throughput increase with N. A common measure of comparison for RRA voice protocols is the maximum N with $P_{drop} \leq 1\%$ (voice capacity). Analytical results obtained for the random access protocols operating at their respective voice capacity are summarized in Table 2. The multiplexing gain is equal to N/F; and with no P_{drop} constraint, the maximum gain is 2.16 (i.e., $I/(Fp_T)$).

Access Protocol	N (terminals)	P_{drop} (percent)	Access Delay (ms)	Throughput (packets/frame)	Multiplexing Gain
Aloha	97	0.963	31.40	42.46	1.94
Two-cell	97	0.905	30.57	42.49	1.94
Ideal	97	0.868	30.07	42.50	1.94

Table 2. Voice performance for the access protocols operating at voice capacity.

In Figure 2, we observe that for N values below about 97 the aloha access algorithm consistently yields slightly higher P_{drop} values than the two-cell and ideal algorithms; and, that for $N > 97$ all of the algorithms produce similar P_{drop} values. Voice packet dropping is due to delays encountered by a terminal that is in either the contender or queued state. We show the fraction of voice packets dropped from terminals in the contender and queued states for the aloha and two-cell random access algorithms in Figures 3 and 4, respectively. The analytical results were obtained using the definitions in Section 3. With ideal access, any voice packet dropping occurs in the queued state.

The contender and queued lines intersect at about $N = 88$ and $N = 84$ for the aloha and two-cell access algorithms, respectively. This indicates that packet dropping due to contention is more significant for aloha than for the two-cell algorithm. At steady state, the expected number of voice terminals entering talkspurt at the start of a frame is equal to Np_Sp_{ST}. Thus, the expected number of new contenders per frame is much less than one (e.g., from 0.26 to 0.38, for N values from 70 to 100). For the algorithms in Section 2, when there is exactly one contending voice terminal: it successfully transmits its request 90 percent of the time with aloha (due to the probabilistic first time transmission rule and the use six mini-slots per voice request interval); while it always succeeds with the two-cell. Therefore, we conclude that at lower loads (e.g., N < 82) packet dropping depends on the random access algorithm; while at higher loads, the choice of random access algorithm is less crucial since packet dropping is mainly due to the frame being full (i.e., all of the information slots are allocated to voice terminals).

It is well known that the steady state voice packet dropping probability may not be a sufficient measure of voice performance when considering quality of service issues or access control strategies. Table 3 contains simulation results for the steady state packet dropping distribution per talkspurt for each access protocol operating at voice capacity. Comparing the aloha values to the two-cell and ideal values suggests that the probability of dropping zero or one voice packet per talkspurt is due to the efficiency of the random access algorithm; while, dropping more than one is due to the frame being full. The results indicate that a new talkspurt has an almost ten percent chance of dropping more than four packets. This is because the average talkspurt duration is on the order of 100 frames. Our simulations indicate a positive probability for run lengths of up to approximately 53 dropped packets.

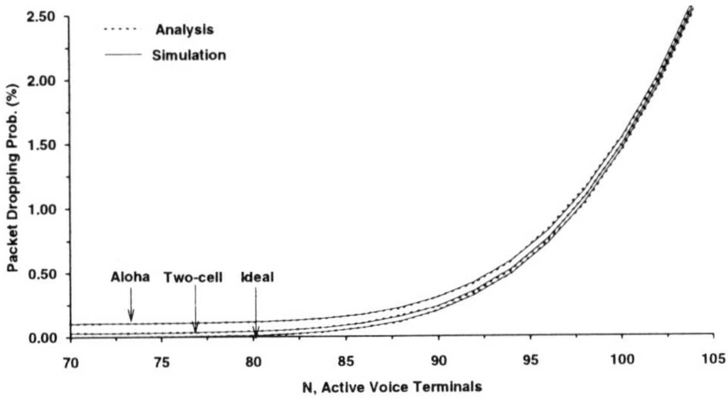

Figure 2. Steady state P_{drop} versus N.

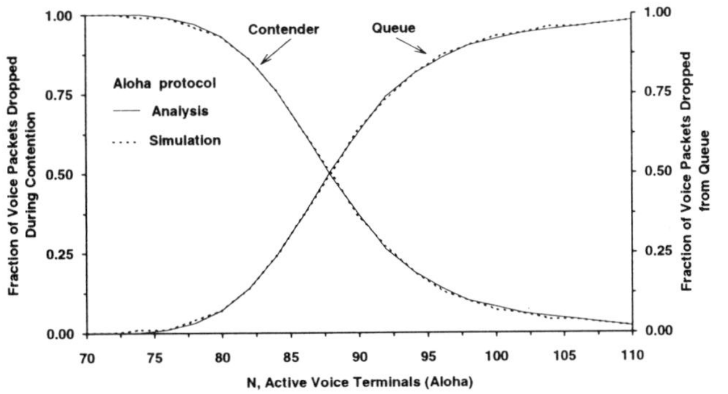

Figure 3. Fraction of packets dropped from contender/queued state (aloha).

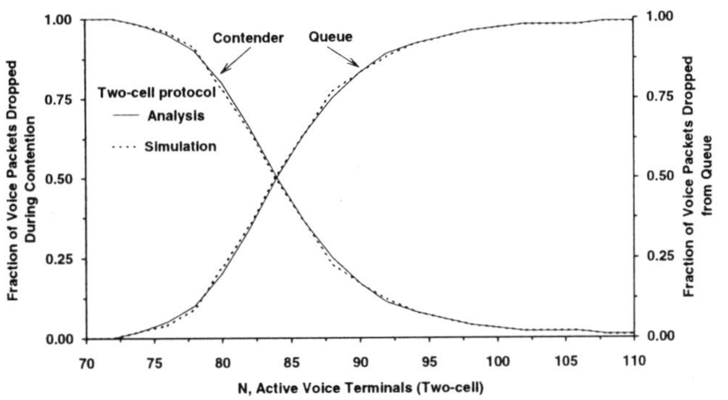

Figure 4. Fraction of packets dropped from contender/queued state (two-cell).

Packets Dropped per Talkspurt	Aloha $N = 97$	Two-cell $N = 97$	Ideal $N = 97$
0	0.76	0.82	0.86
1	0.10	0.04	0.02
2	0.03	0.02	0.02
3	0.01	0.02	0.01
4	0.01	0.01	0.01
> 4	0.09	0.09	0.08

Table 3. Distribution of the voice packets dropped /talkspurt.

Although not included, our results show the voice throughput η, to be nearly linear until P_{drop} starts to exceed one percent; and, that it is accurately described by $\eta = Np_T(1\text{-}P_{drop}/100)$, [4], regardless of the access protocol. In Figure 5, the mean voice access delay curves show similar trends to those in Figure 2. For the same reasons, the aloha algorithm produces slightly higher mean access delay values up until about $N > 97$. We observe that the mean access delay increases above the voice limit delay of 24 ms when N (P_{drop}) is approximately equal to 92 (0.42), 94 (0.52) and 94 (0.49) for the aloha, two-cell and ideal access protocols, respectively. These values indicate that the queuing delay component unduly influences the mean access delay value.

5.1 Voice-Data Integration

Data traffic will gain in importance as small, portable, and inexpensive computing devices proliferate. One difficulty in designing an algorithm for voice-data integration is determining how to characterize and model the user side data traffic in the emerging PCS environment. Besides assuming that voice is of higher priority than data traffic, our RRA scheme is based on the following. We assume that data applications with special needs, such as high speed or minimal variance between packet deliveries, will be accommodated with reservations and a suitable priority designation. We surmise that a typical PCS data user might do things like read/compose short e-mail messages, respond to paging type messages, transfer files between the mobile and a fixed computer, and query some form of database. Finally, we observe that the data may consist of mouse events or text based characters.

The following performance measures are considered for the data traffic: wait delay is the time between the message arrival and the end of the first data packet transmission into a reserved slot; message delay is the time between the message arrival and the end of the last data packet transmission into a reserved slot; and, throughput is the proportion of time slots that successfully carry data packets. We assume that data messages are generated by a theoretically infinite number of data terminals and that the aggregate arrivals are Poisson distributed with mean, λ messages/frame. Additionally, we assume that the message length varies according to a geometric distribution with parameter q and mean B = 1/q. Because the duration of an information slot is fixed (and fairly short), an arriving data message may result in the formation of multiple packets. Thus, expressing B as packets/message, the

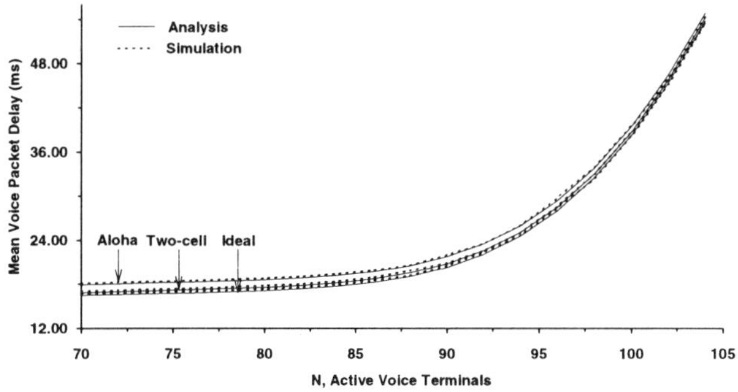

Figure 5. Mean voice access delay versus N.

steady state data rate is equal to λB packets/frame. Data traffic will likely be more"bursty" than our model indicates; however, we submit that the model is commonly used, and thus reasonable for initial investigations.

In this work, we set the parameter q equal to 1/8 (thus, B = 8). This implies that the average data message (as described above) is short (i.e., besides user data, the packet will contain overhead bits for forward error correction and from the upper layers). The maximum throughput of the two-cell stack algorithm (0.429, [9]) together with six mini-slots per frame means that the maximum data throughput is about 20.6 packets/frame (i.e., (0.429)6B). With no voice traffic, our simulations show that up until about 2.5 message arrivals per frame (about 20.0 packets/frame), the average data message delays are below 200 ms.

The data traffic has no effect on voice traffic performance, since the voice and data requests are transmitted separately and since the base station preempts data reservations in favor of voice. Two costs to this approach are the fixed overhead of dedicating a portion of the frame to data requests; and, data throughput is severely limited at high voice loading. For instance, since η = 42.5 packets/frame for the ideal protocol operating at voice capacity the maximum data throughput is limited to about 5.5 packets/frame (i.e., less than the mean message length that we have assumed here). In Table 4, N = 86 and the steady state P_{drop}, voice access delay and throughput are approximately 0.07, 18 ms and 38 packets/frame, respectively. Since the voice access is ideal, the P_{drop} value is due to the frame being full (i.e., there will be consecutive frames when no reservations are received by data terminals). The results indicate that the average data message delays are below 200 ms for message arrival rates up until about 0.75.

Arrivals (msg/fr)	Wait Delay (ms)	Message Delay (ms)	Throughput (pack/fr)
0.25	41.72 ± 2.00	127.88 ± 2.19	2.00 ± 0.01
0.50	76.50 ± 3.16	162.80 ± 3.31	3.99 ± 0.01
0.75	141.61 ± 5.24	227.86 ± 5.28	5.93 ± 0.01

Table 4. Data performance; ideal voice access, N = 86, and B = 8.

6. CONCLUSION

We have proposed and evaluated an RRA scheme for integrating voice and data traffic in outdoor microcellular environments. We developed an approximate Markov model to investigate voice traffic performance measures and investigated several algorithms for the transmission of voice requests. Analytical results were presented for the steady state voice capacity, packet dropping probability, access delay and throughput. Simulations were employed to study the voice packet dropping distribution per talkspurt and to explore preliminary voice-data integration considerations. Our results indicate that this is a promising scheme for providing voice-data integration in outdoor microcellular environments.

7. REFERENCES

[1] D. C. Cox, "Wireless Personal Communications: What is it?", *IEEE Pers. Commun.,* Vol. 2, No. 2, Apr. 1995, pp. 20-35.

[2] V. O. K. Li and X. Qui, "Personal Communication Systems", *Proc. IEEE*, Vol. 83, No. 9, Sep. 1995, pp. 1210-1243.

[3] S. Nanda, D. J. Goodman and U. Timor, "Performance of PRMA: A Packet Voice Protocol for Cellular Systems", *IEEE Trans. Veh. Technol.*, Vol. 40, Aug. 1991, pp. 584-598.

[4] S. Jangi and L. Merakos, "Performance Analysis of Reservation Random Access Protocols for Wireless Access Networks", *IEEE Trans. Commun.*, Vol. COM-42, No. 2-4, pt 2, Feb-Apr. 1994, pp. 1223-1234.

[5] J. M. DeVile, "A Reservation Based Multiple Access Scheme for a Future Universal Mobile Telecommunications System", *Proc. of the 7th IEE Conf. on Mobile and Pers. Commun.*, Brighton, UK, Dec. 1993, pp. 210-215.

[6] J. Dunlop, J. Irvine, D. Robertson and P. Cosimini, "Performance of a Statistically Multiplexed Access Mechanism for a TDMA Radio Interface", *IEEE Pers. Commun.*, Vol. 2, No. 3, Jun. 1995, pp. 56-64.

[7] N. M. Mitrou, T. D. Orinos and E. N. Protonotarios, "A Reservation Multiple Access Protocol for Microcellular Mobile Communication Systems", *IEEE Trans. Veh. Technol.*, Vol. 39, No. 4, Aug. 1990, pp. 340-351.

[8] D. Bertsekas and R. Gallager, *Data Networks*, 2nd Ed., Prentice Hall Inc., 1992.

[9] M. Paterakis, and P. Papantoni-Kazakos, "A Simple Window Random Access Algorithm with Advantageous Properties", *IEEE Trans. Inform. Theory*, Vol. IT-35, No. 5, Sep. 1989, pp. 1124-1130.

[10] J. E. Wieselthier and A. Ephremides, "A New Class of Protocols for Multiple Access in Satellite Networks", IEEE Trans. Automatic Control, Vol. AC-25, No. 5, Oct. 1980, pp. 865-879

DESIGN AND EVALUATION OF PAGING STRATEGIES FOR PERSONAL COMMUNICATIONS

David Goodman, P. Krishnan*, Binay Sugla*

*WINLAB, Rutgers University; *Bell Laboratories, New Jersey; USA*[1]

ABSTRACT

The mobility of phones in a cellular or Personal Communication Services (PCS) environment introduces the problem of efficiently locating the called phone. In this paper, we present an analysis of the delay and number of messages transmitted in different sequential and parallel search strategies, considering for the first time the issue of queuing on radio paging channels. Our analysis shows that parallel search may not reduce the time to find a mobile phone if the parameters of the system are unfavorable. We also develop an efficient algorithm for searching with minimum expected number of messages when the location of the phone is given by a probability vector.

1 INTRODUCTION

An important problem with mobile phones is that the location of the called phone is unknown and must be determined before the call is completed. Several schemes have been employed and suggested for this problem. A centralized paging scheme in which the called phone number is broadcast is inefficient in the use of radio bandwidth. In the more recent search techniques, the basic unit of paging is the cell level. The problem now is to design an efficient search algorithm such that the relevant costs are minimized.

A number of research papers address facets of the mobile location problem. The problem is complex because there are several performance criteria including costs of reporting, recording, and retrieving the locations of mobile phones, costs of searching for mobile phones when calls arrive, the probability of a successful search, and delays in finding phones. Tracking and search costs include radio channel occupancy, transmissions in fixed networks, and database transactions. The overall complexity is also important because a scheme that requires a highly distributed real-time system may introduce problems of its

[1]Email: dgoodman@winlab.rutgers.edu, pk@research.att.com, sugla@research.att.com

own, particularly with respect to reliability. Previous papers [1, 2, 3, 4, 6, 8, 9] address network architecture issues, database structures, and tradeoffs between registration and paging costs.

This paper focuses on the search process. We assume that the system has accurate knowledge that a phone is in a "location area" which consists of a collection of cells. We then examine sequential search strategies for determining the cell in which the phone is located. The search procedure consists of sending paging messages to a group of cells in the location area and waiting for a response from the phone. If no response arrives within a fixed waiting period, the network sends paging messages in another group of cells, and again waits for a response. The procedure continues until the phone responds to a paging message. The quality criteria that we examine are the search delay and the total number of messages transmitted.

An important issue is the queuing delay of paging messages on radio channels. Our probabilistic analysis reveals that parallel search may not reduce the time to find a mobile phone if the parameters of the system are unfavorable. We also develop an efficient algorithm for paging to minimize the total number of messages when we know the probability of finding a phone in any specific cell.

Rose and Yates inform us that they have also independently come up with a similar queuing delay analysis [10]. Our goal in this paper is to study the design implications of our analysis taking into account both the delay incurred and number of messages transmitted. The delay analysis focuses on uniformly distributed traffic and includes the effect of a timeout at each cell. In [10], the authors analyze delay as a function of the mean of the ordered distribution that describes the uncertainty about the location of the mobile phone.

In [7], Madhavapeddy et al. propose methods to empirically compute the probability of a mobile phone being in any paging cell using the registration history. They also develop algorithms to minimize the expected number of messages while searching for a mobile phone; we compare and contrast our algorithms with the ones in [7] in Section 4. In [11], the authors deal with minimizing expected number of messages for different types of distribution.

The rest of the paper is organized as follows. In Section 2 we present our model. In Section 3, we introduce and present an analysis of sequential and parallel schemes used for locating a mobile phone. The simple analysis in Section 3.1 raises important questions about the issue of queuing delays. In Section 3.2, we analyze the effect of queuing on radio paging channels, assuming a Poisson call-arrival. In Section 4, we develop an efficient paging strategy to minimize the total number of messages sent in locating the mobile phone when the possible locations of the phone are specified by a probability vector. We conclude in Section 5 with directions for future research.

2 THE MODEL

We denote the number of base stations by N_B. The called telephone can be in any one of the N_B cells served by the base stations. In the basic search technique, the search is divided into *stages*, where in the jth stage, k_j stations that have not searched for the phone in earlier stages query for the phone; the search stops when the phone is found, or all stations have searched for the phone. The stations that page in a stage form a *paging group*. The maximum number of paging groups is denoted by N_G. The case when $N_G = N_B$ is called *polling*, and the case when $N_G = 1$ is called *flooding*.

Let p_i, for $1 \leq i \leq N_B$, be the probability that the phone is in the service area of base station i, let S be the random variable denoting the number of stages in the search for the phone, and let M be the random variable denoting the total number of query messages sent in locating the phone. We define the *uniform case* as a situation in which a mobile phone is equally likely to be in any cell in the location area. Furthermore, we divide the location area into N_G equal-sized paging groups, each containing k cells. In terms of our notation, we have for the uniform case that $p_i = 1/N_B$, $\forall 1 \leq i \leq N_B$, and $k_j = k = N_B/N_G$, $\forall 1 \leq j \leq N_G$.

Two important quantities we study in this paper are the *delay* and the *number of messages transmitted* in searching for a mobile phone. In this paper we assume that the number of messages transmitted is equal to the number of base stations that page before the phone is located. In general, the delay (or time taken in locating the phone) is a function of the number of paging groups, the delay in the wireless channel, the timeout variable (i.e., the amount of time a station waits before deciding that the phone is not present in its paging area), the delay in the wired network, and the computational time at each base station and the switching network. Typically, the delay in the wired network and the computational delays are comparatively negligible.

3 ANALYSIS OF SEARCH TECHNIQUES

In this section, we study the uniform case in detail. This corresponds to the situation when $p_i = 1/N_B$, for all i.

3.1 Basic Sequential and Parallel Search

The primary quantities of interest in analyzing a search strategy are the expected and maximum number of stages (denoted by $E(S)$ and $\max(S)$ respectively), and the expected and maximum number of query messages (denoted by $E(M)$ and $\max(M)$ respectively).

Method	$E(S)$	$\max(S)$	$E(M)$	$\max(M)$
General k	$\frac{N_B+k}{2k}$	$\frac{N_B}{k}$	$\frac{N_B+k}{2}$	N_B
Polling ($k=1$)	$\frac{N_B+1}{2}$	N_B	$\frac{N_B+1}{2}$	N_B
Flooding ($k=N_B$)	1	1	N_B	N_B

Table 1 The expected and maximum number of search stages and messages for the uniform case when k stations page in a stage, and the phone is equally likely to be found by any of N_B stations.

It is obvious that $\max(S) = N_G = N_B/k$, and $\max(M) = N_B$. Clearly, $E(S) = \sum_{j=1}^{N_B/k} j \cdot (k/N_B)$, since with probability k/N_B, we will have j stages. Hence, $E(S) = (N_B + k)/2k = (N_G + 1)/2$. Since in each stage k messages are sent, $E(M) = k \cdot E(S) = (N_B + k)/2$.

Table 1 summarizes this information. The tradeoffs are easy to notice: if the delay is proportional to the number of stages, flooding finds the phone quickly with more messages, while polling takes longer but sends fewer messages.

3.2 Analysis with Queuing Delays

The tradeoff observed from the analysis in Section 3.1 assumes that the delay in finding a phone is proportional to the number of stages. However, in practice, paging messages will get queued at a station before being broadcast, because of the volume of calls. This implies that the response time delay associated with finding a mobile phone is not proportional to the number of stages anymore. We now analyze search techniques taking into account the effect of queuing delays at the stations.

In Theorem 1, we determine an expression for the expected queue length at a station. In Theorem 2, we derive an expression for the expected delay in paging based on our expression for the expected queue length. We then use this expression for the expected delay to derive in Section 3.3 an interesting result that affects the design of paging systems.

To obtain our expression for the queue length, we make some simplifying assumptions about the system, which are reasonable in our current setting.

Assumption 1 We assume the uniform case as described in Section 2. We assume that the entire delay bottleneck in any stage is in the radio channels. Each channel is an $M/M/1$ queue with average capacity of μ messages per unit time (i.e., each channel services broadcast requests at an average rate of μ messages per unit time). We assume that call arrivals are Poisson with aggregate arrival rate of ϕ calls per unit time. We assume that the k stations in each paging group are chosen at random from stations that have not yet paged. In other words, the paging groups are not fixed.[2] Assuming that the call arrivals are uniform over the N_B base stations, the arrival rate on each paging channel is $\phi S/N_G$. The average arrival rate on each paging channel is $\lambda(N_G) = \phi E(S)/N_G$. From Table 1, $\lambda(N_G) = \phi \cdot (N_G + 1)/2N_G$. (Notice that in flooding $\lambda(N_G) = \lambda(1) = \phi$, while in polling $\lambda(N_G) = \lambda(N_B) = \phi \cdot (N_B + 1)/2N_B$.)

We denote by $\rho(N_G)$ the quantity $\lambda(N_G)/\mu$. For notational convenience, we denote $\lambda(N_G)$ by λ and $\rho(N_G)$ by ρ; it should be borne in mind that for any fixed ϕ, the quantities λ and ρ are functions of N_G. □

Let Q_i be the random variable denoting the queue length at base station (paging station) i. The delay induced by a queue with ℓ outstanding paging requests is ℓ/μ. Assume without loss of generality that in the jth stage we request stations $1, 2, \ldots, k$ to broadcast paging messages. Intuitively, the delay D_j in the jth stage is closely related to the *maximum queue length* amongst the k base stations that page in that stage; i.e., D_j, for $1 \leq j < S$ is related to the random variable $G_j = \max(Q_1, Q_2, \ldots, Q_k)$. To get an expression for the expected delay, we would like to derive an expression for $E(G_j)$. We proceed to get an expression for $\Pr(G_j = \ell)$ from an expression for $\Pr(Q_i \leq \ell)$ as follows.

By M/M/1 queuing theory, the queue length Q_i at base station i is a random variable with geometric distribution having parameter $\rho = \lambda/\mu$; i.e., $\forall i$, $\Pr(Q_i = \ell) = (1 - \rho)\rho^\ell$. Hence, $\Pr(Q_i \leq \ell) = 1 - \rho^{\ell+1}$. By assumption, the Q_i are i.i.d. Therefore,

$$\Pr(G_j \leq \ell) = (\Pr(Q_i \leq \ell))^k = \left(1 - \rho^{\ell+1}\right)^k. \tag{1.1}$$

Further,

$$E(Q_i) = \frac{\rho}{1 - \rho}. \tag{1.2}$$

Clearly, for all $\ell > 0$, $\Pr(G_j = \ell) = \Pr(G_j \leq \ell) - \Pr(G_j \leq \ell - 1)$. Substituting for $\Pr(G_j \leq \ell)$ from (1.1) and simplifying yields

$$\Pr(G_j = \ell) = \sum_{m=1}^{k} \binom{k}{m} (-1)^m \rho^{m\ell} \left(\rho^m - 1\right), \qquad \ell > 0. \tag{1.3}$$

Now, $E(G_j) = \sum_{\ell > 0} \ell \cdot \Pr(G_j = \ell)$. Substituting for $\Pr(G_j = \ell)$ from (1.3), interchanging the order of summation, and simplifying yields the following theorem.

[2] We make this assumption for mathematical simplicity. In practice, the paging groups may be fixed before-hand, which introduces a mathematical dependency between the queue lengths at different stations.

Theorem 1 *Under the conditions listed in Assumption 1, $E(G_j)$, the average maximum queue length amongst the k stations that page in the jth stage is given by*

$$E(G_j) = \sum_{m=1}^{k} \binom{k}{m} (-1)^{m-1} \frac{\rho^m}{1-\rho^m}. \tag{1.4}$$

We are now ready to derive an expression for $E(D)$, the expected delay in locating a mobile phone. We denote the expected delay of the jth stage by $E(D_j)$. Since the k stations to page in the jth stage are chosen at random, it implies that $E(G_j)$ is the same for all j, where $1 \leq j < S$; with a slight abuse of notation, we will denote this average by $E(G)$. Similarly, by Assumption 1, since the Q_i's are i.i.d., we will use the notation $E(Q)$ to denote the expected queue length at any base station.

When a call arrives at a switch, the system sends messages to the paging group. The request eventually finds its way to the front of the queue and is broadcast. The station either gets a positive acknowledgement (say, immediately), or waits a fixed amount of time W before it times out and assumes the mobile phone is not found. We now compute the delay for the search strategy.

Let us assume the mobile phone is eventually found. (The analysis for the case when the phone cannot be located is similar, and is omitted.) If the mobile phone is found in the sth stage, the expected delay in this sth stage is $E(D_s) = E(Q)/\mu$, where $E(Q) = E(Q_i)$ is as in (1.2). (If station i locates the mobile phone, the expected delay is $E(Q_i)/\mu = E(Q)/\mu$.) For an unsuccessful stage j, the expected delay is $E(D_j) = E(G)/\mu + W$, since it takes $E(G)/\mu$ time to exhaust the queue, and the last station that broadcasts the request waits W units of time before timing out. Hence the average delay is

$$E(D) = \frac{E(Q)}{\mu} + (E(S) - 1) \cdot \left(\frac{E(G)}{\mu} + W\right),$$

where S is the number of stages. Substituting for $E(S)$ from Table 1, and for $E(Q)$ from (1.2) we get the following theorem.

Theorem 2 *For any given load $\rho = \rho(N_G) = \lambda(N_G)/\mu$, under the conditions of Assumption 1, the expected delay $E(D)$ in locating a mobile phone in the presence of queuing delays is given by*

$$E(D) = \frac{\rho}{\mu(1-\rho)} + \left(\frac{N_G - 1}{2}\right) \cdot \left(\frac{E(G)}{\mu} + W\right),$$

when the phone is eventually found, where N_G is the number of paging groups, $E(G) = E(G_j)$ is as in Theorem 1, and W is the maximum waiting time for a response from the phone.

Theorem 2 gives us an analytical handle for the paging delay in the presence of queuing delays. To better understand the design implications of this analysis, we now study how the average delay varies with the number of search stages and the system loading factor $\rho(1) = \phi/\mu$.

3.3 Design Implications of Queuing Analysis

In this section we look at numerical results based on the analysis in Section 3.2. For convenience, we study the quantity $E(D) \times \mu$ (rather than just $E(D)$); multiplying by μ only scales the units. Our delay is therefore measured in the number of messages; we call our unit of delay a *clock*, where one clock is the time required to send one message on the paging channel. Similarly the *waiting timeout* $w = W \cdot \mu$ is the number of clocks spent waiting for a response.

In Figure 1, for a fixed load $\phi/\mu < 1$, we plot the average delay as a function of the number of paging groups N_G. The graphs show that when the paging channel is not heavily loaded ($\phi < 0.95$, for example), flooding results in the shortest delay. However, when paging channels face congestion, flooding places a high volume of messages on the paging channels. This results in very high queuing delays. In going from flooding to a two-stage search, we add to the average waiting time. However, this is more than compensated by the reduction in queuing time. Adding further stages beyond two is generally counter-productive. The increased waiting time is more significant than further reductions in queuing. Given that $E(M) = (N_B + (N_B/N_G))/2$ (from Table 1), perhaps $N_G = 4$ is a good value for the number of paging groups—this reduces the expected number of messages (relative to $N_G = 1$) from N_B to $0.625 N_B$ in addition to providing small delay.

Figure 2a graphs the average delay as a function of channel load ϕ/μ. The graph illustrates dramatically that while the delay blows up for single stage paging, it remains manageable for two or more stages. Although not obvious, this is not surprising, since in a one stage search, the paging channels get inundated with messages which blows up the delay, while having two search stages reduces the messages more than linearly, as given by the expression for $E(G)$ in Theorem 1.

An alternative way of looking at the problem is to see what is the maximum single stage loading (ϕ/μ) that can be supported for a specific value of expected delay for different values of N_G. Figure 2b plots the minimum ϕ/μ such that the average delay exceeds 200 vs. N_G. As we see from the graph the sustainable load increases with N_G, but beyond a point, the delay introduced by multiple stages predominates.

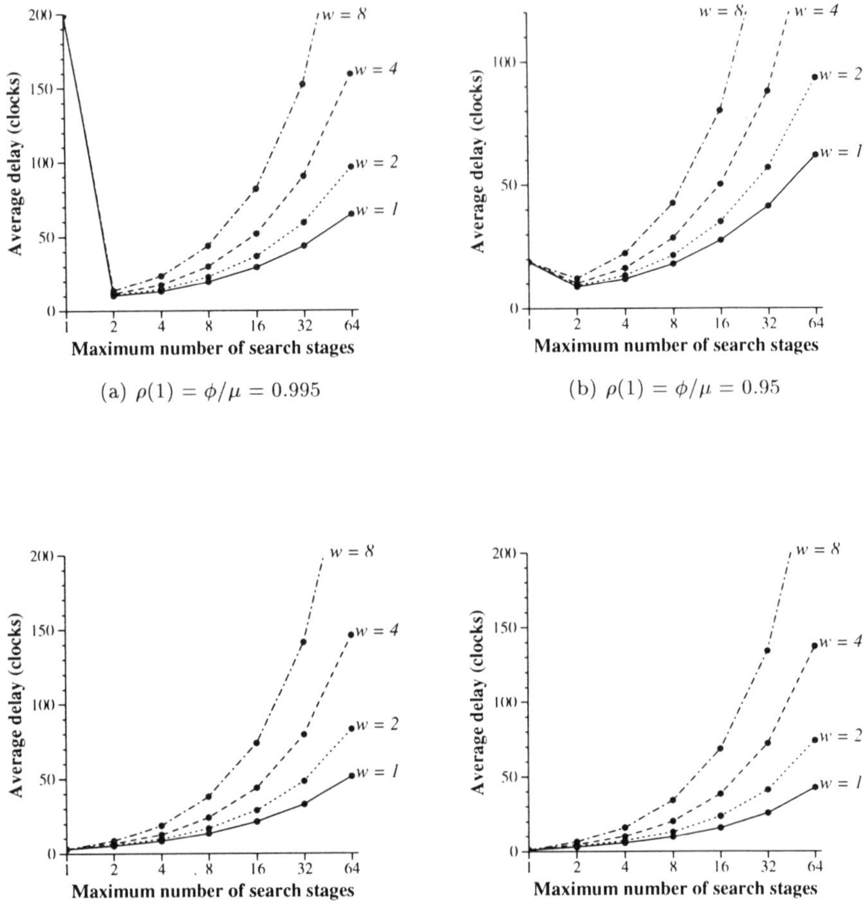

(a) $\rho(1) = \phi/\mu = 0.995$

(b) $\rho(1) = \phi/\mu = 0.95$

(c) $\rho(1) = \phi/\mu = 0.75$

(d) $\rho(1) = \phi/\mu = 0.5$

Figure 1 Average delay vs. the maximum number of paging groups N_G for four different values of load $\rho(1) = \phi/\mu$. The number of base stations is $N_B = 64$. The four different curves in each graph correspond to different values of the waiting time, w. The x axis is logarithmic to better illustrate the interesting portions of the graph corresponding to small N_G. Portions of the graph corresponding to a high delay have been cropped. As seen from (a) and to a lesser extent from (b), the delay reduces dramatically in going from one to two paging groups in the presence of substantial load.

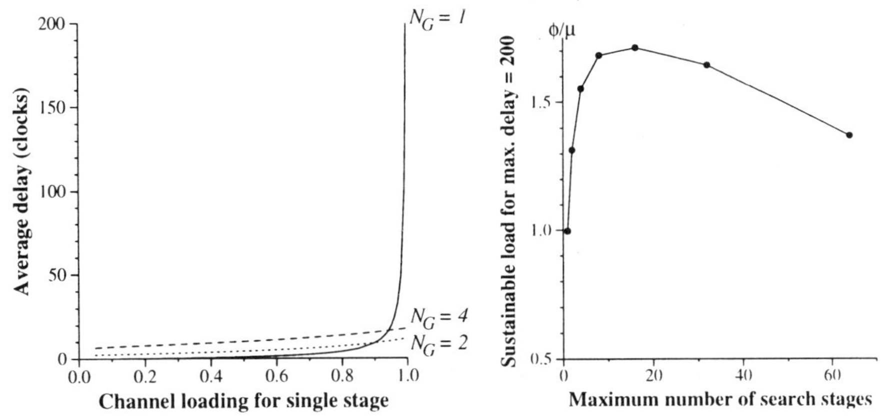

(a) Average delay vs. ϕ/μ (b) ϕ/μ vs. N_G for fixed delay

Figure 2 Average delay vs. ϕ/μ (a), and the minimum single channel loading ϕ/μ such that delay exceeds 200 plotted against N_G (b). The waiting timeout $w = 4$.

4 MINIMIZING $E(M)$ GIVEN $\vec{P}$

In this section we consider the general (non-uniform) case when the phone is not necessarily equally likely to be found in any of the N_B paging cells. Instead, we are given a probability vector $\vec{p} = (p_1, p_2, \ldots, p_{N_B})$ such that p_i is the probability that the phone is in the paging area of base station i. (These probabilities could be estimated in practice by using the technique from [7], for example.) In this scenario, our goal is to find an efficient algorithm that groups the N_B stations into at most N_G paging groups to minimize $E(M)$, the average number of broadcast messages.[3]

For this problem, we study two cases: We first study the special case when $N_G = 2$ in detail and develop an algorithm *Divide2* that partitions the N_G stations optimally and efficiently into two paging groups. Algorithm *Divide2* is based on "Algorithm 2" from [7], with a couple of simple enhancements. The $N_G = 2$ case is particularly interesting since our delay analysis from Section 3.3 suggests that we gain a lot in going from $N_G = 1$ to $N_G = 2$ with respect to queuing delay at high loads. It is also of great interest to practitioners given its simplicity. We then develop a polynomial time algorithm *Group* that minimizes the expected number of messages for the general case when N_G is arbitrary.

[3]The queuing delays would depend on modeling how the probability vectors over different calls interact. We do not tackle the queuing delay problem in this general setting.

Without loss of generality, assume $p_1 \geq p_2 \geq p_3 \geq \cdots \geq p_{N_B}$. *Since the p_i's are not necessarily equal, each paging group need not have $k = N_B/N_G$ elements.* Let paging set B_j be the set of k_j base stations that page in the jth stage. We denote $\sum_{i \in B_j} p_i$ by P_j. By the definition of expectation, it follows that

$$E(M) = \sum_{j=1}^{N_G} P_j \cdot \left(\sum_{m=1}^{j} k_m \right). \tag{1.5}$$

We make some interesting observations that form the basis for our algorithms *Divide2* and *Group*. In Lemma 1, we show that there is a configuration that minimizes $E(M)$ such that for any $1 \leq i \leq N_B - 1$, stations i and $i+1$ are in the same paging set B_j or in consecutive paging sets (i.e., station i is in set B_j and station $i+1$ is in set B_{j+1}). In other words, we will never have a station ℓ page later than station i if $p_\ell > p_i$. (This fact is also noted in [7, 11].) In Lemma 2, we show that in the configuration that minimizes $E(M)$, there are indeed N_G paging sets (and not $N'_G < N_G$ paging sets). For brevity, the proofs are omitted here, but can be found in the full version of the paper [5].

Lemma 1 *Let the sets $B_1, B_2, \ldots B_{N_G}$ be a partition of the N_B stations into N_G sets such that $E(M)$ is minimized. If station $i \in B_j$, then no station ℓ such that $p_\ell > p_i$ can be in set B_n, where $n > j$.*

Lemma 2 *Let the sets $B_1, B_2, \ldots B_{N_G}$ be a partition of the N_B stations into $N_G \leq N_B$ sets such that $E(M)$ is minimized. Then $k_j > 0$, for $1 \leq j \leq N_G$.*

Notice from Lemmas 1 and 2 that by setting $N_G = N_B$, we essentially want to sort the probability vector; hence the lowerbound of $\Omega(N_B \log N_B)$ on sorting N_B numbers (in the comparison model) holds for our problem too.

4.1 When $N_G = 2$: Algorithm *Divide2*

When the number of paging groups is 2, we want to divide the N_G stations into two sets B_1 and B_2 so that $E(M)$ is minimized. Based on Lemmas 1 and 2, we see that we can first sort the N_B probabilities, and "test" the $N_B - 1$ possible positions to determine which position gives us two sets that minimize $E(M)$. We can do this by iteratively calculating the value of $E(M)$ for each k, for $1 \leq k \leq N_B - 1$, when the first k stations are in set B_1, and choose the k that minimizes $E(M)$.

"Algorithm 2" from [7] computes B_1 and B_2 by observing the following property: If B_1 has the first $k-1$ stations, and $P = \sum_{i=1}^{k-1} p_i$, by adding the kth station to set B_1, we have $\Delta E(M)$, the change in $E(M)$, as

$$\Delta E(M) = P - (N_B - k) \cdot p_k. \tag{1.6}$$

We need to increase k only as long as $\Delta E(M) < 0$.

Since $E(M)$ when taken as a function of k, the number of elements in set B_1, is convex, the above technique provides us with an optimal partitioning of the stations into two paging groups.

We make two simple observations below that can be used in conjunction with "Algorithm 2" from [7] to get *Divide2*.

1. In our iterative process of computing the sets B_1 and B_2, let $|B_1| = k-1$, and let $P = \sum_{i=1}^{k-1} p_i$ be the sum of the probabilities of the $k-1$ elements in B_1. *If $P + p_k \leq 1/2$, the kth station can be added to set B_1.* This is because $p_k \leq p_{k-1} \leq P/(k-1)$, since $\vec{p}$ is sorted. By virtue of $\vec{p}$ being sorted, and $P+p_k \leq 1/2$, we have $p_k \cdot (N_B - k) \geq \sum_{i=k+1}^{N_B} p_i \geq 1/2$. Hence, $\Delta E(M) \leq 0$.
2. The convexity of $E(M)$ as a function of k combined with our knowledge that $\vec{p}$ is sorted implies that *in the optimal breakup, $k \leq (N_B+1)/2$.* This is because we know that $p_k \leq P_{k-1}/(k-1)$; from (1.6) it follows that $\Delta E(M) \geq 0$ for $k \geq (N_B+1)/2$.

Notice that Observation 2 does not improve the time complexity of the strategy, but helps us in getting a better running-time bound: The algorithm takes at most $(N_B+1)/2$ iterations after sorting the vector $\vec{p}$. An interesting related question is: How unbalanced in terms of probability can sets B_1 and B_2 be in the optimal partitioning? We discuss this issue briefly in Appendix A.

4.2 Algorithm *Group*

In this section we look at the case when N_G is arbitrary. An obvious generalization of algorithm *Divide2* to the case when $N_G > 2$ is a greedy method that moves elements between adjacent sets as long $E(M)$ decreases; an orderly way to do this is via "Algorithm 3" from [7]. Unfortunately, this greedy method is not always optimal; in particular, we can get "stuck" in a local minimum as shown in Appendix A.

We can divide the N_B elements into N_G groups in polynomial time using the following dynamic programming method. Let $T(i,j,k)$ be the minimum expected number of messages possible by concentrating only on stations $\{i, i+1, \ldots, j\}$ and partitioning this set of stations into $k \geq 1$ paging groups, for $j \geq i+k-1$. Algorithm *Group* wants to find the partition corresponding to $T(1, N_B, N_G)$. Clearly, $T(i,j,1)$ can be computed, for all i,j. Iteratively compute $T(i,j,k)$ as follows: from Lemmas 1 and 2, the best way to partition $(i, i+1, \ldots, j)$ into k sets is to find a station u such that $\{i, i+1, \ldots, u\}$ is the first set and

$\{u+1, u+2, \ldots, j\}$ is divided into $k-1$ sets. In particular, by using the formula for $E(M)$ given in (1.5) we get

$$T(i,j,m+1) = \min_{i \leq u \leq j-m} \left(T(i,u,1) + T(u+1,j,m) + (u-i+1) \sum_{q=u+1}^{j} p_q \right).$$

Algorithm *Group* sorts the N_B probabilities, if required, and precomputes for all $1 \leq i, j \leq N_B$ the partial sums $\sum_{u=i}^{j} p_u$. It then computes $T(i,j,k)$ for all $1 \leq i \leq N_B - N_G$, $i \leq j \leq N_B$, and $1 \leq k \leq N_G$ in $O(N_B^2 N_G + N_B \log N_B)$ time. (Precomputing the $\sum_{u=i}^{j} p_u$'s helps in determining $T(i,j,k+1)$ iteratively in $O(1)$ time.) The partition corresponding to each $T(i,j,k)$ can be maintained implicitly by storing the corresponding value of u with each $T(i,j,k)$.

Can the Gap be Closed? There is a gap between our upperbound of $O(N_B^2 N_G)$ and the sorting lowerbound (as mentioned at the end of Section 4) of $\Omega(N_B \log N_B)$ we have for this problem. Certain optimizations can be done for specific values of N_G; for example, in Section 4.1 we looked at the important case when $N_G = 2$. The following lemma helps in cutting down the computation for general N_G by a constant factor. (The big-oh bound still remains $O(N_B^2 N_G)$.) A proof sketch is given in Appendix A.

Lemma 3 *Amongst all possible partitionings of the stations into paging groups that minimize $E(M)$, there is one partitioning into sets B_1, $B_2, \ldots B_{N_G}$ such that $\forall j > 1$, $k_{j-1} \leq k_j$.*

5 CONCLUSIONS

In this paper we have looked at important issues in searching for a mobile phone. We have looked at the question of queuing delays and derived the result that parallel search may not reduce the delay in searching for a phone if the system is heavily loaded. In particular, our analysis sheds light on important design issues in paging in cellular telephone networks. We find that at high loads, the delay decreases significantly in going from one to two paging groups. When considered along with the number of messages sent, perhaps 2–4 paging groups is a good value; e.g., having four paging groups reduces the expected total number of messages (relative to flooding) from N_B to $0.625 N_B$. The minimum is $0.5 N_B$, achieved with flooding which leads to a high delay. It would be interesting to study the impact of queuing delay on other results in the literature.

We have also presented an efficient method to group stations into paging groups when a probability vector defining the likelihood of finding a phone in any paging area is given, and the goal is to minimize the expected number of messages. Our algorithm *Divide2* optimally and efficiently partitions the states into $N_G = 2$ paging groups, an important case as suggested by our delay analysis. For general N_G, our algorithm *Group* takes worst case time proportional to $O(N_B^2 N_G + N_B \log N_B)$ to determine the optimal partitioning. It would be interesting to close the gap between our known lowerbound of $O(N_B \log N_B)$ and the upperbound of *Group*. Another open issue is to derive an appropriate model for analyzing the effect of queuing delays when the probability vector is given. Some discussion related to these open issues appears in the Appendix.

Acknowledgements. We would like to thank Ming Kao for discussions related to some of the material in Section 4.2.

REFERENCES

[1] B. Awerbuch, D. Peleg. "Concurrent Online Tracking of Mobile Users," Proceedings of the 1991 ACM SIGCOMM Conference, pp. 221–233.

[2] A. Bar-Noy, I. Kessler, M. Sidi. "Mobile Users: To Update or not to Update," Wireless Networks, 1(1), 1995.

[3] A. Bar-Noy, I. Kessler. "Tracking Mobile Users in Wireless Networks," INFOCOMM '93, pp. 1232–1239.

[4] S. T. S. Chia. "Location Registration and Paging in a Third Generation Mobile System," BT Technology Journal, 9(4), October 1991, pp. 61–68.

[5] D. Goodman, P. Krishnan, B. Sugla. "Minimizing Queuing Delays and Number of Messages in Mobile Phone Location," AT&T Bell Laboratories Technical Report BL0112500-951005-03TM.

[6] R. H. Katz. "Adaptation and Mobility in Wireless Systems," R. H. Katz, IEEE Personal Communications, First Quarter 1994, pp. 6–17.

[7] S. Madhavapeddy, K. Basu, A. Roberts. "Adaptive Paging Algorithms for Cellular Systems," Proceedings of the Fifth WINLAB Workshop on Third Generation Wireless Information Networks, April 1995, pp. 347–361.

[8] K. S. Meir-Hellstern, E. Alonso, and D. O'Neil. "The Use of SS7 and GSM to Support High Density Personal Communications," Proceedings of the International Conference on Communications (ICC), 1992.

[9] S. Mohan, R. Jain. "Two User Location Strategies for Personal Communication Services," IEEE Personal Communications, First Quarter 1994, pp. 42–50.

[10] C. Rose and R. Yates. "Ensemble Polling Strategies for Increased Paging Capacity in Mobile Communication Networks," manuscript.

[11] C. Rose and R. Yates. "Minimizing the Average Cost of Paging Under Delay Constraints," ACM Wireless Networks, 1(2), pp. 211–219, 1995.

A APPENDIX

In this appendix, we present some issues related to possible future work, and a proof sketch for Lemma 3. For more details, please see [5].

Minimizing $E(M)$ for $N_G = 2$. We continue here with the question we posed at the end of Section 4.1: Can $\Delta E(M)$ be negative if $P_{k-1} > 1/2$? The following example shows that the answer is "yes." Consider the probability vector $(5/8, 1/3, 1/240, 1/240, \ldots, 1/240)$, for example. Although $5/8 > 1/2$, the optimal partitioning is: $\{5/8, 1/3\}$, $\{1/240, \ldots, 1/240\}$.

The Greedy Strategy for General N_G. In order to get a better-than-$O(N_B^2 N_G)$ optimal strategy, an intuitive idea (e.g., "Algorithm 3" from [7]) is a simple greedy strategy. If we start with a partitioning of the N_B stations into N_G sets obeying the property stated before Lemma 1 (that for any $1 \leq i \leq N_B - 1$, stations i and $i+1$ are in the same paging set B_j or in consecutive paging sets), and move stations between adjacent sets if $E(M)$ decreases (i.e., greedily), will we reach a configuration that minimizes $E(M)$?

The following example shows that we may not. For our counter-example, we use $N_B = 23$ stations with $\vec{p} = (1/3, 49/150, 1/60, 1/60, \ldots, 1/60, 1/150)$. Let $N_G = 3$. The partitioning $\{1,2\}, \{3,4,\ldots,13\}, \{14,15,\ldots 23\}$, is optimal in this case, with $E(M) = 1096/150$. If we start with a partition of stations into sets $\{1\}, \{2\}, \{3,4,\ldots,23\}$ and use the greedy method (or "Algorithm 3" from [7]), we will end up with a final partition of stations into sets $B_1 = \{1\}$, $B_2 = \{2,3\}$, and $B_3 = \{4,5,\ldots,23\}$ with $E(M) = 1320/150$, which we know is not optimal. That is, we get "stuck" in a local minimum. The basic problem here is that to get out of the local minimum, we need to make two changes *simultaneously*; i.e., move station 2 to set B_1 and station 4 to set B_2.

Proof of Lemma 3. (Sketch) Lemma 1 implies that there is a configuration that minimizes $E(M)$ such that for any $1 \leq i \leq N_B - 1$, stations i and $i+1$ are in the same paging set B_j or in consecutive paging sets. Consider such an optimal partitioning. Let i be the station with the smallest probability in set B_{j-1}. It can be verified that we can move station i from B_{j-1} to B_j without increasing $E(M)$ if

$$p_i \cdot (k_j + 1) \leq P_{j-1}. \qquad (1.7)$$

If i is the station with smallest probability p_i amongst stations in B_{j-1}, it implies that every station in B_j has an associated probability at most p_i. If in the optimal partitioning $k_{j-1} \geq k_j + 1$, then $p_i \cdot (k_{j-1}) \geq p_i \cdot (k_j + 1)$. Since $P_{j-1} \geq p_i \cdot k_{j-1}$, it follows that $p_i \cdot (k_j + 1) \leq P_{j-1}$. From (1.7), station i can be moved from B_{j-1} to B_j. This argument can be repeated until the desired configuration is attained.

DATABASE ARCHITECTURES AND LOCATION STRATEGIES FOR MOBILITY MANAGEMENT IN MOBILE RADIO SYSTEMS

Nébil Tabbane, ISG Tunis

Sami Tabbane, ESPTT

Route de Raoued, Km 3,5 - 2083 Cité El Ghazala - TUNISIA

Abstract - With the increasing number of mobile service users, cellular systems must offer higher traffic capacities. Therefore, cells sizes are reduced which increases the mobility signalling load. Part of this signalling is due to handovers ("radio mobility") and the other part to mobility management ("network mobility"). In this paper we combine location strategies and database architectures to evaluate their impact on user mobility management cost. Two location management strategies are studied: the classical strategy used in GSM, and the alternative strategy based on users' mobility patterns. The centralized, distributed and hybrid types of databases architectures are considered here and combined with the two location strategies. Evaluation of different scenarios is done analytically. We show that, for the values used here, the hybrid database architecture combined with the alternative strategy allows to save more than 50% of the costs involved by the classical strategy combined with the centralized database architecture (which last combination is used in most of current cellular systems).

I. INTRODUCTION

The demand in mobile services is showing a quasi-exponential growth since the early nineties. To satisfy and comply with the demand of the users and the quality of service, cells are made smaller to increase network capacity and quality. One of the main drawbacks when reducing the cell size is a substancial increase of the mobility signalling load: first, handover signalling and second, signalling due to location updates and paging procedures. Cellular systems development will be limited if their capacity does not grow in the same way. This condition partly relies on the optimisation of the radio interface. This means that signalling, which is a system information that is invisible to the users, must be reduced as much as possible.

In this study we focus on the location management process in order to identify how to reduce its costs in terms of signalling load. The location management relies first on the location strategy and second on the network architecture, and more precisely on the « *mobility* » database architecture (e.g. HLR and VLRs structuring). Thus, in this paper, we consider two location strategies combined with three database architectures.

Schematically, a cellular network, say GSM [5], is composed of cells gathered into location areas. When a mobile station (MS) crosses a location area border, it generates a location update which consists in updating the *Home Location Register* (HLR) and the *Visitor Location Register* (VLR), the main two databases types of the system. When the networks has to find the mobile (when the user receives an incoming call typically), it pages it in the location area where the mobile has processed its last location update. This process, used in most present cellular systems, is called here *Classical Stategy* (CS).

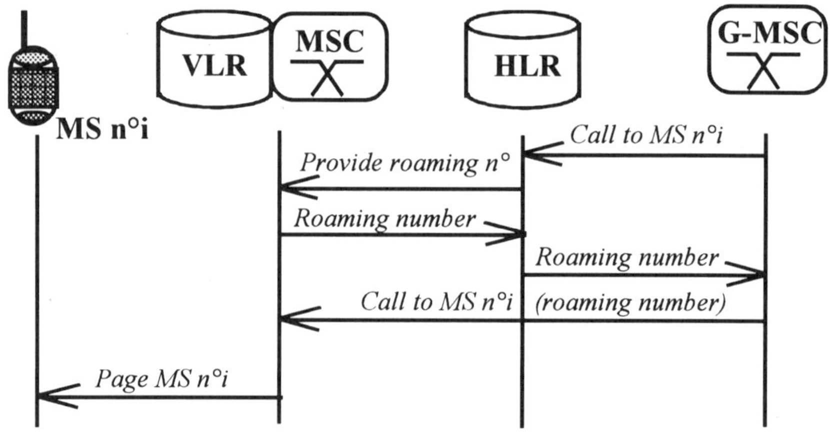

Figure 1 Example of mobile paging (GSM-like)

Because CS consumes an important amount of radio bandwidth and observing that users show repetitive mobility patterns, we have defined the *Alternative Strategy* (AS) [1]. The main goal of AS is to reduce the traffic due to mobility management by taking advantage of the users' highly predictable patterns, thus, reducing the location updating cost. In this strategy, we consider that the system handles a *profile*, for each user, where are recorded his most probable mobility patterns. The structure and contents of this profile are as follows.

- For an individual user and to each period of time $[t_i, t_j]$, corresponds a set of location areas, *k*, which can be optimized, in the following way:

  - (a_f, α_f) with $1 \leq f \leq k$ where:
    - a_f is the location area the user can be located in,
    - α_f is the probability that the user is located in a_f,
  - with $\alpha_1 > \alpha_2 > ... > \alpha_k$ and $\sum_{i=1}^{k} \alpha_i < 1$

The profile of each user is stored in the HLR. The process follows the following algorithm.

> If a user is called at time t_g (with $t_i \leq t_g < t_j$), the system will page him over the location area a_1, if not found in this location area, the system will page him over the location area a_2 ... until the location area a_k (in case when the user has not been found in the previous areas a_i).

It is of prime importance to insure that the system always keeps track of the user's location. Thus, the following procedure is provided.

When the subscriber moves away from the recorded zone $\{a_1, ..., a_k\}$ during the corresponding period $[t_i, t_j[$, the terminal processes a voluntary registration by pointing out its presence to the nearest *Base Station* (BS).

Using successively CS and AS location strategies, we study the database architecture influence on the generated signalling burden. We present three database architectures: *centralized* (*CDB*) - as the one used in GSM - *distributed* (*DDB*) and *hybrid* (*HDB*).

The cost of each strategy is evaluated in terms of messages generated by the roaming of one given subscriber. Each interface is considered for these evaluations; nevertheless, as stretched before, attention must be focused on the number of messages carried on the radio interface.

II. CENTRALIZED DATABASE ARCHITECTURE

The centralized database architecture includes a central database called HLR where subscriber's data as well as its current location (i.e. its current VLR number) are stored. This architecture includes also several VLRs which contain the data of the subscribers located under their service area. For each user, the VLR handles a copy of his data stored in the HLR and his location area identifier. On Figure 2, we have reproduced the CDB hierarchy.

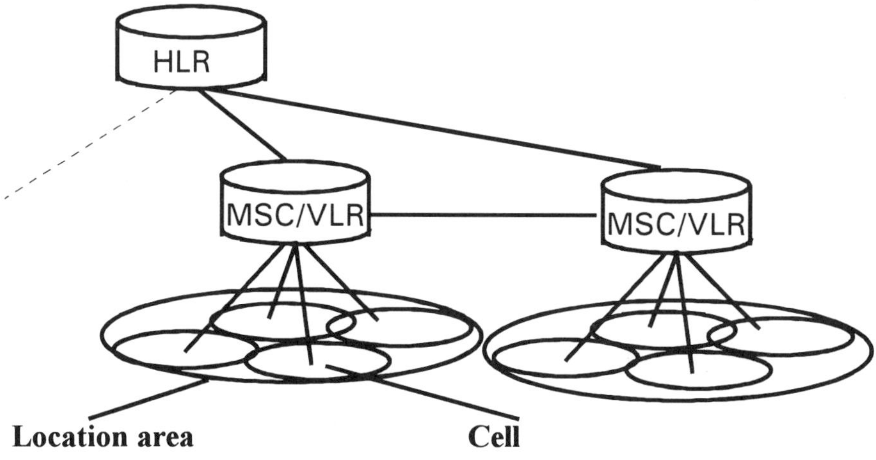

Figure 2 Centralized Database Architecture

II.A. Classical Strategy

The Classical Strategy defined in the introduction is the location strategy widely used in current cellular systems such as GSM.

a. Location Updating Cost

In CS, a location update is generated each time the user crosses the boundary of a location area. This event occurs at a rate equal to $\frac{8\,\mathrm{V}}{3\pi\sqrt{\mathrm{N}}\,\mathrm{R}}$ where V is the mean user's velocity, N is the number of cells per location area and R is the cell radius [1, 2, 4].

We introduce the following additional parameters:

β_1: probability that an intra-VLR location update (i.e. the mobile moves from a location area to a new one without leaving its VLR service area) occurs (*case 1*),

β_2: probability that an inter-VLR using *Temporary Mobile Subscriber Identity* (TMSI) occurs (*case 21*),

β_3: probability that an inter-VLR using *International Mobile Subscriber Identity* (IMSI) occurs (*case 22*). In *case 21* the new VLR of the MS has to request the user's data from the old VLR instead of requesting it from the HLR in *case 22*. *Case 22* shows a higher number of transmitted messages than *case 21*.

$\mathrm{Nbl}_{\mathrm{cost,case}i}(\mathrm{int})$: number of messages generated by a location update on the interface *int* in the case *i* (i=1, 21 or 22).

Using this parameters, the cost of this procedure (expressed per time unit) is:

$$\frac{8\,\mathrm{V}}{3\pi\sqrt{\mathrm{N}}.\mathrm{R}}.\left[\beta_1.\mathrm{Nbl}_{\mathrm{cost,case1}}(\mathrm{int})+\beta_2.\mathrm{Nbl}_{\mathrm{cost.case21}}(\mathrm{int})+\beta_3.\mathrm{Nbl}_{\mathrm{cost,case22}}(\mathrm{int})\right]$$

b. Paging Cost

We suppose that a successful paging (*case 1*) occurs when the mobile acknowledges the first paging message. The paging is said to have failed when the mobile did not answer (*case 2*) after the transmission of two paging messages. We define $\mathrm{Nbp}_{\mathrm{cost}i}(\mathrm{int})$ as the cost (in number of messages) generated by a paging procedure on the interface *int* in the case *i* (i=1 or 2), λ_{t1} as the incoming call rate per user (in calls/hour) and λ_{t2} as the unsuccessful call rate per user (in calls/hour). The paging cost (per time unit) is then equal to [4]:

$$N\left[\lambda_{t1}.\mathrm{Nbp}_{\mathrm{cost1}}(\mathrm{int})+\lambda_{t2}.\mathrm{Nbp}_{\mathrm{cost2}}(\mathrm{int})\right]$$

II.B. Alternative Strategy

AS relies on user's profile. This profile is used to decide whether or not to generate a location update and where to page the user in case of mobile terminating call.

a. Location Updating Cost

The probability that in the time slot $[t_j, t_{j+1}[$ the user is situated in a location area recorded in his profile is $\sum_{i=1}^{k} \alpha_i$ and the probability that the user is not in a location area recorded in his profile is $(1-\sum_{i=1}^{k} \alpha_i)$. The cost generated in the first case includes the cost of one location update (which occurs when the mobile enters a location area of his profile and remains in his profile) which occurs with the probability $\sum_{i=1}^{k} \alpha_i$. The cost generated in the second case is the same as in the CS case. Using the parameters introduced before, the cost of AS location updating procedure, computed per time unit, is the following:

$$\left\{\sum_{i=1}^{k} \alpha_i [\beta_1.\mathrm{Nbl}_{cost,case1}(\mathrm{int}) + \beta_2.\mathrm{Nbl}_{cost,case21}(\mathrm{int}) + \beta_3.\mathrm{Nbl}_{cost,case22}(\mathrm{int})] + \frac{8\mathrm{V}}{3\pi\sqrt{\mathrm{N}}.\mathrm{R}}.(t_{j+1}-t_j).(1-\sum_{i=1}^{k}\alpha_i).[\beta_1.\mathrm{Nbl}_{cost,case1}(\mathrm{int}) + \beta_2.\mathrm{Nbl}_{cost,case21}(\mathrm{int}) + \beta_3.\mathrm{Nbl}_{cost,case22}(\mathrm{int})]\right\} / (t_{j+1}-t_j)$$

b. Paging Cost

To compute the paging cost, we consider two cases: either the user is located in a location area which is not recorded in his profile, or he is located in a location area of his profile. In the latter case the user will be paged first in a_1 (which paging will occur with probability α_1), then in a_2 if not found in a_1 (which paging will occur with the probability $\alpha_1+\alpha_2$), ...

The paging cost (per time unit) is then given by the following formula:

$$\left\{N.\left[\lambda_{t1}.Nbp_{cost1}(\mathrm{int}) + \lambda_{t2}.Nbp_{cost2}(\mathrm{int})\right]\right\}.\left\{(1-\sum_{i=1}^{k}\alpha_i) + \sum_{i=1}^{k} \mathrm{i}.\alpha_i\right\}$$

III. Distributed Database

In [7, 8, 9] different database architectures are investigated. We first consider the distributed database architecture which allows to manage the coverage area in a distributed manner (each network entity will manage a small service area).

To implement a distributed database different sizes of location areas are defined so that they are overlapping (see Figure 3).

We define *m* as the number of levels in the architecture. At each level *i* corresponds several VLR_i which manage the user's located in location areas of type ZL_i. The highest level (the HLR one) corresponds to $i=m$ and the lowest corresponds to $i=1$.

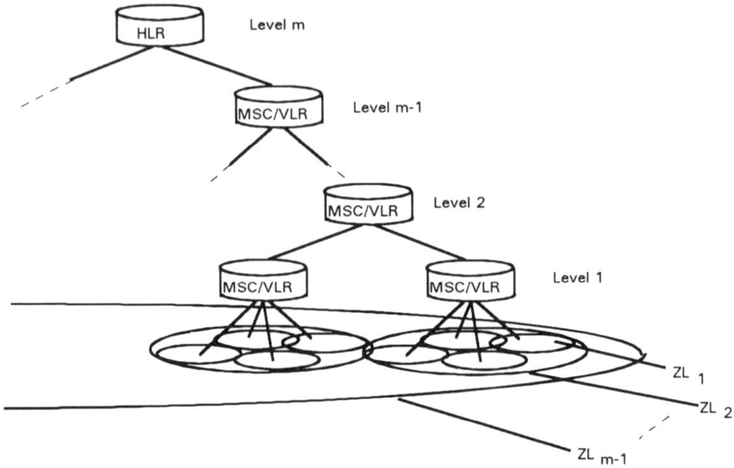

Figure 3 Distributed Database Architecture

III.A. Classical Strategy

a. Location Updating Cost

Let us give a simple example of use of this architecture. We take the case of an intra VLR_{i+1} location update. When the mobile crosses a ZL_i area border and remains in the same ZL_{i+1} area, the location update process involves all VLR_j (inter-VLR_j location update) for j=1, ...i+1. So that levels i+2 to m are not concerned by this location updating.

We define δ_i as the probability that a mobile generates a location update at level i, without moving between location areas at level (i+1). Note that $\Sigma\delta_i=1$. The probabilities of generating intra-VLR, inter-VLR with IMSI or TMSI are denoted β_{iD} (instead of β_i previously) as their value differ from the CDB case. The analytic formula for the location updating cost (per time unit) is determined from the centralized database case by introducing the δ_is. Note that β_{1D} does not appear because included in the δ_is. According to these assumptions and parameters, the location updating cost becomes:

$$\frac{8V}{3\pi\sqrt{N}.R}\cdot\left\{\left(\sum_{i=1}^{m-2}\delta_i\right).\mathrm{Nbl}_{cost,case1}(\mathrm{int})+\left(\sum_{i=2}^{m-1}\delta_i.(i-1)\right).\left[\beta_{2D}.\mathrm{Nbl}_{cost,case21}(\mathrm{int})+\beta_{3D}.\mathrm{Nbl}_{cost,case22}(\mathrm{int})\right]\right\}$$

b. Paging Cost

The paging process is carried on in the same way as in the CDB case. The only difference is that the HLR requests from the involved VLR at level (m-1) the roaming number of the user. In the general case (i=1,.., m-2), a VLR at level (i+1) requests the mobile roaming number from a VLR at level i. These requests generate messages on the corresponding interfaces and therefore the

the Nbp_{costi} values differ from that of the CDB case. Therefore, the paging cost formula remains the same as in the centralized database case.

III.B. Alternative Strategy

a) Location Updating Cost

The location update process is the same as in the CDB case, only the computation of the cost differs. Using the parameters introduced in the CS case, the cost of the location update is computed per time slot $(t_{j+1} - t_j)$.

$$\left[(\sum_{i=1}^{k}\alpha_i)+\frac{8V}{3\pi\sqrt{N}.R}.(t_{j+1}-t_j).(1-\sum_{i=1}^{k}\alpha_i)\right]$$
$$.\left\{\left(\sum_{i=1}^{m-2}\delta_i\right).Nbl_{cost,case1}(int)+\left(\sum_{i=2}^{m-1}\delta_i.(i-1)\right).\left[\beta_{2D}.Nbl_{cost,case21}(int)+\beta_{3D}.Nbl_{cost,case22}(int)\right]\right\}/(t_{j+1}-t_j)$$

b) Paging Cost

The paging is processed the same as in the centralized database case. The cost formula remains identical apart from the Nbp_{costi} which values are different from the CDB case.

IV. HYBRID DATABASE ARCHITECTURE

Compared with the DDB architecture, the hybrid database architecture only includes two location area levels: ZL_0 (smaller type) and ZL_1 (larger type, which kind includes several ZL_0). Then, a location area can be seen as including several paging areas. We distinguish two VLR types: VLR_Ls and VLR_Cs. VLR_Ls are local databases which control several ZL_0 location areas. VLR_Cs are central databases which control ZL_1 type location areas. Four levels are distinguished: level 4 (HLR), level 3 (VLR_c), level 2 (VLR_L) and level 1 (cells).

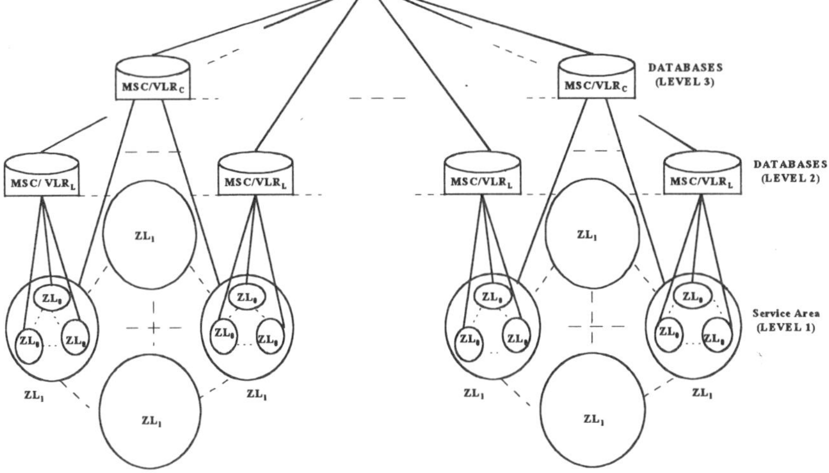

Figure 4 Hybrid Database Architecture

IV.A. Classical Strategy

We first introduce the new parameter denoted δ_{hybi} as the probability that a mobile generates a location update at level i, without moving between location areas at level (i+1). The δ_{hybi} values differ from that of δ_i defined in the DDB case. Similarly, we define β_{2H} as the probability that an inter-VLR using TMSI occurs (*case 21*) and β_{3H} as the probability that an inter-VLR using IMSI occurs (*case 22*).

a) Location Updating Cost

The location updating process is carried on the same as in the DDB architecture case. Using the new parameters and the formula determined in the DDB case, the location updating cost (per time unit) becomes:

$$\frac{8\,\mathrm{V}}{3\pi\sqrt{\mathrm{N}}.\mathrm{R}}\cdot\left\{\left(\delta_{\mathrm{hyb}1}+\delta_{\mathrm{hyb}2}\right).\mathrm{Nbl}_{\mathrm{cost,case1}}(\mathrm{int})+\left(\delta_{\mathrm{hyb}2}+\delta_{\mathrm{hyb}3}.2\right).\left[\beta_{2\mathrm{H}}.\mathrm{Nbl}_{\mathrm{cost,case21}}(\mathrm{int})+\beta_{3\mathrm{H}}.\mathrm{Nbl}_{\mathrm{cost,case22}}(\mathrm{int})\right]\right\}$$

b) Paging Cost

The paging process remains the same as in the DDB architecture. To page a user located in a ZL_0 type location area, the HLR determines his VLR_c in order to find out the VLR_L. Therefore, the formula remains the same, with a change only in the $Nbp_{cost\,i}$ values.

IV.B. Alternative Strategy

a) Location Updating Cost

As the HDB can be considered as a particular case of the DDB, the location updating process remains the same as in the DDB case. In the present case, the only difference is that we only have two databases levels. Using the parameters introduced in the CS cost evaluation, the AS location updating cost is :

$$\left[(\sum_{i=1}^{k}\alpha_i)+\frac{8\,\mathrm{V}}{3\pi\sqrt{\mathrm{N}}.\mathrm{R}}.(\mathrm{t}_{j+1}-\mathrm{t}_j).(1-\sum_{i=1}^{k}\alpha_i)\right]/(\mathrm{t}_{j+1}-\mathrm{t}_j)$$

$$.\left\{\left(\delta_{\mathrm{hyb}1}+\delta_{\mathrm{hyb}2}\right).\mathrm{Nbl}_{\mathrm{cost,case1}}(\mathrm{int})+\left(\delta_{\mathrm{hyb}2}+\delta_{\mathrm{hyb}3}.2\right).\left[\beta_{2\mathrm{H}}.\mathrm{Nbl}_{\mathrm{cost,case21}}(\mathrm{int})+\beta_{3\mathrm{H}}.\mathrm{Nbl}_{\mathrm{cost,case22}}(\mathrm{int})\right]\right\}$$

b) Paging Cost

The paging process also remains the same as in the DDB case with a change only in the $Nbp_{cost\,i}$ values.

V. RESULTS

V.1. Hypotheses

The values of the parameters taken in these evaluations are the following:

R=1 km (cell radius), N=7 (cell number in a location area), N_c=1 (number of location areas per MSC/VLR), V=12 km/hour (mobile mean speed),

λ_{t1}=0.6 call/hour [3], $\lambda_{t2}=\lambda_{t1}/100$ call/hour and $\Sigma\alpha_i$=0.75 (α_1=0.50, α_2=0.15, α_3=0.10).

CDB values :

β_1=0.29 (intra-VLR location update probability), β_2=0.41 (probability of inter-VLR location update with TMSI), β_3=0.3 (probability of inter-VLR location update with IMSI).

DDB values :

$\beta_{2D}=\beta_{3D}$=0.5 (probabilities of inter-VLR location update with TMSI and IMSI). Each VLR_1 controls 4 ZL_1, a VLR_2 controls 2 ZL_2 and a VLR_3 controls 3 ZL_3 in the distributed database architecture. δ_1=0.78, $\delta_2=\delta_4$=0 and δ_3=0.22.

We have taken the numbers of cells per location area respectively of level 1, 2 and 3 to be 7, 28 and 49.

HDB values :

$\beta_{2H}=\beta_{3H}$=0.5. A VLR_L controls 4 ZL_0 (ZL_0 a location area with 7 cells) and a VLR_C controls 3 ZL_1 (ZL_1 location area with 49 cells). $\delta_{hyb\,1}$=0.78, δ_{hyb2}=0.22 and $\delta_{hyb\,3}$=0.

We define the A interface as the BSC-MSC interface, B is the MSC-VLR interface, C_i is the VLR_i-VLR_{i+1} interface, D is the HLR-VLR interface, E is the MSC-MSC interface.

V.2. Database Architecture Comparison

On the following figures, we have plotted the location update and paging costs versus the interface and for each database architecture. Figure 5 shows that using the CS, the hybrid architecture gives rise to the lowest costs for most of the interfaces.

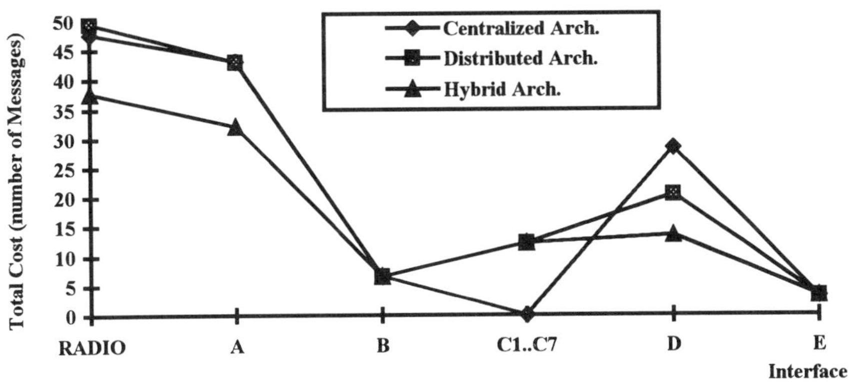

Figure 5 The Classical Strategy performance

On Figure 6, the costs generated using the AS are plotted. As in CS case, the hybrid database architecture gives the best results for most of the interfaces, especially for the radio interface which is the most critical one.

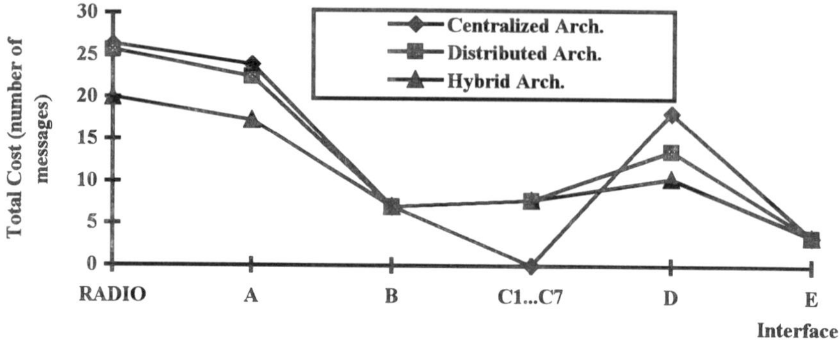

Figure 6 The Alternative Strategy performance

The results obtained show that the hybrid database architecture allows to reduce the mobility management costs compared to the centralized database case. This result is of prime importance as the additional infrastructure required to implement a HDB architecture is not very important. Using the HDB structure, potential savings on the radio interface are around 20% using both strategies, compared to the CDB case currently implemented in analog and digital cellular systems. The values used to compute these evaluations are a medium case. Precise statistics on users mobility and traffic behaviour will allow the choice of the best architecture.

V.3. Strategies Comparison

In this section, we have plotted the costs generated by each strategy versus the interface in the centralized and distributed database architectures. On Figure 7, we consider the centralized database architecture case.

It is clear that AS gives the best results for all the interfaces. CS generates almost twice as much messages as AS on the radio interface.

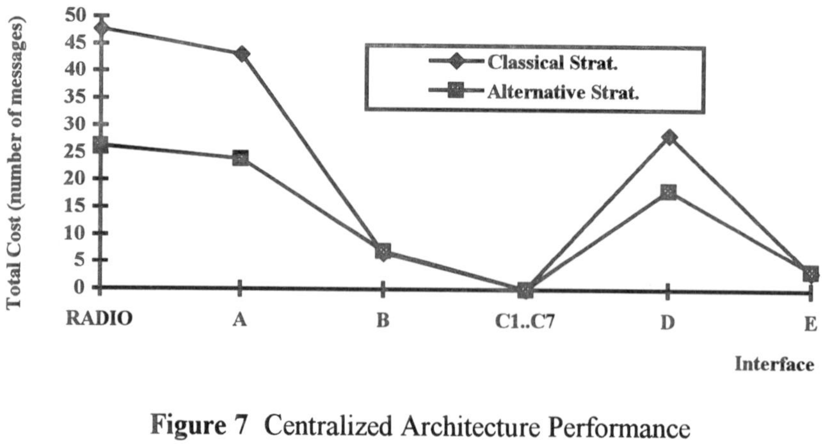

Figure 7 Centralized Architecture Performance

Results obtained with the distributed architecture (see Figure 8) are similar, showing that AS gives rise to lower costs than CS. Here again the ratio between the costs generated by the two strategies is around two for the radio interface.

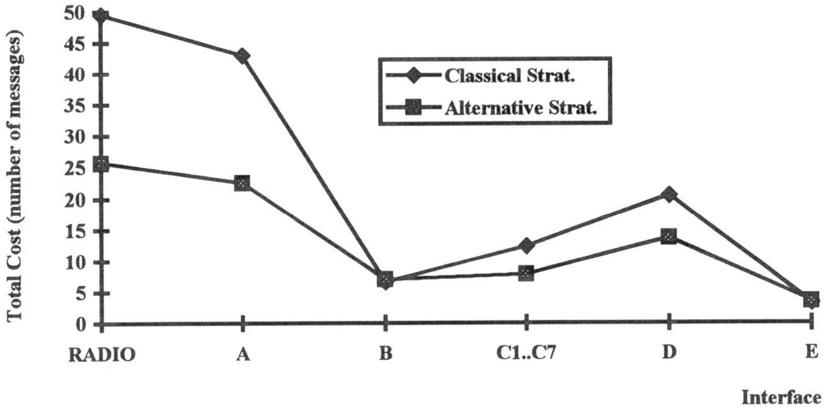

Figure 8 Distributed Architecture Performance

Comparison between the two location strategies points out a big advantage of AS over CS. The main reason is because this strategy allows to save location updates when mobility patterns are predictable. Values taken to produce these evaluations are that of medium predictable mobility behaviour. In the case when the user shows unpredictable patterns, AS may generates the same amount of messages as CS.

VI. CONCLUSIONS

Mobility management is one of the most critical function of cellular networks. It mainly consists on *radio mobility* (handover) and *network mobility* (combination of the location management and the paging procedures). With the increasing demand in mobile services which leads to a cell size reduction, network mobility importance increases. To improve the cellular networks capacity, the signalling load, especially carried on the radio interface, must be reduced as far as possible.

In this paper, we investigated two means for reducing this burden. We compare the classical and alternative location strategies for three different database architectures (centralized, distributed and hybrid). The Alternative Strategy, which takes advantage of user's predictible mobility patterns by building corresponding profiles, allows to save a great amount of signalling over all the interfaces and especially over the air interface, compared to the Classical Strategy presently used in the GSM system.

Regarding to the database architectures, results show that the hybrid database architecture gives rise to the lowest costs in terms of signalling messages transmitted over the air interface. This architecture is of special interest because it only involves a single additional database and switching level in the network architecture and equipement. Combining the Alternative Strategy and the hybrid database architecture allows to save more than 50% signalling compared to the Classical Strategy combined with the centralized database architecture (which scheme is implemented in most current cellular networks).

As our evaluations have been based on medium values (user mobility and traffic parameters), statistics about the cellular population behaviour should allow the operator to choose the adequate location strategy and database architecture in order to increase its network capacity.

References

[1] S. Tabbane, « An Alternative Strategy for Location Tracking, » *IEEE Journal on Selected Areas in Communications*, Vol. 13, NO. 5, June 1995.

[2] E. Alonso, K. Meier Hellstern, G. Pollini, « Influence of Cell Geometry on Handover and Registration Rates in Cellular and Universal Personal Telecommunications Networks, » Rutgers University, *WINLAB-TR-31*, May 1992.

[3] CCIR Report M/8 (mod. F), « Future Public Land Mobile Telecommunications System, » *Doc. 8/1014-E*, December 1989.

[4] R. Thomas, H. Gilbert, G. Mazziotto, « Influence of the Moving of the Mobile Stations on the Performance of a Radio Mobile Cellular Network, » *Proc. Third Nordic Seminar on Digital Land Mobile Radio Communications*, Copenhagen, Denmark, Sept.12-15, 1988.

[5] X. Lagrange, P. Godlewski, S. Tabbane, « Les réseaux GSM/DCS, » *Hermès*, Novembre 1995.

[6] David C. C. Wang, « A Survey of Number Mobility Techniques for PCS, » in *Proc. IEEE Int. Conf. Comm.*, April 1994.

[7] John Zhonghe Wang, « The Hierarchical Structure of Tracing Strategy for Universal Personal Communication Systems, » *Applied Research*, Motorola Inc., June 23 1992.

[8] John Z. Wang, « A Fully Distributed Location Registration Strategy for Universel Personal Communication Systems, » *IEEE Journal on Selected Areas in Communications*, Vol.11, No.6, Aug. 1993.

[9] T. Hattori, H. Sekiguchi, K. Kohiyama, and H. Yamamoto, « Personal Communication Concept and architecture, » in *Proc. IEEE Int. Conf. Comm.*, pp. 335.4.1 - 335.4.7, Apr. 16 - 19, 1990.

INFINITE SERVER TRAFFIC MODELS FOR CDMA CELLULAR MOBILE NETWORKS

Jamie Evans and David Everitt

Department of Electrical and Electronic Engineering
University of Melbourne, Parkville 3052, Victoria, Australia

ABSTRACT

The aim of this paper is to contribute to the understanding of the teletraffic behaviour of CDMA cellular mobile networks. In particular we examine a technique to assess the reverse link traffic capacity and its sensitivity to various propagation and system parameters.

1 INTRODUCTION

Code division multiple access (CDMA) is an alternative multiple access strategy to frequency division and time division multiple access. Provided the synchronisation and power control problems can be overcome, CDMA is a very attractive technique for wireless communications. Its advantages over other multiple access schemes include higher spectral reuse efficiency, greater immunity to multipath fading, gradual overload capability, simple exploitation of sectorisation and voice inactivity and more robust handoff procedures [1].

As early as 1978 a CDMA system had been proposed for mobile communications [2], however interest was limited until Qualcomm demonstrated the feasibility of implementing such a system in the late 1980s [3]. Since then there has been an explosion in CDMA research mainly concentrating on the design and performance analysis of receivers, coding and modulation techniques and power control algorithms. Yet to be properly examined is the traffic behaviour of a cellular mobile network employing CDMA. Important issues of traffic capacity, call admission control, analysis of soft handoff and the effects of gradual overload and imperfect power control on traffic behaviour, are not well understood. The limited amount of published research in these areas is quite surprising considering their tremendous significance to the successful operation of a network.

The goal of this paper is to contribute to the development of a deeper understanding of the traffic behaviour of CDMA mobile cellular networks through the determination of analytic tools for performance analysis and design of these networks. Such an understanding is vital to sensible network operation under the stochastically varying loads that characterise teletraffic.

The paper is organised as follows. In Section 2 the system structure and propagation models used throughout the paper are introduced. Section 3 examines methods of quantifying the interference produced by mobiles in other cells of the network. The few attempts at traffic modelling of CDMA cellular networks that are scattered through the literature are based on modelling each cell as an independent $M/G/\infty$ queue. These attempts are reviewed in Section 4 before a new model based on this assumption is presented in Section 5. From the network operators point of view the model corresponds to a system where no calls are blocked and no calls are terminated prematurely. From a mathematical point of view the number of users in each cell becomes a Poisson random variable and the total interference can be modelled as a compound Poisson sum. Methods for approximating tail probabilities associated with these sums are discussed in Section 6. These methods lead to a very efficient technique for assessing the reverse link traffic capacity of CDMA cellular mobile networks and for investigating its sensitivity to propagation and system parameters.

2 SYSTEM AND PROPAGATION ISSUES

Throughout this work we consider the standard, uniform, hexagonal layout with a base station at the centre of every cell. The forward and reverse links use disjoint frequency bands and can thus be analysed independently. We only consider the reverse link and in the sequel, all mention of path loss, signal to interference ratio (SIR) and capacity refers to the reverse link. We also note that we are solely concerned with direct sequence CDMA systems.

Unless otherwise stated, a mobile connects to the base station that offers the least path loss at any given time. The chosen base station employs power control to maintain the received signal power at one unit. Finally it is assumed that all systems are interference limited and that background noise is negligible. In real systems the background noise provides the reference from which absolute signal powers are set.

The simplest model for the mobile radio channel is a propagation loss inversely proportional to the distance between the transmitter and the receiver raised to an exponent. If the transmitter and receiver are separated by d units, then the received power is given by

$$P_R(d) = P_T \; P_0 d^{-\gamma} \tag{1.1}$$

where P_T is the transmit power and P_0 and γ are independent of distance. P_0 is a function of carrier frequency, antenna heights and antenna gains, and we assume it is constant for all paths between a mobile and a base station. γ is the path loss exponent (PLE) which varies with antenna heights and is typically in the range two to six.

3 OTHER CELL INTERFERENCE

3.1 Introduction and Review

There are two important sources of variation in the interference received at a target base station (BS) from an active mobile station (MS) in another cell, the location of the MS within the cell and lognormal shadowing. Variation in the total interference is also dependent on the number of active MSs in the system, however this is a traffic issue which is dealt with in later sections.

In all the papers we review there is no modelling of multipath fading. It is generally assumed that the use of techniques such as interleaving, diversity reception and equalization, as well as the employment of a RAKE receiver, greatly mitigate fast fading. At any rate it is reasonable to assume that the effects of the fast fading are encapsulated in the E_b/I_0 requirements of the system. This means that the propagation models used centre on distance driven path loss like (1.1) and the inclusion of lognormal shadowing. Another common theme is the assumption that an MS is power controlled by the BS to which it is connected to be received at a fixed signal power. Power control errors are usually not modelled.

The first paper to give an analysis of other-cell interference in spread spectrum mobile systems was [2]. Although the authors study a frequency hopped system their interference analysis in terms of SIR is applicable to direct sequence systems. The propagation model of (1.1) is assumed, which coupled with perfect power control leads to a simple expression for the interference at the desired BS from an MS with known position in the network. The total interference from a cell other than the desired one is calculated by integrating the above interference expression mixed with a continuous and uniform user density over a circular region approximating an hexagonal cell. An analytic result is only possible for restricted values of the path loss exponent (PLE) and so the authors use numerical integration to calculate the interference levels. The overall other-cell interference results after summing the contributions from all interfering cells apart from the desired one. This paper does not deal with the randomness of the user locations and is equivalent to calculating expected values when each user is independently and uniformly distributed over the cell of concern.

In [4] a very similar analysis to the above is presented with the exception that the fixing of the PLE at 4 leads to analytic expressions for the interference from the circular cells. This is extended to an analytic result for the variance in [5].

An extension of [2] which includes the effects of shadowing and voice activity monitoring is found in [6]. A standard hexagonal cellular layout is assumed with a propagation model that includes lognormal shadowing taken to be independent on distinct paths.

The total interference at a target BS is examined assuming that there are an equal number of users per cell, N, spread evenly and continuously over each cell. MSs are initially assumed to connect to the BS offering the least path loss. If this BS is the target then the interference is the fixed, constant power specified by the power control, otherwise the interference is a lognormal random variable (rv) with mean dependent on the position of the MS. To simplify the analysis, an MS decides between the closest BS (not including the target) and the target BS only. An expression for the interference dependent upon the MS position is then multiplied by the user density (not a probability density function) and integrated over the network to give the total other-cell interference. This total interference is a rv due to the lognormal shadowing and in the paper its mean and variance are calculated numerically as functions of the user density. Approximating the probability density function (pdf) as a Gaussian, the other-cell interference is fully characterised.

An extension of the reverse link analysis of [6] is discussed in [7]. Firstly, the propagation model is extended to take into consideration the dependence of the shadowing from an MS to different BSs. Secondly, rather than choosing between the target BS and the closest BS, an MS can connect to any of the nearest M BSs. This involves a fairly straightforward extension of the analysis in [6] although the computational complexity increases considerably to the extent that only mean values for the interference are calculated. The results show a dramatic drop in the mean other-cell interference from $M = 1$ to $M = 2$ for typical values of the shadowing variance, while the improvement is small for $M > 3$.

In [8] there is no modelling of shadowing but more detailed and accurate versions of (1.1) are employed. The analysis assumes a circular target cell plus wedge shaped adjacent cells, this geometry allowing a fairly simple investigation into the sensitivity of other-cell interference to user density profile variation. Power control errors are not examined and the results are all numerical.

In all of the above treatments, the randomness in user location within a cell has not been dealt with. Rather some (usually uniform) continuous user density has been assumed and its product with an interference function integrated over the network. An alternative approach is to look at the interference as a function of a random position vector as in [9] where the MS location is assumed uniformly distributed over each cell. In this paper however, only the mean and variance of the interference are required since a Gaussian approximation is used. This means the treatment is identical to [6] and it is only because of the slightly different angle taken that it is mentioned here.

Section 3.2 discusses a novel method of characterising other-cell interference in CDMA cellular networks. As with the above work, Raleigh fading is not studied and perfect power control of received signal strength is assumed. To begin we work with the propagation model of (1.1). Given uniformly distributed users and circular approximation of the hexagonal cells, the distribution function (df) of the interference from a MS in another cell is calculated analytically as a function of the PLE and the location of the cell. The calculation of this df compares to [4, 5] which assume a similar geometry and propagation model yet only derive an expression for the mean and variance respectively for a fixed PLE of 4. The results can be extended to include lognormal shadowing [10] however in this case the df must be calculated numerically.

3.2 Deterministic Path Loss

Consider the situation where an MS is located at (x, y) within a hexagonal cell of the standard two dimensional layout. The MS is connected to the BS with coordinates $(0,0)$ and causes interference to the BS at $(-a, 0)$. Based on the power control assumptions and (1.1), the interference is

$$I(x,y) = \left(\frac{x^2 + y^2}{(x-a)^2 + y^2} \right)^{\gamma/2} \tag{1.2}$$

and we would like to be able to calculate the df of the rv $I(X, Y)$ given the joint df $F_{X,Y}$ of the rvs X and Y.

This is in general a complicated problem, so to simplify matters let us assume the joint density of X and Y is uniform over the hexagon. Due to the large number of possible orientations of the hexagonal cell and to the dependence of X and Y, the analysis remains exceedingly tedious. These problems can be eradicated by approximating the hexagonal cells by circles of radius b. The orientation difficulty clearly vanishes and by having a and b as parameters, a great deal of flexibility results.

If the MS is at location (r, ϕ) in polar coordinates and is connected to the BS at the origin, the interference caused at the BS with location (a, π) is given by

$$I(r,\phi) = \left(\frac{r}{\sqrt{r^2 + a^2 + 2ar\cos\phi}} \right)^{\gamma}$$

and the problem now is to calculate the df of the rv $I(R, \Phi)$ where R and Φ are independent random variable with readily calculated distributions.

The derivation is carried out in [10] and leads to the df of the interference received at (a, π) from an MS that has a uniformly distributed location within the circle of radius b and centre the origin. The df is given below.

$$F_I^{a,b}(z) = \begin{cases} 0, & z < 0 \\ g_1(z), & 0 \le z < \left(\frac{a}{b}+1\right)^{-\gamma} \\ g_2(z), & \left(\frac{a}{b}+1\right)^{-\gamma} \le z < \left(\frac{a}{b}-1\right)^{-\gamma}, z \neq 1 \\ g_3(z), & 1 < \left(\frac{a}{b}-1\right)^{-\gamma}, z = 1 \\ 1, & \left(\frac{a}{b}-1\right)^{-\gamma} \le z \end{cases} \tag{1.3}$$

where

$$g_1(z) = \frac{a^2 z^{2/\gamma}}{b^2 \left(z^{2/\gamma} - 1\right)^2}$$

$$
\begin{aligned}
g_2(z) &= \frac{1}{\pi}\arccos\left(h_1(z)\right) + \frac{1}{\pi}g_1(z)\left[\pi - 2z^{2/\gamma}h_1(z)h_2(z)\right. \\
&\quad \left. - \frac{a}{b}z^{2/\gamma}h_2(z) - \arccos\left(h_1(z)\right) - \arcsin\left(z^{1/\gamma}h_2(z)\right)\right] \\
g_3(z) &= \frac{1}{\pi}\arccos\left(-\frac{a}{2b}\right) + \frac{a}{4\pi b^2}\sqrt{4b^2 - a^2}
\end{aligned}
$$

and

$$
\begin{aligned}
h_1(z) &= \frac{-a^2 - b^2 + b^2 z^{-2/\gamma}}{2ab} \\
h_2(z) &= \sqrt{1 - h_1^2(z)}
\end{aligned}
$$

In all future numerical calculations, b is chosen such that the areas of the hexagonal cell and approximating circle are equal.

4 THE $M/G/\infty$ APPROXIMATION

4.1 Introduction and Review

In this section we present a technique that allows the teletraffic capacity of CDMA cellular networks to be estimated. The simplicity of the technique arises from modelling each cell of the network as an independent $M/G/\infty$ queue and consequently the theory of this section provides no input to the understanding of how calls should be admitted to the system. A more advanced network model which does impact on call admission control schemes is presented in [11].

We begin with a review of several papers [12, 13, 14, 9] that employ the $M/G/\infty$ approximation and compare them to the model presented in this paper and in [15]. Note that in these papers the generally distributed holding times are replaced by negative exponentially distributed holding times giving $M/M/\infty$ queues. However all their results apply in the general case since it is only the stationary distribution of the number in the system that is used.

The first of the above papers to appear was [12]. The paper looks only at the limiting reverse link and has as its aims the development of a model that deals with variability in the number of users per cell, voice activity and variable E_b/I_0 requirements. Concentrating on a single cell (or sector) the authors assume that no new call requests are denied and as such model the cell as an $M/M/\infty$ queue. The number of users in the cell is thus modelled as an rv with a Poisson distribution having mean equal to the cell offered traffic. Voice activity is included simply by assuming each mobile is gated on with probability ρ and off with probability $1-\rho$. Voice activity is thus modelled by a Bernoulli rv ν. Based on fitting of empirical data on the received E_b/I_0 values required to maintain frame error rates below 1%, the authors model the E_b/I_0 required by each mobile as a lognormal rv. Since the data involves actual received values required for acceptable performance it must include the effects of imperfect power control and varying propagation conditions.

With the above aims achieved the authors essentially define blocking to occur when the instantaneous E_b/I_0 and bit rate requirements of all users cannot be met. This leads to the relation

$$P_{blocking} = P\left(\sum_{i=1}^{k} \nu_i (E_b/I_0)_i \geq \text{constant}\right) \tag{1.4}$$

where k is a Poisson rv, the ν_i are Bernoulli rvs, the $(E_b/I_0)_i$ are lognormal rvs, and all rvs are assumed independent. This probability is bounded for a range of cell offered traffics using a modified Chernoff bound and compared to results from a Gaussian approximation and simulation. Based on this comparison and the relative ease with which it is calculated, the Gaussian approximation is used in the extension to multiple cells.

The first assumption of the extension is that the number of users in each cell is the same. The second assumption is that the k users in every outer cell produce a combined interference equivalent to kf users in the inner cell where f is an expected outer-cell interference fraction obtained from [16]. Accepting the assumptions the multiple cell case reduces to the single cell problem with an equivalent number of active mobiles of $(1+f)k$ where as before k is a Poisson rv. To apply the Gaussian approximation the mean and variance of the sum in (1.4) need only be scaled by $(1+f)$.

In [13] a computationally intensive procedure is presented for the evaluation of the teletraffic capacity of both forward and reverse links in a CDMA cellular mobile system. Each cell is modelled as an independent $M/M/\infty$ queue and the quality of service (QoS) criterion evaluated is the outage probability or the probability that the SIR of a link is below a certain threshold. A uniform hexagonal layout, a uniform density for the mobile location within each cell, and a propagation model including lognormal shadowing are other features of the model presented.

In [14] a teletraffic model of the reverse link is considered. The assumptions include uniform hexagonal layout, equal traffic offered to every cell, uniform density for mobile locations within cell, two layers of interfering cells considered, deterministic propagation loss only, and perfect power control of received signal strength. Despite the initial discussion of a model including a finite number of modems, trunk reservation for handovers, and mobility the subsequent analysis does not allow for mobility and assumes traffic levels which reduce the new call blocking probability to a negligible figure. Thus the system is actually modelled as a network of independent $M/M/\infty$ queues. Once more the QoS measure concerns the probability of the SIR being below a given threshold which with the assumption that each mobile is received at a fixed power level involves calculating the probability that the interference gets too large. The total interference is calculated as the sum of inner-cell interference and outer-cell interference. In line with the $M/M/\infty$ assumption the contribution from within the desired cell is taken as a Poisson rv with mean equal to the cell offered traffic. The outer-cell interference is approximated as a Gaussian rv with mean and variance obtained via simulation. A numerical convolution of the Gaussian and Poisson dfs then leads to the df for the total interference. The traffic capacity corresponding to two QoS values is presented as a function of the path loss exponent of the propagation model.

Many of the assumptions of [9] are as in [14]. Only the reverse link is considered, the QoS is based on a minimum SIR requirement, perfect power control of received signal strength is assumed, and each cell acts as an independent $M/M/\infty$ queue. The internal and external interference are again treated separately, the internal a Poisson rv and the external a Gaussian rv. The mean and variance of the external interference are calculated by analytical and numerical methods based on the treatment in [6] which includes lognormal shadowing in the propagation model. The blocking or outage probability is then given in the form of a convolution as in [14].

The analysis of reverse link traffic capacity for CDMA cellular mobile networks developed in this paper and in [15] shares many of the features of the above papers. In particular we employ the independent $M/G/\infty$ queue model for each cell, the service requirement is in terms of SIR, and each mobile is power controlled to a fixed and equal power. In the most general development [15] arbitrary network layouts, user distributions, and traffic profiles are allowed, and lognormal shadowing is included in the propagation model. If however, a symmetric structure is imposed, the calculation of the service measure reduces to evaluating the probability of a compound Poisson sum exceeding a certain threshold. If the propagation model does not include shadowing, an analytic expression is available for the distribution of the random summands. The service measure is approximated using a standard Gaussian approximation and bounded with the Chernoff bound.

4.2 Model Description

In this section we discuss the assumptions leading to, and justification for, modelling the CDMA cellular network as a collection of independent $M/G/\infty$ queues. The assumptions of the traffic model we will use are as follows:

- the call initiation processes in each cell are modelled as independent Poisson streams,
- all arriving calls are accepted into the network and remain in the network for the full call duration (no blocked or dropped calls),
- call durations are generally distributed and independent of the arrival processes and other holding times,
- mobility is not modelled and thus the mobile is associated with the cell of its call initiation for the duration of the call.

The first and third points are standard assumptions from teletraffic engineering that have been employed for several decades to model the stochastic nature of call arrivals to telephone exchanges and their circuit holding times. The second assumption is reasonable for systems operating with CDMA since there is no theoretical hard limit on the number of quasi-orthogonal codes available to assign to users. From a more practical point of view it is reasonable to assume that there are enough codes available so that the new call blocking probability is negligible for moderate offered traffic. The final assumption is a good approximation when the cell size is large compared to the distance a typical mobile will travel during a call. In particular the trend toward hand held mobiles drastically reduces the need to include mobility in traffic models.

The assumptions imply that each cell of our network behaves like an independent $M/G/\infty$ queue [17]. This is one of the most basic queueing models and has a particularly simple form for the steady state distribution of the number of active calls. If the mean time between call arrivals is $1/\lambda$ seconds and the mean call holding time is $1/\mu$ seconds then the traffic to the system is $A = \lambda/\mu$ Erlangs. Let N be the rv representing the number of active calls in the system at steady state. Then N has the Poisson distribution

$$P(N = n) = e^{-A}\frac{A^n}{n!}.$$

The independence of each cell in the network implies that the joint steady state distribution for the number of active calls in each cell is simply a product of Poisson distributions.

Before proceeding to the next section we make one final point. In the rest of this paper it is assumed that the traffic offered to every cell of the network is equal. We emphasise that this equality applies to the parameters of a stochastic model and is distinctly different to assuming an equal static load in every cell. This along with the infinite, symmetrical, cellular layouts and uniform user distributions that we have assumed allows all calculations to be performed for one cell of the network only. The extension of this work to asymmetrical layouts, offered traffic, and user distributions is straightforward from a theoretical point of view [15] and is not included here.

5 BLOCKING PROBABILITY

In this section we develop a simple expression for a QoS indicator which we call the blocking probability. Calculation of the blocking probability reduces to the evaluation of the probability that a compound Poisson rv exceeds a given threshold. The analysis of such an expression is left to the following section.

5.1 Definition

In the absence of new call blocking we use as a performance measure the probability that a mobile achieves an insufficient SIR. Although this measure is more closely related to outage probability than blocking probability we choose the latter name to reflect the traffic modelling orientation.

To calculate the blocking probability, B, we must determine the stationary probability that an arbitrary mobile anywhere in the cellular network receives a reverse link SIR that is insufficient for acceptable QoS. If certain symmetries exist then B will be the same for mobiles at any point in the network and we may just as well consider calls that are connected to a particular BS. Associate with this target BS and its cell the index 1.

Because of the standard power control assumptions, mobiles are received at BS_1 with one unit signal power. We can thus easily translate the SIR requirement into a constraint on the total interference at BS_1. That is,

$$B = P(I > \Gamma) \tag{1.5}$$

where I is a rv representing the total power received at an arbitrary BS in the network. Γ is a measure of the *size* of the CDMA system and is related to the system bandwidth (W Hz), the data rate (R bps), and the required bit energy to interference density ratio (E_b/I_0) by [6]: $\Gamma = (W/R)/(E_b/I_0)$.

5.2 Interference

Suppose that there are $M - 1$ cells apart from cell_1 that generate significant interference at the target BS with labels $2, \ldots, M$. Assume that the interference rvs for mobiles in cell_i are iid with df F_i and the interference rvs from different cells are independent. Given N_i calls in cell_i the total power received at BS_1 is given by

$$I = \sum_{i=1}^{M} \sum_{j=1}^{N_i} I_{ij}$$

where I_{ij} is the interference from the jth mobile in cell_i and the N_i are independent Poisson rvs with mean A.

It is readily shown using characteristic functions that

$$I \doteq \sum_{j=1}^{N} I_j \tag{1.6}$$

where N is a Poisson rv with parameter MA and I_j is an rv with df $M^{-1} \sum_{i=1}^{M} F_i$ being a finite mixture of the original dfs. The symbol $\doteq$ indicates equality in distribution. In (1.6) the interfering cells are combined and the total traffic into the conglomeration considered.

Combining (1.6) and (1.5) we arrive at a simple expression for the blocking probability in the network

$$B = P\left(\sum_{j=1}^{N} I_j > \Gamma\right). \tag{1.7}$$

We now present an example to illustrate and clarify the ideas of the last sections.

5.3 Example

Consider the standard two dimensional layout with A Erlangs of traffic offerered to each cell. Assume the mobile locations within each cell are iid rvs uniformly distributed over each cell and that the propagation environment is governed by (1.1). The interference resulting at some target BS from a mobile randomly located in a cell of the network is characterised by the approximate df of (1.3). If only the target cell and the first two surrounding rings of cells are taken to contribute significantly to the total interference at the target BS we have $I = \sum_{j=1}^{N} I_j$ where N is a Poisson rv with mean $19A$ and the I_j are a sequence of iid rvs with df

$$F(z) = 19^{-1}(u(z-1) + 6F_I^{1,b}(z) + 6F_I^{\sqrt{3},b}(z) + 6F_I^{2,b}(z)). \tag{1.8}$$

In the above $u(z)$ is the unit step function and $F_I^{a,b}(z)$ are given in (1.3).

To calculate the blocking probability as a function of the offered traffic per cell we are faced with evaluating (1.7). Two methods of approximating this probability are described in the next section.

6 APPROXIMATIONS

In this section we consider techniques for approximating

$$P\left(\sum_{j=1}^{N_A} X_j \geq \Gamma\right)$$

where $X_1, X_2, \ldots$ are iid rvs and N_A is a Poisson rv with mean A that is independent of the X_j.

Let us define $S_{N_A} = \sum_{j=1}^{N_A} X_j$ and denote the mean and variance of S_{N_A} by $E[S_{N_A}] = \mu_S$ and $\text{var}(S_{N_A}) = \sigma_S^2$ respectively. Then if $E[X_1] = \mu_X$ and $\text{var}(X_1) = \sigma_X^2$,

$$\begin{aligned} \mu_S &= A\mu_X \\ \sigma_S^2 &= A(\sigma_X^2 + \mu_X^2). \end{aligned}$$

6.1 Normal Approximation

The normal approximation is

$$\frac{S_{N_A} - \mu_S}{\sigma_S} \doteq \mathcal{N}(0,1)$$

where $\mathcal{N}(0,1)$ is a zero mean, unit variance, normal rv. We thus have

$$P\left(\sum_{j=1}^{N_A} X_j \geq \Gamma\right) = P\left(S_{N_A} \geq \Gamma\right) \approx 0.5 - \text{erf}\left(\frac{\Gamma - \mu_S}{\sigma_S}\right)$$

where

$$\mathrm{erf}(x) = \frac{1}{\sqrt{2\pi}} \int_0^x e^{-t^2/2}\, dt.$$

To examine the asymptotic behaviour of this approximation as $A \to \infty$ we first assume that A takes non-negative integral values only. Since the sum of Poisson rvs is also Poisson, we have

$$S_{N_A} = \sum_{j=1}^{N_A} X_j \doteq \sum_{j=1}^{A} Y_j$$

where $Y_1, Y_2, \ldots, Y_A$ are iid rvs and

$$\begin{aligned} Y_1 &\doteq \sum_{j=1}^{N_1} X_j \\ E[Y_1] &= \mu_Y = \mu_X \\ \mathrm{var}(Y_1) &= \sigma_Y^2 = \sigma_X^2 + \mu_X^2. \end{aligned}$$

The sum with a random number of summands has been transformed into a standard deterministic sum of iid rvs for which we can apply the central limit theorem (CLT) in its simplest form. The CLT [18] states that provided μ_Y is finite and σ_Y^2 is positive and finite,

$$\frac{S_{N_A} - A\mu_Y}{\sigma_Y\sqrt{A}} = \frac{S_{N_A} - \mu_S}{\sigma_S} \Rightarrow \mathcal{N}(0,1) \text{ as } A \to \infty.$$

The symbol $\Rightarrow$ refers to convergence in distribution.

This result is readily extended to the case of real A [10].

6.2 Large Deviations Bound

We now give an upper bound on the blocking probability using the Chernoff bound. The asymptotic behaviour of this bound is discussed in the context of elementary large deviations theory.

Consider first the case when A takes on positive integer values and the Poisson sum can be rewritten as the deterministic sum $S_{N_A} \doteq \sum_{j=1}^{A} Y_j$. The large deviation rate function is defined by

$$I(t) = \sup_{\theta} [\theta t - M_Y(\theta)]$$

where θ is real and

$$M_Y(\theta) = \log E\left[e^{\theta Y_1}\right] = e^{M_X(\theta)} - 1$$

is the log moment generating function (LMGF) of the Y_j which is related as shown to the LMGF of the X_j, $M_X(\theta)$.

Provided $M_Y(\theta) < \infty$ for all θ and that Y_1 is not a bounded random variable in the sense that $P(Y_1 \in (a,b)) < 1$ for all finite a and b, then from Cramer's Theorem [19]

$$\lim_{A\to\infty} \frac{1}{A} \log P\left(\frac{1}{A} S_{N_A} \geq \eta\right) = -I(\eta) \tag{1.9}$$

for $\eta > \mu_Y$. Moreover for all positive integral A

$$\frac{1}{A} \log P\left(\frac{1}{A} S_{N_A} \geq \eta\right) \leq -I(\eta).$$

The above bound is commonly called the Chernoff bound and is directly applicable for any positive real value of A. To extend the limit result of (1.9) to the case when A is real is more involved but is readily accomplished either by modified use of Cramer's Theorem or by direct application of the more powerful Gartner-Ellis Theorem.

Applying the above to our problem we have

$$\begin{aligned} \frac{1}{A} \log P\left(S_{N_A} \geq \Gamma\right) &\leq -I\left(\frac{\Gamma}{A}\right) \\ &= \inf_\theta \left[M_Y(\theta) - \theta\frac{\Gamma}{A}\right] \\ &= \inf_\theta \left[e^{M_X(\theta)} - 1 - \theta\frac{\Gamma}{A}\right] \end{aligned}$$

and from Cramer's Theorem the bound becomes tight as $A \to \infty$ with Γ/A held constant.

7 CONCLUSION

In this paper we have presented an analysis for the reverse link traffic capacity of CDMA cellular mobile networks. The key assumption was that each cell can meaningfully be modelled as an independent $M/G/\infty$ queue. After discussing the justification for, and consequences of the $M/G/\infty$ model, an expression for blocking probability was developed in terms of a compound Poisson random variable. Two techniques were then applied to approximate the blocking probability along along with corresponding asymptotic results.

Acknowledgements

This work was partially supported by the Australian Telecommunications and Electronics Research Board, the Australian Research Council and a Telstra Research Laboratories Postgraduate Fellowship.

REFERENCES

[1] W. Lee, "Overview of cellular CDMA," *IEEE Trans. Veh. Technol.*, vol. 40, pp. 291–302, May 1991.

[2] G. Cooper and R. Nettleton, "A spread spectrum technique for high capacity mobile communications," *IEEE Trans. Veh. Technol.*, vol. VT-27, pp. 264–275, Nov. 1978.

[3] R. Padovani, "Reverse link performance of IS-95 based cellular systems," *IEEE Personal Commun.*, vol. 1, pp. 28–34, Third Quarter 1994.

[4] K. I. Kim, "CDMA cellular engineering issues," *IEEE Trans. Veh. Technol.*, vol. 42, pp. 345–350, Aug. 1993.

[5] J. Lee and L. Miller, "On the Erlang capacity of CDMA cellular systems," in *Proc. GLOBECOM'95, Singapore*, pp. 1877–1883, Nov. 1995.

[6] K. Gilhousen, I. Jacobs, R. Padovani, A. Viterbi, L. Weaver, and C. Wheatley, "On the capacity of a cellular CDMA system," *IEEE Trans. Veh. Technol.*, vol. 40, pp. 303–312, May 1991.

[7] A. J. Viterbi, A. M. Viterbi, K. S. Gilhousen, and E. Zehavi, "Soft handoff extends CDMA cell coverage and increases reverse link capacity," *IEEE J. Select. Areas Commun.*, vol. 12, pp. 1281–1288, Oct. 1994.

[8] T. S. Rappaport and L. B. Milstein, "Effects of radio propagation path loss on DS-CDMA cellular frequency reuse efficiency for the reverse channel," *IEEE Trans. Veh. Technol.*, vol. 41, pp. 231–242, Aug. 1992.

[9] M. Frullone, G. Riva, P. Grazioso, and M. Missiroli, "Comparisons of multiple access schemes for personal communication systems in a mixed cellular environment," *IEEE Trans. Veh. Technol.*, vol. 43, pp. 99–109, Feb. 1994.

[10] J. Evans and D. Everitt, "Non-blocking traffic models of CDMA cellular mobile networks," 1996. Submitted.

[11] J. Evans and D. Everitt, "Effective bandwidth based admission control for multiple service CDMA cellular networks," 1996. Submitted.

[12] A. Viterbi and A. Viterbi, "Erlang capacity of a power controlled CDMA system," *IEEE J. Select. Areas Commun.*, vol. 11, pp. 892–900, Aug. 1993.

[13] A. Fapojuwo, "Radio capacity of direct sequence code division multiple access mobile radio systems," *IEE Proceedings-I*, vol. 140, pp. 402–408, Oct. 1993.

[14] W. Lavery and D. Everitt, "Analysis of the uplink teletraffic behaviour in CDMA cellular systems," in *Proc. IEEE VTC'94, Stockholm, Sweden*, pp. 868–872, June 1994.

[15] J. Evans and D. Everitt, "Analysis of reverse link traffic capacity for cellular mobile communication networks employing code division multiple access," in *Proc. Australian Telecommunication Networks and Applications Conference, Melbourne, Australia*, pp. 775–780, Dec. 1994.

[16] A. Viterbi, A. Viterbi, and E. Zehavi, "Other cell interference in cellular power controlled CDMA," *IEEE Trans. Commun.*, vol. 42, pp. 1501–1504, Feb/Mar/Apr 1994.

[17] L. Kleinrock, *Queueing Systems Volume 1: Theory.* New York: Wiley, 1975.

[18] P. Billingsley, *Probability and Measure.* New York: John Wiley and Sons, second ed., 1986.

[19] J. Bucklew, *Large Deviations Techniques in Decision, Simulation, and Estimation.* New York: Wiley, 1990.

Modeling Priority Traffic and Hot Spots in Wireless PCS Systems

Maria Luisa Merani
Dipartimento di Scienze dell'Ingegneria - University of Modena - Italy

Abstract

In this paper we investigate a possible technique for handling the reverse link (mobile-to-base) traffic inside the cell of a wireless PCS system, assigning different priority classes to new calls and hand-offs. For the corresponding queueing scheme we demonstrate how we can achieve the complete steady-state solution, exploiting a matrix analytic technique. Further and more importantly, through this approach we show how to efficiently model system behavior when critical traffic conditions - hot spots - are present in the mobile radio network, consequently determining the traffic quantities which directly affect transmission design. Lastly, we face the issue of tying the design of a CDMA transmission system to traffic constraints: to this regard we demonstrate how traffic parameters influence this system performance both in ordinary – and in critical – traffic conditions .

1 Introduction

Wireless Personal Communications envision seamless access, featuring wire-like quality and smooth integration with fixed networks. Within this framework an issue critical to PCS system design is modeling and managing wireless traffic: numerous contributions have been recently furnished in such a field, see e.g. [1, 4], and some papers began to deal with the problem of interfacing bounds dictated by traffic conditions with constraints obeying to transmission design [5], [6].

In the present work we handle the reverse link (mobile-to-base) traffic inside the cell of a radio network introducing different priorities for new calls and hand-offs. The objective is to efficiently deploy the pool of available resources, be it the codes of a DS-CDMA implementation or the time slots within the narrow band frequency channels of a TDMA system. More precisely, we refer to a queueing

*Work financially supported by M.U.R.S.T. and by C.N.R., within the framework of the WWLAN strategic project (Italy).

scheme with m available resources and two priority classes. New calls and hand-offs are modeled through two distinct Poisson streams with different arrival rates; the service distribution is exponential for both type of calls. New and handed-off calls have free access to the resources as long as some of them are free. When all the resources are busy, a queue forms at the base station; new calls waiting in queue are allowed into service following a FCFS modality only when no handed-off calls have a service request pending. Our choice is alternative to the idea of partitioning the resources into two distinct pools, with one of them exclusively dedicated to hand-offs [6]; it is also alternative with respect to a scheme which contemplates a cut-off mechanism, [1]. In [7] we previously reported the probability of a call being delayed in our scheme, and compared it with the corresponding probability of being delayed/blocked for the schemes cited above. Bypassing the complete solution of the examined Markov chain, in that circumstance we were also able to determine the trade-off between network and transmission design of both TDMA and CDMA transmission systems, when no traffic upsurges within the mobile radio network were accounted for.

In the present contribution however, we point out the analytic procedure which leads to the complete steady-state solution of the examined Markov chain. To this regard we exploit a matrix geometric approach, underlining its computational efficiency. Further and more importantly, we outline that this methodology opens the way to the modeling of critical traffic conditions within the wireless system - hot spots -. We are therefore allowed to analytically investigate system behavior in the presence of sudden traffic upsurges, so frequent in a wireless scenario. In these new operating conditions we determine those traffic quantities whose knowledge is vital for transmission design, and subsequently tie the transmission system engineering to the presence of traffic build-ups. Finally, through the numerical results we quantitatively specify how CDMA system performance be deeply influenced by traffic parameters both in ordinary – and in critical – traffic conditions.

The paper is organized as follows: in Section 2 we recall how to model the two-priority-class system and how to determine the queueing quantities of interest to interface traffic and transmission design. In Section 3 we analytically obtain the steady-state complete solution through the matrix geometric approach. In Section 4 we show how to render a hot spot situation and how to characterize system behavior in such a context. In Section 5 we merge hot spot modeling and transmission design. The Numerical Results are comprised in Section 6 and the conclusions follow in Section 7.

2 Queueing Model

Our aim is to handle the reverse link (i.e., mobile-to-base) traffic in a cell of a wireless cellular PCS system introducing different priority classes for new calls and hand-offs: in order to pursue the corresponding performance analysis, we will take as our starting point the hypothesis that both type of calls can be modeled through independent Poisson processes of parameter λ_N and λ_H, respectively. As regards the assumption of independence between the two Poisson streams, we observe that it does not truly reflect the real situation of a wireless network: however, we deem the coupling between the two stochastic processes be loose enough to allow for this *analytically tractable* model. For what pertains the Poisson assumption, we observe that it has been widely validated, see e.g. [9], and used [1], [6]. We also make the hypothesis that the time spent in service in each cell by a generic mobile radio user is drawn from an exponential, i.e., memoryless, distribution of parameter μ, regardless of the corresponding call being new or handed-off.

The reference cell of our model is represented by a multiple resource service facility (be it the pool of available codes of a DS-CDMA system, the frequency carriers of a FH-CDMA implementation or the time slots within the narrow-band frequency channels of a TDMA scheme). New calls and hand-offs can freely access all the resources, as long as some of them are available. A queue is allowed to form at the facility when all the m resources are busy and arriving calls join it following a head-of-the-line priority discipline: since we attribute hand-offs higher priority with respect to new calls originated within the cell, queued hand-offs will receive service on a FCFS basis *before* new calls that are waiting in queue. The latter ones will therefore go into service following a FCFS modality only when no more hand-offs are waiting for service in queue. Service is assumed non-preemptive: indeed, it would be meaningless to drop a new call while it is in progress to make room for a handed-off call, because once a new call is being served, it is as though it acquired the same priority level as a hand-off (they actually become indistinguishable). We will denote the state of the system in stationary conditions by the triplet (n, h, k), with (n, h) respectively furnishing the number of new calls and the number of hand-offs waiting in queue, and k the number of busy servers. $P(n, h, k)$ will correspondingly represent the probability of being in state (n, h, k). In Fig.1 we report the state diagram for the case of m available resources in the single cell, when an infinite queue is allowed to form at the service facility.

As one target of this paper is to highlight the trade-off between traffic and transmission design aspects, our first step will be to focus on those queueing parameters whose impact on transmission dimensioning is more relevant: we will correspondingly draw from the previous state diagram the needed information without completely solving the Markov chain. To this regard, we first of all observe that if we consider as our state variable the number of users – calls – in system (meant as servers plus queue), neglecting the distinct description of the number of hand-offs and new calls waiting for service in queue, then the corresponding state diagram is the one of an M/M/m system with coefficient of utilization $\rho = \frac{\lambda_H + \lambda_N}{m\mu}$. It immediately follows from this observation that the state probabilities for this well known system, which we name P_t, $t \geq 0$, do coincide with $P(0, 0, t)$ of the original chain, $\forall t$, $t \leq m$, i.e.,

$$P(0,0,0) = P_0 = \left[\sum_{k=0}^{m-1} \frac{(m\rho)^k}{k!} + \frac{m^m \rho^m}{m!} \frac{1}{1-\rho} \right]^{-1} \tag{1}$$

and

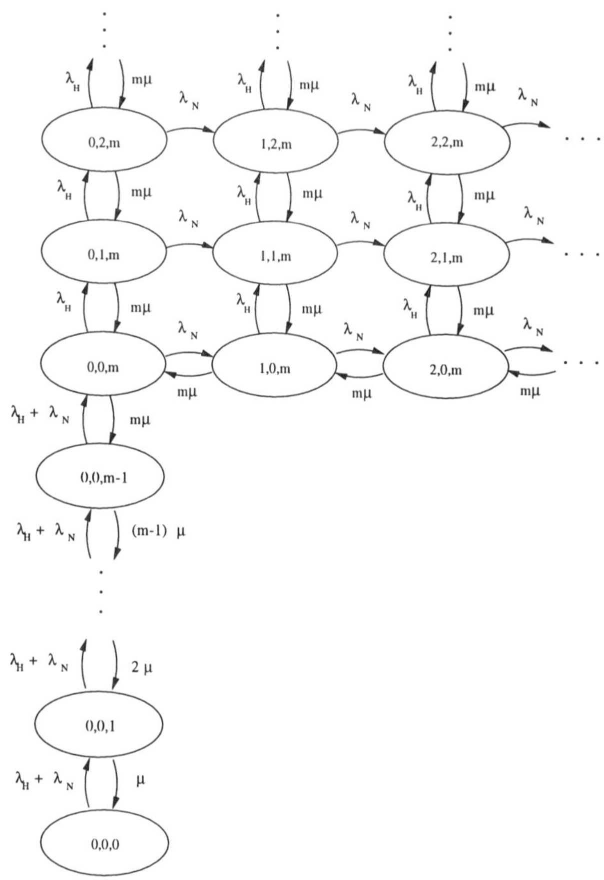

Figure 1: The state diagram of the examined queueing system.

$$P(0,0,t) = P_t = \frac{\left(\frac{\lambda_H+\lambda_N}{\mu}\right)^t}{t!} P(0,0,0) \quad , \quad \text{for } t \leq m\,. \tag{2}$$

As regards the probability of having all the resources busy, we observe that it coincides with the probability of a call to be placed in queue, regardless of it being new or handed-off. In both queueing systems it is expressed by:

$$P_{delay} = \sum_{t=m}^{\infty} P_t = P(0,0,0)\frac{m^m}{m!}\frac{\rho^m}{1-\rho}\,. \tag{3}$$

Hence, we finally write the probability of t among the m available resources being busy as:

$$Pr\{t\; resources\; are\; busy\} = \begin{cases} P(0,0,t) & , \quad 0 \leq t \leq m-1 \\ P_{delay} & , \quad t = m. \end{cases} \tag{4}$$

3 The Complete Solution

Let us now look for the complete steady-state solution, in order to better characterize both high and low priority calls. In order to do so, let us give a finite dimension to the hand-off queue, say r_1: this considerably facilitates the analytic approach; more importantly, it represents a physically reasonable hypothesis, as the blocking of high priority calls when the corresponding queue grows too much reflects their realistic inability to bear too long a delay. Let us then imagine to terminate the upper part of the state diagram of Fig.1 with the (r_1+1)-th row, which corresponds to index h in (n,h,m) taking on a finite set of values, ranging from 0 to r_1. Further, let us group the state probabilities $P(0,0,k)$, $0 \leq k \leq (m-1)$ into a vector Π_0 of size m, defined as

$$\Pi_0 = [P_{000}\, P_{001}\, \ldots P_{00m-1}] \;, \tag{5}$$

where with short notation P_{00k}, $0 \leq k \leq (m-1)$, represents the $P(0,0,k)$ state probability; let us also group the state probabilities $P(n,h,m)$ of the upper part of the diagram, which neglecting index $k=m$ we will refer to as P_{nh} from now onward, into vectors as well, with the generic vector Π_{n+1} defined as

$$\Pi_{n+1} = [P_{n0}\, P_{n1}\, \ldots P_{nr_1}] \quad , \quad n = 0, 1, \ldots \quad . \tag{6}$$

Finally, let us consider the whole state probability vector Π, obtained through the following ordering of the Π_n vectors, $(n = 0, 1, \ldots)$:

$$\Pi = [\Pi_0\, \Pi_1\, \Pi_2\, \ldots] \;, \tag{7}$$

and let us name $\tilde{Q}$ the corresponding transition probability matrix, which naturally obeys the $\Pi\tilde{Q} = 0$ equation. Let us now look for a matrix-geometric solution for the considered Markov chain, taking advantage of the distinctive structure of the upper part of the state diagram, which reflects into the $\tilde{Q}$ structure itself. The following interpretation applies to the upper part of the Markov chain: it is as though we had an $M/M/m$ queue with (r_1+1) phases [8], one for each of the rows of the diagram shown in Fig.1. Phase (row) 0 has $\lambda_0 = \lambda_N$ and $\mu_0 = m\mu$, whereas phase (row) i $(i = 1, 2, \ldots, r_1)$ is characterized by $\lambda_i = \lambda_N$ and $\mu_i = 0$. The A matrix which describes the transitions from one phase to the others is tridiagonal and is $(r_1+1) \times (r_1+1)$ (recall that r_1 is the dimension of the hand-off queue); since we move from row i to row $i+1$ with rate λ_H and from row i to row $i-1$ with rate $m\mu$, A has the following structure:

$$A = \begin{bmatrix} -\lambda_H & \lambda_H & 0 & & \cdots & 0 \\ m\mu & -(\lambda_H + m\mu) & \lambda_H & & \cdots & 0 \\ 0 & m\mu & -(\lambda_H + m\mu) & & \cdots & 0 \\ 0 & 0 & m\mu & & \cdots & 0 \\ & & & \ddots & & \\ \vdots & \vdots & \vdots & & -(\lambda_H + m\mu) & \lambda_H \\ 0 & 0 & 0 & & m\mu & -m\mu \end{bmatrix} \tag{8}$$

However, we cannot forget the lower part of the original Markov chain, the one described by the Π_0 vector: what we will do is to interpret its states as determining part of the odd boundary behavior of the global $\tilde{Q}$ matrix. Having defined Π as in (7), it is possible to demonstrate that $\tilde{Q}$ takes on the following structure:

$$\tilde{Q} = \begin{bmatrix} B_{00} & B_{01} & 0 & 0 & 0 & \dots \\ B_{10} & B_{11} & \Delta\lambda_N & 0 & 0 & \dots \\ B_{20} & B_{21} & A - \Delta\lambda_N - \Delta\mu & \Delta\lambda_N & 0 & \dots \\ B_{30} & B_{31} & \Delta\mu & A - \Delta\lambda_N - \Delta\mu & \Delta\lambda_N & \dots \\ \cdot & \cdot & 0 & \Delta\mu & A - \Delta\lambda_N - \Delta\mu & \dots \\ \cdot & \cdot & \cdot & 0 & \Delta\mu & \dots \\ \vdots & \vdots & \vdots & \vdots & \vdots & \vdots \end{bmatrix}, \tag{9}$$

where A is the matrix defined in (8), $\Delta\lambda_N$ is a diagonal $(r_1+1)\times(r_1+1)$ matrix, defined as $\Delta\lambda_N = \lambda_N I$, where I represents an $(r_1+1)\times(r_1+1)$ identity matrix and $\Delta\mu$ is a $(r_1+1)\times(r_1+1)$ matrix with only one non-null element, namely, $\Delta\mu[1,1] = m\mu$. B_{00} is a tridiagonal $m\times m$ matrix:

$$B_{00} = \begin{bmatrix} -(\lambda_H+\lambda_N) & \lambda_H+\lambda_N & 0 & \dots & 0 \\ \mu & -(\lambda_H+\lambda_N+\mu) & \lambda_H+\lambda_N & \dots & \cdot \\ 0 & 2\mu & -(\lambda_H+\lambda_N+2\mu) & & \vdots \\ \cdot & 0 & 3\mu & \ddots & 0 \\ \vdots & \vdots & \vdots & & \lambda_H+\lambda_N \\ 0 & 0 & 0 & \dots & -(\lambda_H+\lambda_N+(m-1)\mu) \end{bmatrix} \tag{10}$$

B_{01} is another very special $m\times(r_1+1)$ matrix with one non-null element, element $B_{01}[r_1+1,1] = \lambda_H+\lambda_N$; the same can be said of B_{10}, which is an $(r_1+1)\times m$ matrix whose only non-null element is $B_{10}[1,m] = m\mu$, whereas B_{11} is an $(r_1+1)\times(r_1+1)$ matrix given by: $B_{11} = A - \Delta\lambda - \Delta\mu$. B_{21} is an $(r_1+1)\times(r_1+1)$ matrix, $B_{21} = \Delta\mu$, and finally we have the remaining matrices $B_{20} = B_{30} = \ldots = 0$, $B_{31} = B_{41} = \ldots = 0$.

As a consequence of the $\tilde{Q}$ distinctive structure we just described, the analytic solution of the examined Markov chain is given by [8]:

$$\Pi_n = \Pi_1 R^{n-1} \;,\; for\;\; n \geq 2\,, \tag{11}$$

where R is the minimal non-negative solution of the equation [10]:

$$A_0 + RA_1 + R^2A_2 = 0\,, \tag{12}$$

with $A_0 = \Delta\lambda_N$, $A_1 = A - \Delta\lambda_N - \Delta\mu$ and $A_2 = \Delta\mu$.

As regards the Π_0 and Π_1 vectors, i.e., the probabilities for the states of the lower part of the Markov chain and for the ones belonging to the first row of the diagram in Fig.1, they can be obtained from the system [10, 8]:

$$\begin{aligned} [\Pi_0 \quad & \Pi_1]\, B[R] = 0 \\ \Pi_0 e \quad & + \Pi_1 (I - R)^{-1} e = 1\,, \end{aligned} \tag{13}$$

where e represents a unitary column vector, and $B[R]$ can be shown to have the structure:

$$B[R] = \begin{bmatrix} B_{00} & B_{01} \\ B_{10} & B_{11} + RB_{21} \end{bmatrix}. \tag{14}$$

As with any matrix-analytic method, with the proposed matrix geometric approach we gain in computational efficiency with respect to other methodologies available to solve the system global balance equations. More importantly however, we open the way to the inclusion in the model of fluctuations in input traffic of both type of calls: these variations can be expected to occur frequently in a wireless context, caused by traffic upsurges - hot spots - which arise within the cell of the mobile radio system we are examining. Let us then move on, to the characterization of these phenomena.

4 Hot Spot Rendering

The occurrence of sudden increases in the input traffic of a cell of a cellular PCS system has to be regarded as a frequent phenomenon: unexpected changes in users' movements, together with the traditional "rush-hour" condition, represent just two possible causes of such deep variations. It is of great importance to investigate the behavior of the proposed scheme in a scenario which embodies such events, and to dimension system resources so as to cope with them.

The idea we exploit to describe both the ordinary traffic condition and the upsurge traffic situation is to refer to a very simple bi-state ergodic Markov chain, whose behavior we synthesize in Fig.2: the transition rates α_u and β_u that appear in this figure statistically detail the modalities of arising and falling of the variations in the mean traffic values.

In ordinary traffic conditions the arrival rate is λ_N for new calls and λ_H for hand-offs, while the service rate of each exponential server is μ; when in the upsurge condition, the input rates are modified into $\lambda_N + \lambda_{NU}$ and $\lambda_H + \lambda_{HU}$ respectively, whereas no variation is introduced onto the service rate μ. With the aim to graphically understand how things work when hot-spots are

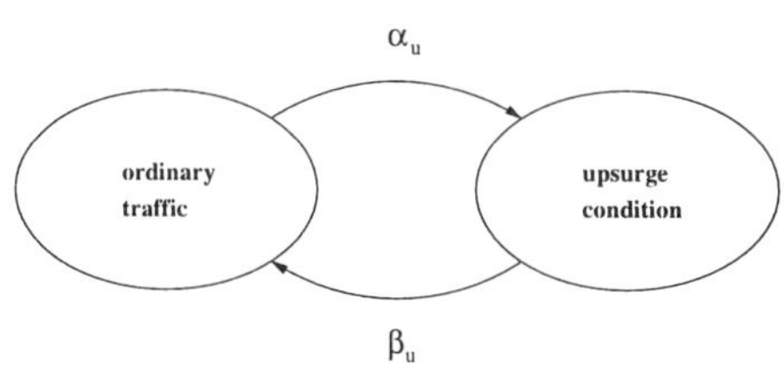

Figure 2: Bistate Markov chain to describe varying traffic conditions.

accounted for, let us refer to the model mentioned in Section 2, which is the one of interest to evaluate the probabilities of having t resources busy ($t \leq m$), hence to determine those traffic parameters which directly influence transmission design. Let us then draw in Fig. 3 the corresponding state diagram, modified accordingly to the presence of the random environment described by the Markov chain of Fig.2.

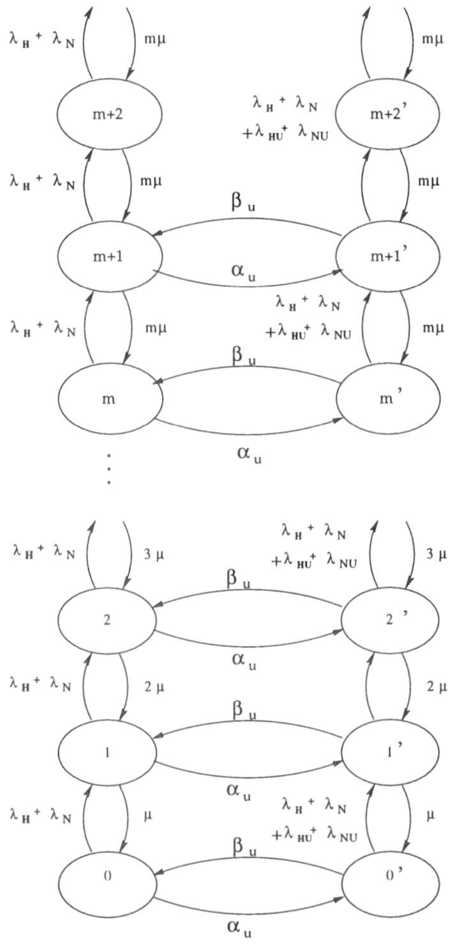

Figure 3: State diagram in the presence of varying traffic conditions.

Note that we have introduced the notation superscript " ' " to distinguish the quantities (states and state probabilities) referring to the upsurge part of the diagram from the corresponding quantities which refer to the ordinary condition part (e.g., $P_t^{'}$ represents the probability of having t resources busy in critical traffic conditions). In accordance with the *modus operandi* presented in last section, we next build the Π vector for this Markov chain, ordering the state probabilities in the following manner:

$$\Pi = [P_0 P_0' P_1 P_1' \ldots P_{m-1} P_{m-1}' P_m P_m' P_{m+1} P_{m+1}' \ldots] \tag{15}$$
$$= [\Pi_0, \Pi_1, \Pi_2 \ldots], \tag{16}$$

where the vector Π_0 is defined as: $\Pi_0 = [P_0 P_0' P_1 P_1' \ldots P_{m-2} P_{m-2}']$,whereas $\Pi_i = [P_{m-2+i} P_{m-2+i}']$, $i = 1, 2, \ldots$. As in the model described in last section, this time too the regular part of the state diagram is the one which describes system behavior when the m servers are busy, and the corresponding state probabilities are grouped into the Π_i vectors with index i, $i \geq 2$.

Given the distinctive architecture this diagram presents, the solving procedure for this Markov chain parallels the one highlighted in the previous section: again, we look for a matrix-geometric solution, and in order to pursue this goal we need to identify the various matrices which appear in the $\tilde{Q}$ matrix of this chain, characterized by the structure already pointed out in (9). This time however matrix $\Delta\lambda_N$ appearing in (9) is replaced by

$$\Delta\lambda = \begin{bmatrix} \lambda_N + \lambda_H & 0 \\ 0 & \lambda_N + \lambda_{NU} + \lambda_H + \lambda_{HU} \end{bmatrix}, \tag{17}$$

matrix A is defined as

$$A = \begin{bmatrix} -\alpha_u & \alpha_u \\ \beta_u & -\beta_u \end{bmatrix}, \tag{18}$$

i.e., it is the matrix governing the transitions from the normal operating conditions to the hot spot situation; matrix $\Delta\mu$ is replaced by $\Delta\mu(m) = m\mu I$, where I is a 2×2 identity matrix.

As regards the boundary behavior of the chain, i.e., the determination of the P_0, P_0', P_1, $P_1' \ldots P_{m-1}, P_{m-1}'$ probabilities appearing in Π_0 and Π_1, the matrices of interest are the following: B_{00}, which is a $2(m-1) \times 2(m-1)$ matrix,

$$B_{00} = \begin{bmatrix} A - \Delta\lambda & \Delta\lambda & \ldots & 0 \\ \Delta\mu(1) & A - \Delta\lambda - \Delta\mu(1) & \ldots & 0 \\ 0 & \Delta\mu(2) & \ldots & 0 \\ \vdots & \vdots & \vdots & \vdots \\ 0 & 0 & \ldots & A - \Delta\lambda - \Delta\mu(m-2) \end{bmatrix}, \tag{19}$$

where $\Delta\mu(i)$ is the diagonal 2×2 matrix defined as

$$\Delta\mu(i) = i\mu I\ , \quad i = 1, 2, \ldots \tag{20}$$

B_{01}, which is $2(m-1) \times 2$:

$$B_{01} = \begin{bmatrix} 0 \\ 0 \\ \vdots \\ \Delta\lambda \end{bmatrix}, \tag{21}$$

B_{10}, which is $2 \times 2(m-1)$:

$$B_{10} = \left[\begin{array}{cccc} 0 & 0 & \dots & \Delta\mu(m-1) \end{array} \right], \tag{22}$$

the 2×2 B_{11} matrix:

$$B_{11} = A - \Delta\lambda - \Delta\mu(m-1), \tag{23}$$

and finally the 2×2 B_{21} matrix:

$$B_{21} = \Delta\mu(m), \tag{24}$$

with all the remaining B_{ij} matrices appearing in (9) identically equal to zero. We finally have that the solution to this chain is still formally expressed by (11) and (13), provided that the matrices A_0, A_1 and A_2 in (12) are equal to $\Delta\lambda$, $A - \Delta\lambda - \Delta\mu(m)$ and $\Delta\mu(m)$ respectively, and that the boundary matrices in (14) are the ones defined in (19) through (24).

Note that any work-conserving queueing system with arrival rates $\lambda = \lambda_H + \lambda_N$ and $\lambda + \lambda_U = \lambda_H + \lambda_N + \lambda_{HU} + \lambda_{NU}$ in the two phases, has the solution furnished in this section, regardless of the queueing discipline (e.g. LIFO, random order). It follows that since the probabilities of having $0, 1, \dots, m$ resources busy do not vary for varying queueing discipline, transmission design is unaffected by this last choice.

5 Hot Spots and Transmission Design

In order to demonstrate how transmission design is shaped by traffic parameters in the presence of hot spots, we next consider a power controlled DS-CDMA scheme which operates inside the single cell of a wireless network, employing the queueing scheme with two priority classes we investigated.

As regards transmission design, we focus onto a single merit figure, namely, the outage probability P_{out}, defined as the probability that the bit error probability a generic user experiences on the reverse link (mobile-to-base direction) is above a predefined value P_{e0}. As in [7] we neglect intracell interference and assume that system performance be limited by intercell interference. We then transpose the P_e condition onto the r.v. "carrier-to-interference ratio" C/I: hence, $P_{out} = Pr\{C/I \leq (C/I)_0\}$, where $(C/I)_0$ represents the C/I value which corresponds to the P_{e0} threshold. Note however that due to the varying traffic conditions we described in the previous sections, the number N of simultaneous active interferers which determine the actual distribution of the r.v. I is a random variable too: if we therefore wish to compute the outage probability accounting for the fact that the N interfering units are not active all the time, we have to pursue the evaluations reported below.

Given that $P_{out}|N$ represents the outage probability conditioned to the presence of N active interferers, in ordinary traffic conditions the reference user will experience an outage given by:

$$P_{out|ord} = \frac{\sum_{N=1}^{m-2} P_{out}|N \cdot P_{N+1} + P_{out}|N = (m-1) \cdot P_{delayord}}{P_{ord} - P_0}, \qquad (25)$$

where $P_{ord} = \frac{\beta_u}{\alpha_u+\beta_u}$ gives the probability of being in ordinary traffic conditions, and $P_{delayord} = P_{ord} - \sum_{i=0}^{m-1} P_i$ represents the probability of having all the m resources – codes – busy in ordinary traffic conditions.

In critical traffic conditions, the reference user will correspondingly experience an outage expressed by

$$P_{out|crit} = \frac{\sum_{N=1}^{m-2} P_{out}|N \cdot P'_{N+1} + P_{out}|N = (m-1) \cdot P'_{delaycrit}}{P_{crit} - P'_0}, \qquad (26)$$

where $P_{crit} = \frac{\alpha_u}{\alpha_u+\beta_u}$ furnishes the probability of being in the upsurge condition, while $P_{delaycrit} = P_{crit} - \sum_{i=0}^{m-1} P'_i$ is the probability of having all m resources busy in critical traffic conditions.

We can also evaluate an overall value of the outage probability, unconditioning with respect to the two possible situations – ordinary and upsurge phase –, obtaining:

$$P_{out} = P_{out|ord} \cdot \frac{P_{ord} - P_0}{1 - P_0 - P'_0} + P_{out|crit} \cdot \frac{P_{crit} - P'_0}{1 - P_0 - P'_0}, \qquad (27)$$

which we also restate as

$$P_{out} = \frac{1}{1 - P_0 - P'_0} \cdot \left(\sum_{N=1}^{m-2} P_{out}|N \cdot \left(P_{N+1} + P'_{N+1}\right) + \right.$$
$$\left. P_{out}|N = (m-1) \cdot \left(P_{delayord} + P'_{delaycrit}\right) \right). \qquad (28)$$

In next section we will numerically show the behavior of these quantities for different choices of the parameters describing the examined Markov chain.

6 Numerical Results

In our numerical evaluations we refer to a DS-CDMA system which utilizes $m = 18$ PN sequences and features a processing gain $G = 127$, setting the threshold P_{e0} to 10^{-3}. We account only for the effects of an imperfection in the power control loops [5, 7]: as in [5] we suppose the corresponding error be Gaussian once expressed in dB, unbiased and have a standard deviation $\sigma_\Delta = 2$ dB. It follows that the r.v. C appearing in the carrier-to-interference ratio C/I is lognormally distributed; the r.v. I representing the interference, $I = \sum_{i=1}^{N} I_i$, is a sum of i.d. lognormal random variables, where I_i represents the contribution of the i-th interferer among the N that are active.

In Fig.4 we first report the behavior of both $P_{out|ord}$ and $P_{out|crit}$ as a function of the λ_u/λ ratio, with $\lambda_u = \lambda_{HU} + \lambda_{NU}$, and $\lambda = \lambda_N + \lambda_H$, when $\lambda = 0.04$ and $\mu = (180)^{-1}$. Solid lines refer to the following choice: $(\alpha_u, \beta_u) = (10^{-3}, 10^{-2})$, dashed lines refer to $(\alpha_u, \beta_u) = (10^{-4}, 10^{-3})$. For each (α_u, β_u) choice the upper curve gives $P_{out|crit}$, the lower curve gives $P_{out|ord}$. It is important to highlight that in the range of interest for the outage probability values, (10^{-2} or less), the $P_{out|ord}$ and $P_{out|crit}$ curves lie far apart, regardless of the (α_u, β_u) pair examined. Note also that when $\lambda_u/\lambda = 1$ we have an overall coefficient of utilization $\rho = 0.44$, whereas for $\lambda_u/\lambda = 10$ we have $\rho = 0.8$. Unfortunately, we can never utilize the m resources so intensively, since it is evident that a heavily loaded CDMA system performs poorly from the transmission viewpoint. Futhermore, let us note that for both the considered (α_u, β_u) pairs we have $P_{ord} = 0.9$ and $P_{crit} = 0.1$: however, the mean sojourn times in each phase are different. More precisely, solid lines refer to an average time spent in ordinary traffic conditions $\bar{t}_{ord} \simeq 16.6$ min and an average time spent in critical traffic conditions $\bar{t}_{crit} \simeq 1.67$ min, dashed lines refer to $\bar{t}_{ord} \simeq 2.8$ hours and $\bar{t}_{crit} \simeq 16.6$ min.

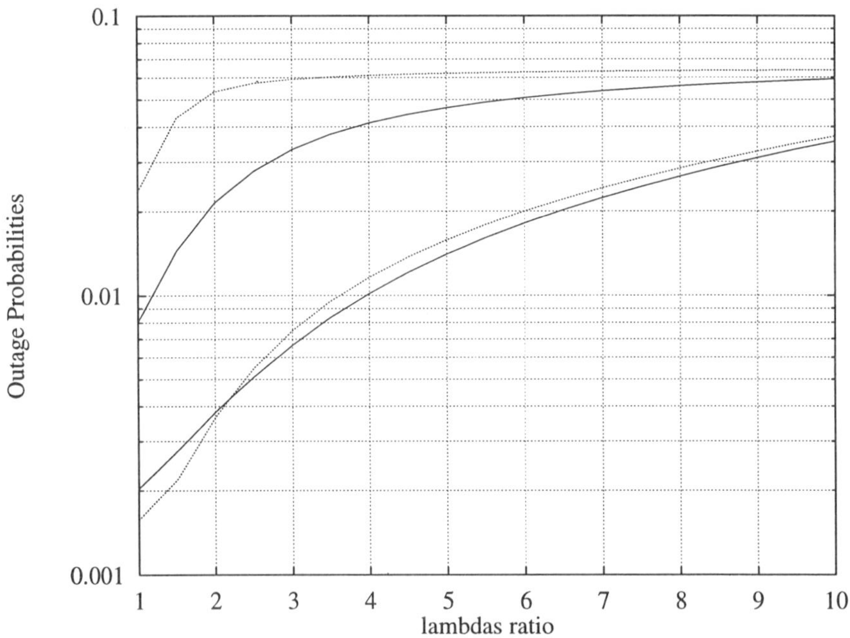

Figure 4: $P_{out|ord}$ and $P_{out|crit}$ as a function of the λ_U/λ ratio for the DS-CDMA system considered. $P_{ord} = 0.9$, $P_{crit} = 0.1$.

In Fig.5 we also report $P_{out|ord}$ and $P_{out|crit}$ when $P_{ord} = P_{crit} = 0.5$, for $(\alpha_u, \beta_u) = (10^{-2}, 10^{-2})$ (solid lines) and for $(\alpha_u, \beta_u) = (10^{-3}, 10^{-3})$ (dashed lines). Again, let us point out how deeply service quality as perceived by the generic network user can vary: if he finds himself in a hot spot, the outage probability he experiences may be subject to a sharp increase with respect to the corresponding value in ordinary traffic conditions!!!

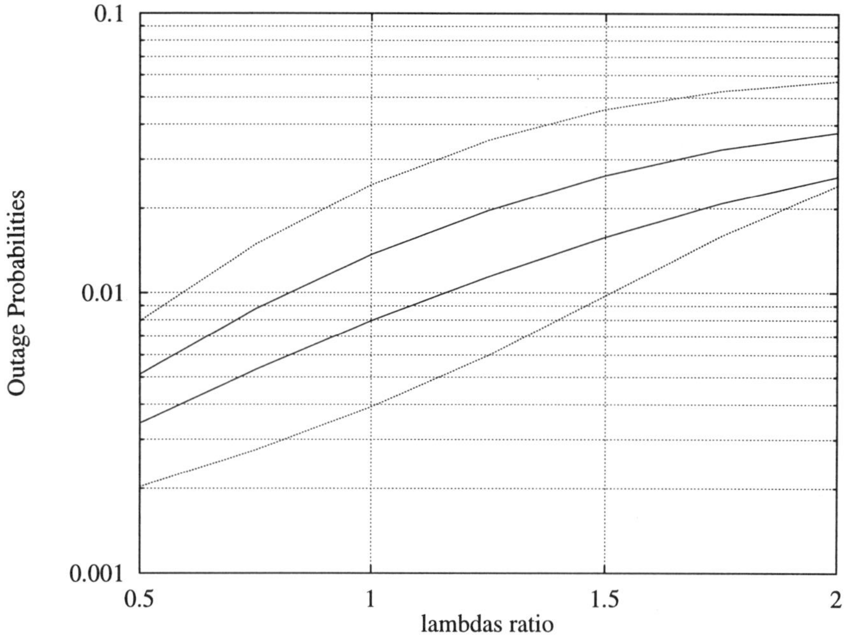

Figure 5: $P_{out|ord}$ and $P_{out|crit}$ as a function of the λ_U/λ ratio for the DS-CDMA system considered. $P_{ord} = 0.5$, $P_{crit} = 0.5$.

7 Conclusions

In this paper we investigated a possible technique to handle the mobile-to-base traffic inside the cell of a wireless PCS system, assigning different priority classes to new calls and hand-offs. For the corresponding queueing scheme we demonstrated how we can achieve the complete steady-state solution, exploiting a matrix analytic technique. This approach further allowed us to efficiently model the behavior of the examined system in the presence of critical traffic conditions inside the cell. Lastly, we demonstrated how traffic parameters influence the design of CDMA systems both in ordinary - and critical - traffic conditions, numerically determining how sensitive system performance is to the duration and intensity of traffic build-ups.

Acknowledgments
The author would very much like to thank Prof. Gianni Immovilli and Dr. Franco Callegati for several helpful discussions and comments.

References

[1] R.Guerin, "Queueing-Blocking System with Two Arrival Streams and Guard Channels", IEEE Trans. on Comm., Vol.36, no.2, February 1988, pp.153-163.

[2] F.T.Pavlidou, "Two Dimensional Traffic Models for Cellular Mobile Systems", IEEE Trans. on Comm., Vol.42, no.2/3/4, February/March/April 1994, pp.1505-1511.

[3] J.Keilson, O.C.Ibe, "Cutoff Priority Scheduling in Mobile Cellular Communication Systems", IEEE Trans. on Comm., Vol.43, no.2/3/4, February/March/April 1995, pp.1038-1045.

[4] K.K.Leung, W.A.Massey, W.Whitt, "Traffic Models for Wireless Communication Networks", IEEE Journ. on Select. Areas in Comm., Vol.12, no.8, October 1994, pp.1353-1364.

[5] A.M.Viterbi, A.J.Viterbi, "Erlang Capacity of a Power-Controlled CDMA System", IEEE Journ. on Select. Areas in Comm., Vol.11, no.6, pp.892-900, August 1993, pp.892-900.

[6] D.Everitt, "Analytic Traffic Models of CDMA Cellular Networks", Conference Record of ITC, Adelaide, Australia, 1994, pp.349-356.

[7] G.Immovilli, M.L.Merani, "Priority Traffic in Wireless PCS Systems: Transmission and Traffic Parameters Trade-Off", accepted for oral presentation at IEEE VTC '96, Atlanta (GA), USA, April 28-May 1, 1996.

[8] M.F.Neuts, "Matrix Geometric Solutions in Stochastic Models: An Algorithmic Approach", John-Hopkins, Baltimore, MD, 1981.

[9] R.Vijayan, J.M.Holtzman, "Foundations of Level Crossing Analysis of Hand-Off Algorithms", Conference Record of IEEE ICC '93, Geneva, Switzerland, April 1993, pp.935-939.

[10] K.Kant, "Introduction to Computer System Performance Evaluation", McGraw-Hill, 1992.

A SATELLITE-AUGMENTED CELLULAR NETWORK CONCEPT

Deepak Ayyagari, Anthony Ephremides

University of Maryland at College Park, College Park, MD 20740

ABSTRACT

Satellite systems have been used recently to extend geographical coverage of cellular service; however, the additional satellite capacity may also be used to off-load congestion within the area covered by the terrestrial network.

In this paper we investigate the performance (in terms of call-blocking probability) of such a hybrid satellite-terrestrial system. The footprints of a satellite with multiple spot beams may be viewed as the highest layer overlaying a multilayered cellular system. While such hierarchical schemes have been well studied, the effects of the large size of the footprints and the different reuse constraints on satellite bandwidth as compared to conventional terrestrial cellular networks, require different trade-offs that are investigated here.

We present a model for a 1-dimensional (e.g highway) cellular system that is augmented by multiple spot beams and we obtain results for the call blocking probability. We then use an elaborate simulation program to study the call blocking probability and handoff blocking rate in a planar cellular network supported by satellite spot beam coverage. The effect of different hierarchical positions of the satellite system and the cellular system in the integrated system are also investigated (i.e., we consider the relative priority assigned to the cellular and satellite systems by the policy used to direct new call requests and handoffs).

1 INTRODUCTION

Satellite and cellular networks are being developed, both as stand-alone systems and as integrated (or hybrid) networks to support the surging demand for cellular services. One such hybrid system seeks to use the satellite capacity to service mobile users leaving the cellular area. This additional satellite capacity could possibly be used to offload cellular congestion also.

An integrated system, where the satellite system has multiple spot beams overlaying a terrestrial cellular network, may be viewed as a two-tier hierarchical system. Hierarchical cellular systems with multiple overlays have been well studied [1, 2]. All the layers in such systems are assumed to have the same propagation characteristics and similar re-use constraints. A recent study [3] analyzes the performance of an integrated satellite-cellular system. According to the model proposed in this study, the satellite footprints are treated as the largest macrocells in a hierarchical cellular structure and each layer is analyzed separately.

In this paper, we model and analyze the satellite and cellular components as an integrated system rather than as one of distinct layers as in [3]. The expectation that a hybrid system may provide superior performance is based on the different satellite propagation characteristics that may permit a much denser pattern of reuse of bandwidth than would be possible in the cellular network. The developments in antenna technology and careful evaluation of the propagation effects in satellite systems support this expectation [4, 5]. The main question we are addressing in this paper is the determination of the distribution of the given bandwidth between the satellite system and the cellular system so as to maximize the efficiency of the use of that bandwidth.We also investigate other factors which affect performance These factors are:

1. The relative size of the satellite footprints compared to the size of the cells in the terrestrial cellular network and the different reuse patterns in the cellular and satellite systems.

2. The new-call service procedure (whether, to direct new call requests to the satellite system first and the overflow calls to the cellular network or vice versa) [6, 7].

3. The handover management procedure (whether, to direct handoff requests to the satellite system first and overflow handoff requests to the cellular system or vice versa).

In this paper, we first present a multi-dimensional Markov chain-based model for a one dimensional (e.g.,highway) cellular system augmented by overlapping satellite spot beams. Initially, we do not consider handoffs and make some reasonable assumptions regarding the reuse patterns in the satellite and cellular systems. The call blocking probability is taken as the performance measure. We determine the performance of the integrated system for different distributions of the total bandwidth between the satellite and cellular systems for a fixed load. We then determine the sensitivity of the performance measure to different values of traffic load, given a certain fixed bandwidth allocation to the two systems. This is accomplished analytically. To assess the effects of handoffs and mobility and to apply our approach to more realistic systems, we resort to simulation. To this effect a simulation program has been developed that incorporates mobility models and handoffs. The simulation is used to compute two performance measures (average new call blocking rate and handoff blocking rate)

The rest of the paper is organized as follows: Section 2 defines the analytic model for the one- dimensional hybrid network. Section 3 presents results obtained for a particular configuration of the network. Section 4 describes the simulation model and presents results for simulations of different configurations of a planar network.

2 MODEL DESCRIPTION

The region under consideration (Figure 1) is covered by cells laid out in a contiguous non-overlapping fashion as in a highway cellular system. All cells are of equal size,and so the size of each cell may be normalized to 1. The cells in Figure 1 are numbered 1 to 8. Each cell is excited by a centrally located base station.

The cells are overlaid with the footprints of satellite spot beams, which may overlap. The size of each footprint is equal to an integral number of cell lengths. We intend to use the minimum number of beams required to cover a given area, keeping in mind the reuse constraints of the satellite system. In Figure 1 the beam width is 3 cell lengths. Given this beam width and a minimum reuse distance of 1 cell length for the satellite system, the size of the area where there can be no reuse is at least 4 cell lengths wide. We need 2 beams to cover this area and there is an overlap between the two beams covering this area. The adjoining areas, also 4 cells wide, can now reuse the same frequencies.

The formulation in this model is generally applicable to any beam width and minimum reuse distance when the layout of the footprints is as described.

- In the cellular system, alternate cells can reuse the same set of frequencies for both the uplink and downlink directions.
- In the satellite system, beams one cell length apart can reuse the same frequencies. Thus we see that the beams A1 and A2 use the same set of frequencies for the configuration in Figure 1. This is a very conservative assumption as reuse could possibly be denser (at a closer minimum reuse distance) in the satellite system, while here the reuse distances are the same as the cellular network.

In this model we do not consider handoffs. Therfore, the analysis of each area of size 4 cells (covered by beams A1 and B1 in Figure 1) would be identical. Note that the size of this area would change with the beam width and the minimum reuse distance for the satellite system. New calls arise in the cells and maybe assigned to the satelllite beam(s) covering the corresponding cell or to the base station for that cell (to the cellular network) depending on the policy of assignment. We define the two following policies:

- Cell First Policy (henceforth called CF): The service request (new call or handoff) is first assigned to the cellular system. Overflow requests are assigned to the satellite system. Excess requests are blocked.
- Satellite First Policy (henceforth called SF): The service request (new call or handoff) is assigned to the satellite system first. Overflow requests are serviced by the cellular system. Excess requests are blocked.

We use only the CF policy in the analytical model which does not consider any handoffs. In the simulation that follows, both the policies are investigated and the policies will apply to both new calls and handoffs. Overflow calls from cells within the overlap region of two beams may be directed to either of the two beams with equal probability.

The following assumptions are made regarding the reuse of frequencies:

Finally, we define a parameter that is crucial in the model as well as in the simulation, and call it the Satellite Cellular Bandwidth Ratio (henceforth called SCBR). We assume that the total bandwidth available is fixed and equal to some M duplex channels. A channel is considered to be a unit of bandwidth sufficient to accomodate one call. These M duplex channels can be assigned to the satellite network or the cellular network as S satellite channels and C cellular channels, so that M= S + C. The efficiency of the system depends on how these channels are reused in the cellular and satellite systems. We define SCBR = S/M i.e., the fraction of the total bandwidth allocated to the satellite. Thus SCBR takes values between 0 and 1. All the results that follow will show the performance measures (new call blocking rate and handoff blocking rate) as functions of SCBR under different conditions such as configurations resulting from different beam widths, traffic variations, new call assignment policies and handoff management policies.

We enumerate the cells and the beams in the region under consideration as follows. If there are N cells and P beams covering the same region, the N cells are numbered 1 to N while the beams are numbered $N+1$ to $N+P$. The state of the system can now be defined by the vector $\overline{x} = (x_0, x_1, \cdots, x_N, x_{N+1}, \cdots, x_{N+P})$ where $x_i, i \in \{1, 2, \ldots, N\}$ is the number of calls being served by cell i and $x_i, i \in \{N+1, N+2, \ldots, N+P\}$ is the number of calls being served by beam i.

The call requests arrive as independent Poisson processes in each cell with rate λ_i in cell i. The session duration of a call is exponentially distributed with parameter μ. We model the system as a continuous time Markov chain.

A state is feasible if it satisfies the following constraints: $\forall i = 0, 1, \ldots, N+P$,

$$0 \leq x_i \leq C_i.$$

That is, the number of calls in progress in cell/beam i does not exceed the number of channels C_i assigned to that cell/beam.

Define Z to be the set of feasible states, $p(\overline{x})$ to be the steady state probability of state $\overline{x}$ and $q(\overline{x}, \overline{y})$ as the rate of transitions to state y, given we are currently in state x.

The rate matrix Q is determined by the new call admission policy that is used (either CF or SF). We are concerned only with the CF policy here. This implies that no new arrival can be directed to a satellite beam unless the capacity in the corresponding cell is full. The transition rate matrix may be constructed as follows:

- If $\overline{y} = (x_1, x_2, \cdots, x_{i+1}, \cdots, x_N, \cdots, x_{N+P}), i \leq N$ then $q(\overline{x}, \overline{y}) = \lambda_i$. This corresponds to an arrival in cell i which is assigned a channel in cell i.

- If $\overline{y} = (x_1, x_2, \cdots, x_{i+1}, \cdots, x_N, \cdots, x_{N+P}), i \leq N$ then $q(\overline{x}, \overline{y}) = x_i \mu$. This corresponds to a call departure.

- $\overline{y} = (x_1, x_2, \cdots, x_{i+1}, \cdots, x_N, \cdots, x_{N+P}), N \leq i \leq N + P$ then this corresponds to an arrival being assigned a channel from beam i. Such a transition is forbidden by the CF policy unless the utilisation of at least one of the cells overlaid by the beam is at full capacity. So $q(\overline{x}, \overline{y}) = 0$ unless $x_j = C_j$ for some $j \leq N$ such that beam i covers cell j. Otherwise, $q(\overline{x}, \overline{y}) = \Sigma o_k$, for all cells k such that $x_k = C_k$ (i.e., the cells numbered k have their capacities full) , where o_k is defined as the overflow rate from cell k. Note that $o_k = \lambda_k$ if the cell k is covered by a single beam, and that $o_k = \lambda_k/2$ if the cell k is covered by two beams.

The steady state probabilities for the Markov chain can then be uniquely determined by solving the system of linear equations

$$pQ = 0$$

where p is the steady state probability vector. The matrix Q being singular, we need the boundary condition $pe = 1$ where e is the vector of 1s.

The matrix Q is very sparse. Therefore efficient numerical algorithms may be used to solve the linear equations and compute the steady state probabilities. The probability of blocking is defined as the average fraction of new call requests that cannot be served. We define $PB_i, i \in \{1, 2, \ldots, N\}$ to be the call blocking probability in cell i. Let PB_{av} be the average call blocking probability over the entire region under consideration.

Since the hybrid system comprises satellite beams that overlap, some cells may be overlaid by more than one beam. We therefore define the sets S_i associated with each cell $i, i \in \{1, \ldots, N\}$ as follows:

$$S_i = \{j, j \in \{N, \ldots, N+P\}, \text{ beam } j \text{ overlays cell } i\}$$

A new call originating in cell i is blocked if it cannot be served by the cellular capacity in cell i or by the satellite capacity furnished by the beams in set S_i. This happens if the system is in a state $\overline{x}$ belonging to a set $B_i, i \in \{1, 2, \ldots, N\}$ defined as

$$B_i = \{\overline{x} : x_i \geq C_i, x_j \geq C_j, j \in S_i\}$$

The call blocking probability in cell $i, i \in \{1, 2, \ldots, N\}$ is given by:

$$PB_i = \sum_{s \in B_i} p(s)$$

The blocking probability in the region under consideration as a whole is given by :

$$PB_{av} = \frac{\sum_{i=1}^{N} \lambda_i \times PB_i}{\sum_{i=1}^{N} \lambda_i}$$

3 RESULTS

We consider a simple network example to illustrate our approach and calibrate the results. This network is shown in figure 1.We assume that the system has a total bandwidth of 8 duplex channels i.e., M=8. Alternate cells reuse the same frequencies and alternate beams reuse the same frequencies. If the SCBR were 0 (i.e., all the bandwidth were allocated to the cellular system), the number of channels in each cell would be 4 and number of channels in each beam would be 0. We determine the performance of the system in terms of average call blocking probability as we vary the value of SCBR from 0 to 1(i.e., allocate different portions of the total bandwidth to the satellite system). We assume that the load in each cell is the same, which we fix to be 0.5 erlangs. The results are shown in Figure 2. We see that for the given traffic load, the configuration which yields the lowest average call blocking probability has SCBR = 0.25. The performance of the system for this allocation is better than for the case when SCBR =0 (i.e., totally cellular allocation) or SCBR =1 (i.e., totally satellite allocation). This demonstrates that for a certain fixed load the integrated satellite system performs better than a exclusively cellular system using the same amount of bandwidth.

We also investigate the sensitivity of the blocking probability to changes in load. We consider three different allocations where the SCBR values are 0, 0.5 and 1 respectively.The results are shown in Figure 2. We see that as the load increases, the blocking probability increases for any allocation as expected. But we also note that the configuration that provides best performance or minimum call blocking probability depends on the load. For example, at a load of 1 erlang, the allocation with 2 satellite channels performs best while at a load of 1.75 erlangs the allocation with no satellite channels and 4 cellular channels performs best. This shows that athough allocation of bandwidth as dedicated capacity within cells results in more channels system-wide, the allocation of bandwidth to the satellite as shared capacity and under different reuse considerations provides a better overall performance for lower loads. For higher values of load, the advantages of the satellite are overshadowed by the increase in capacity gained by small cell size in the cellular system.

4 SIMULATION

In this section we describe the simulation studies conducted on the Object Oriented Hybrid Network Simulation (OOHNS) testbed developed at the University of Maryland. The simulation studies become necessary for the following reasons:

- Scale of the Network: The analytical models presented in literature are computationally infeasible for studying large hybrid networks (large number of cells and satellite beams).
- Mobility management: Analysis is simply powerless to track realistic mobility scenarios. This section is organized as follows:
    1. The architecture, parameters and assumptions of the simulation testbed are described.
    2. The various simulation studies and their results are presented.

4.1 Simulation testbed description

The layout of the integrated satellite-cellular network is as follows: the cellular network comprises hexagonal cells controlled by centrally located base stations.The area of coverage is defined by the number of cells and the cell size. The satellite networks comprises multiple spot beams. The foot prints of these beams are assumed to be circular for simplicity. These footprints overlay the same coverage area defined by the cellular network

The testbed is an object-oriented discrete event simulator.The functional units of the simulation belong to four classes of objects: mobile units, base stations, beam objects and switching center.These objects are basically repositories of functions which implement all the actions that are particular to that class.

Mobile units: All mobiles in the system are assigned unique identification numbers. Each mobile unit in the system is concerned with managing its mobility and with the call process at that mobile.The mobility aspect of the mobiles is embedded in two different models,both of which are used in the simulation. We shall refer to them as Mobility Model 1 and Mobility Model 2 .

Model 1: The parameters under this model are maximum speed, minimum speed and the direction. The speed of the mobile is a random variable uniformly distributed between the maximum and minimum speeds. The direction of motion of the mobile is also uniformly distributed between 0 and 360 degrees according to a fixed reference.

Model 2: In this model, each mobile updates its position at fixed intervals of time. At each decision instant, the mobile makes one of two decisions: to stay within the same cell or to move to a neighboring cell. Thus the parameter in this model is the probability distribution that governs the outcome of the decision.

Call procedures: Each mobile exists in one of the following states: call initiate, call receive, call progress, call terminate or on hook. Each mobile unit generates call requests according to a Poisson process. The time interval between call requests is an exponentially distributed parameter called mean interarrival time. If a mobile is in the call progress state and a move event is triggered changing its position, then a handoff procedure may or may not be activated, depending on whether the mobile has moved out of the area of coverage of the base station or the satellite beam supporting the call.

Base stations and Satellite beam : Both these objects perform similar functions. They maintain a set of channels (specified by the number of channels/base and the number of channels/beam)which are assigned and de-assigned to mobile units each time a call request is granted or terminated respectively.

Switching centre: There is only one instance of this object in the simulation. It consists of procedures that govern channel assignment and handoff management. The SF and CF policies for both handover management as well as new call channel assignment are explicitly implemented in these procedures. The switching centre also has routines which compute the performance measures of interest in the system: the overall new call blocking rate and the handoff blocking rate.

There are several assumptions made in the system design of this system, outlined below.

- Reuse patterns in the cellular and satellite system: for the sake of uniformity, we require all base stations to have the same number of channels as well as beams. In the satellite system we specify that non-overlapping but adjacent beams can reuse the same set of frequencies which essentially results in a reuse factor of 2. While common cellular reuse factors for a hexagonal cell plan are 3 or 7, we use a reuse factor of 4 as it is the only feasible reuse factor that provides the same number of channels in all the base stations and the beams for different values of the parameter SCBR. Also, 4 is a very conservative reuse factor as conventional systems use a reuse factor of 7 for the voice channels.
- The new calls are set up at both ends (called party and calling party) entirely on the cellular or the satellite network. However during a handoff, a mobile being served by the cellular network may be handed off to the satellite network or vice versa.
- The propagation effects in a cellular and a satellite system are not explicitly modeled here.
- The capacity of the links between the base stations and the switching centre as well as from the satellite to the hub are assumed to be large enough to handle the control traffic as well as the user traffic.

4.2 Simulation Results

The three configurations of the hybrid network that were analyzed. The number of beams in each configuration was 1, 4 and 8 respectively.The numerical values of all parameters of the simulation that have been outlined above are specified in Table 1. We are concerned mainly with high mobility users in urban/semi-urban environments.

The performance measures of interest, namely the new call blocking rate and the handoff blocking rate, are computed as functions of the parameter SCBR. The simulations were conducted for SCBR values of 0, 0.2, 0.4, 0.6, 0.8 and 1. The objectives of the simulation studies are to determine:

- The effect of reducing beam size on new call blocking and handoff blocking. The Cell First policy is used for both new calls and handoffs. The total bandwidth in the system M= 20 channels. Motion model 1 is used in the simulation.

Number of Beams (configuration-dependent)	1, 4, 8
Number of Base Stations	25
Number of Mobile Units	100
Total System Bandwidth	20
mean interarrival delay	3 min.
mean call duration	3 min.
minimum speed (motion model 1)	10 mph
maximum speed (motion model 1)	50 mph
probability of cell change (motion model 2)	var.
cell radius	2 mi.
beam radius (depending on configuration)	4, 5.25, 10.5 mi.

Table 1 Simulation parameters.

- The effect of different channel allocation and handoff management policies: Here ew have used a layout with 4 spot beams covering the cellular area. The performance measures are computed for different values of SCBR under the SF and the CF policies for handoffs and new call assignments. Motion model 1 is used in the simulation. The total bandwidth in the system $M = 20$ channels.

From Figure 3, we see that for a given beam size and fixed amount of total bandwidth, the minimum call blocking probability is achieved for different values of the parameter SCBR. We see that the best performance is obtained for values of SCBR greater than 0 only when the ratio of footprint-size to cell size is below certain values. When the ratio is greater than these values (not computed here) the gain from the statistical multiplexing and better reuse in the beams is defeated by the capacity gain of smaller cells. Similar trends are evident for handoffs. It must be noted that with a single beam and all- satellite capacity, there are no handoffs.

In the comparison between the channel assignment policies, it is seen from Figure 4 that the CF (Cell First) policy performs better than the SF (Satellite First) policy for all values of SCBR. When the system contains no cellular capacity (SCBR =1) and when the system contains no satellite capacity (SCBR =0) the performance of the two policies is identical. However, when the capacity is predominantly cellular (lower values of SCBR), the SF policy performs much worse than the CF policy. This may be explained by recalling that the satellite capacity in a beam is shared by all the cells covered by the beam. When this capacity is scarce (SCBR is low), the CF policy ensures that these channels are reserved for overflow calls only while the SF policy ensures that these satellite channels are full before the cellular capacity is used. So when congestion results in a particular cell, i.e its cellular channels fill up, then under the CF policy some satellite channels may still be free to off-load the congestion while under SF policy no free channels are available as they are being used also by calls from cells with no congestion . As the number of satellite channels increases, the performance is similar as more channels are now available to service calls from congested areas. The same trends are observed for the handoff blocking rate.

REFERENCES

[1] J.Steele *et al.*, "Systems Aspects of Cellular radio", IEEE Communications Magazine, pp. 80-85, June 1995.

[2] J. Zander, "Generalized Reuse Partitioning in Cellular Mobile Radio", Proc. Vehicular technology Conference, pp. 181-184, May 1993.

[3] Lon-Rong, S. S. Rappaport, "Personal Communication Systems Using Multiple Hierarchical Cellular Overlays", IEEE Journal on Selected Areas in Communication, vol. 13, no. 2, pp. 406-415, Feb. 1995.

[4] W. T. Vogel, "Mobile Satellite System Propagation Measurements", IEEE Transactions on Antennas and Propagation, vol. 38, no. 2, p. 259, Feb. 1990.

[5] L. J. Ippolito, "Propagation Considerations for Emerging Satellite Communication", Proc. IEEE, vol. 81, no. 6, pp. 923-932, June 1993.

[6] R. Beraldi *et al.*, "A Reversible Hierarchical Scheme for Microcellular Systems Overlaying Macrocellular Systems", to appear Infocom 1996.

[7] Hong, S. S. Rappaport, " Traffic Model and Performance Analysisfor Cellular Mobile Radio Telephone Systems With Prioritized and Nonprioritized Handoff Procedures", IEEE Transactions on Vehicular Technology, vol. VT-35, no.3 pp. 77-92, Aug.1986.

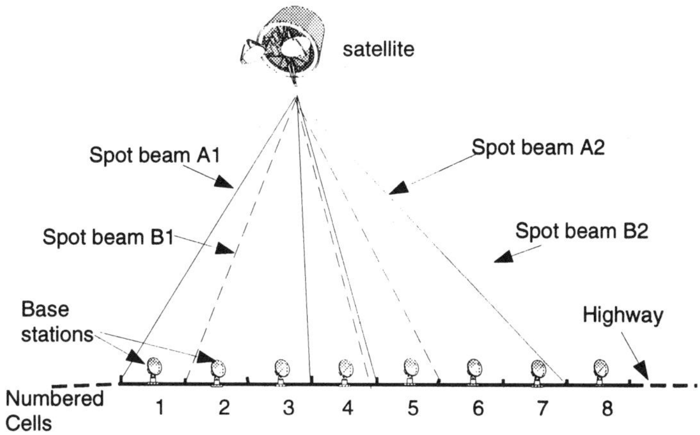

Figure 1 1 dimensional cell plan overlaid with satellite beam footprints. Beam Width = 3 cell lengths. Beam A1,A2 and B1,B2 reuse the same frequencies

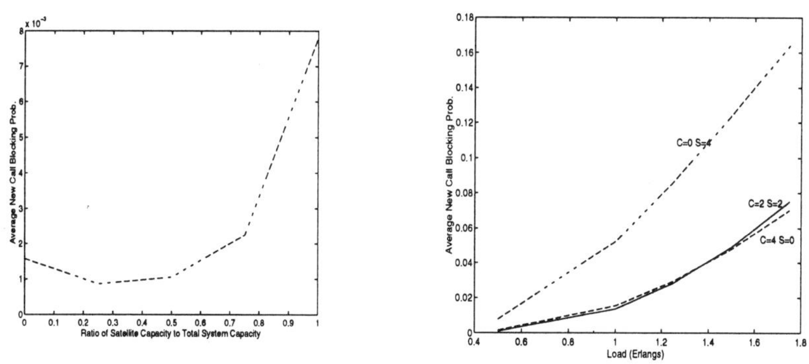

Figure 2 The new call blocking probability for different allocations of satellite capacity (varying SCBR, fixed and varying traffic loads)Total System Capacity M = 8 channels.

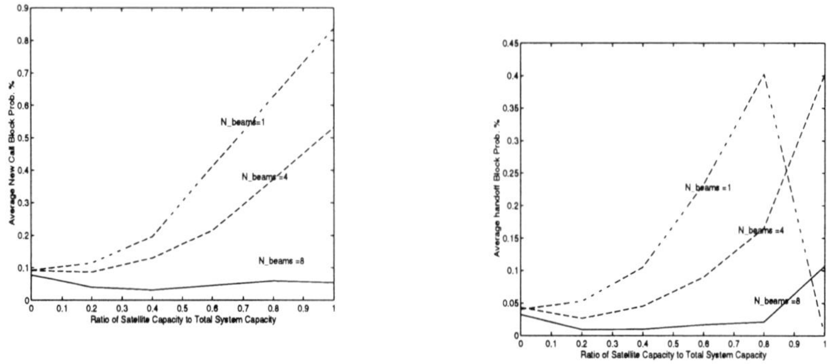

Figure 3 The new call and handoff blocking probabilitiesfor different allocations of satellite capacity (varying SCBR) and different beam widths. Total Bandwidth M=20 channels. Satellite reuse factor=2. Cellular reuse factor =4

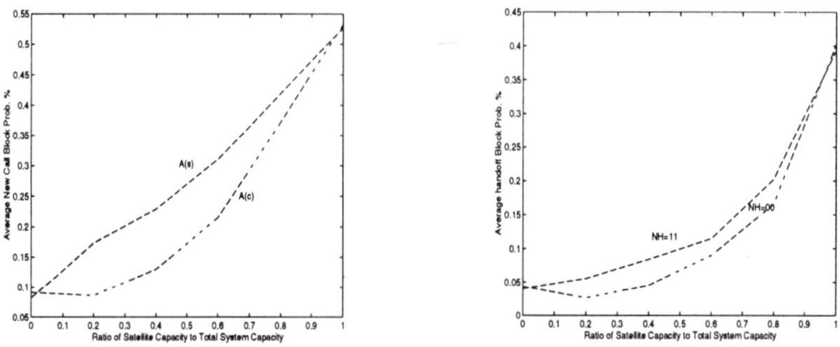

Figure 4 The new call and handoff blocking probabilitiesfor different allocations of satellite capacity (varying SCBR) and policies for new call and handoff management. A(c),NH=00: Cell First policy for new calls and handoffs.A(s),NH=11: Satellite First policy for new calls and handoffs.

Recent Developments of Adaptive Air Interfaces

K. David

DeTeMobil, Deutsche Telekom MobilNet GmbH, P. O. Box 8865, D-48047 Münster, Germany

e-mail: kdavid@ms.detemobil.de

Abstract - Within the European RACE (Research in Advanced Communications for Europe) program several projects have contributed to the development and standardisation of third generation air interfaces. Therefore, several system concepts based on TDMA or CDMA have been developed. For both the selection procedures within the standardisation and a sound comparison of different systems, well defined scenarios and quality of service definitions are a prerequisite.

A major part of this paper concentrates on a description of the ATDMA (Advanced TDMA) system concept and on its evaluation. This project has designed an advanced TDMA radio access scheme for third generation systems. This system concept consists of various new techniques like packet access, novel handover schemes, and link control (a technique adapting the transmission chain to the propagation channel characteristics). Here, an investigation focusing on packet access including capacity issues is given.

I. Introduction

After the major success of the GSM Standard being already chosen by more than 100 mobile operators in more than 60 countries world-wide, the relevant standardisation bodies are working intensively towards the next generation systems, in Europe also called UMTS (Universal Mobile Telecommunications System). In this standardisation work one major step will be the selection of the air interface for the third generation system. The service requirements for these air interfaces have already been formulated and include: the need of a wide range of services, high quality voice, data services up to 2 Mbps, flexibility, and both B-ISDN service and network compatibility. Within the European Union three RACE projects concerning cellular systems have developed such air interfaces and also have been supporting the standardisation work. These projects are ATDMA, CODIT (Code Division Testbed), and MBS (Mobile Broadband Systems).

On the basis of propagation measurements, channel modelling and technique studies the ATDMA system concept has been developed [1]. This concept capitalises on the optimum mixture of baseband processing techniques, and in particular on the ongoing adaptation of each mobile's radio link to meet its quality requirements while minimising the usage of radio resources and will be described in detail in Section III. The multiple access is based on the well-known TDMA technique, like the GSM (Global System for Mobile Communications) or DECT systems.

The CODIT Project has designed a DS-CDMA based air-interface satisfying the requirements above. In order to facilitate flexibility and bit rates up to 2 Mbps the three bandwidths 1, 5, and 20 MHz are used [2]. For each bandwidth different chip rates are foreseen. Both, bandwidth and chiprate can be adapted according to the needs of the service and of the radio channel every 10 ms, the framelength. For variable bit rates between 0.4 and 16 kbps is planned. With the CODIT system concept several problems of DS-CDMA system concepts, especially IS-95, could be solved. Among them, the tight requirements for synchronisation and how to perform a handover between hierarchical-cell structures. The key for solving these issues is "compressed mode handover" a technique developed by CODIT [2]. This is a technique whereby the spreading ratio is reduced and over a few frames a time slotted structure is introduced. At the same time the power is adapted in such a way that the QoS is still met.

For completeness, another area of intensive development for new air-interfaces is mentioned, the work on Mobile Broadband Systems (MBS) [3] allowing for a full integration with B-ISDN. To achieve this, bit rates from 2 Mbps up to 155 Mbps are required. This will allow the wireless extension of ATM-services to the customer. For UMTS Systems, the allocated frequency spectrum is in the range of 2 GHz. Obviously, at 2 GHz there is not enough spectrum available to provide the bit rates above with a high capacity. Indeed, for MBS Systems the 40 or even 60 GHz ranges are considered.

This paper is organized as follows: First, for setting the scene, simulation scenarios and Quality of Service (QoS) issues are presented in Section II. This is also related to the work in ETSI/SMG5 and ITU/TG8/1 concerning the selection procedure of the air interface. Next Section III gives a detailed overview of the ATDMA system concept. This is followed in Section IV by an evaluation of this concept in which especially results of the packet access scheme and capacity are given. Finally, the conclusions will be drawn in Section V.

II. Quality of Service

For both the selection procedures within the standardisation and a sound comparison of different systems, well defined scenarios and quality of service definitions are a prerequisite and will be further elaborated in this Section.

For the comparison of two mobile systems in terms of capacity, especially via simulations, the systems will be loaded for a defined scenario until the QoS can not be met any more, i.e. until the capacity limit is reached. As the simulations show, the QoS definition has a major influence on the results. For the QoS at least two definitions can be used.

II-A. ATDMA QoS Definition (The User Perspective)

For each call, the quality of service is evaluated in terms of a percentage of the call duration, the short term averaged "quality" parameter has fallen below a predefined threshold. A call has met its QoS requirements when this percentage is lower than a predefined threshold.

*Percentage of time (short-term averaged quality < QoS_Quality_Threshold) < Time_Threshold (**1%** or **5%**) for a given Coverage (**99%**) of all calls*

The quality parameter for data services is Bit Error Rate (BER), with QOS_Th equal to $\mathbf{10^{-6}}$. This BER value is directly the output of the Block level look-up table which can be interpreted as the error rate that would experience the call if these conditions (in terms of C/I, speed, environment, and others) would last for the whole call duration. The short-term averaged quality is obtained by averaging Block BER values (in linear scale) over a period of 1 second of activity at the source level.

Speech quality is evaluated in terms of segmented signal to noise ratio (SegSNR) with QOS_Th equal to **4 dB**. SegSNR is defined as the arithmetic mean of the SNR, with SNR being the ratio of the input vocal signal variance to reconstruction error variance. The short-term averaged quality value is obtained by averaging SegSNR values (in dB) over the talk spurt duration (mean value is 1.4 second). Clipping due to PRMA++ access delay is treated as 0 dB SegSNR and is taken into account in the averaging.

II-B. SIG5 QoS definition (the network perspective)

In order to compare ATDMA capacity figures with, for example the CODIT capacity figures, quality had to be evaluated from the network point of view rather than from the user point of view. A new approach has thus been defined in ATDMA for the purpose of SIG5 (Special Interest Group number 5, a joined activity of several RACE projects) work. The system capacity is reached when a certain percentage of quality samples collected during the whole simulation (after the warm-up period) are above QoS thresholds. This is derived from the cumulative distribution function of quality parameter with samples collected among centre cells.

Quality of samples collected among centre cells >
*Service_QoS_Quality_Threshold for a given Coverage **(99%)** of all samples*

The QoS thresholds are identical to those for ATDMA definition (4 dB for SegSNR and 10^{-6} for BER).

II-C. Additional QoS constraints equal for all two definitions

For speech and data services using packet access techniques, an additional condition about packet dropping statistics has to be met :

Percentage of packets dropped (because of access delay) <
*QoS_Dropping_Threshold **(1%)***
*for a given Coverage_Threshold **(99%)** of all calls*

This constraint on clipping results from subjective tests and is evaluated from the user point of view. A call has to fulfill both conditions in order to meet the QoS requirements.

A predefined blocking probability also has to be ensured for the whole system. This is evaluated for each service class independently, as the ratio between admission failures and calls attempts over the whole simulation duration (after the warm-up period):

Admission failures number / Call attempts number <
*Service_QoS_Blocking_Threshold **(1%)***

The QoS equations are evaluated on both uplink and downlink for each call not rejected at admission. The maximum system capacity is reached when **99 %** of calls of each service class have met their QoS requirements. In case of service mix, the capacity is reached as soon as one service class exceeds the QoS requirements.

II-D. Test Scenarios

Besides the QoS definition, the definition of the environments and scenarios is an important topic. In line with the developments in the standardisation bodies (SMG5 and ITU TG8/1) representative environments have been selected. These encompass indoor office environments, vehicular environment, outdoor environment among others. One example scenario, which has also been used for simulating some of the results presented in Section IV is briefly given here, see Figure 1 [4]:

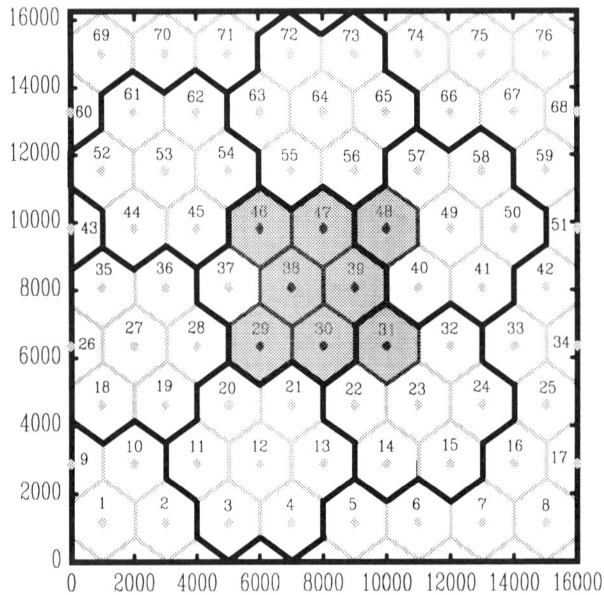

Figure 1: Drawing of a typical Macro-Cell environment with cluster size 7

- An open area of 16.0 km • 16.15 km with 76 base stations (BSs), deployed according to a hexagonal shape with a BS separation of 2 km is used. Each BS uses omnidirectional antennas and has one carrier with 14 traffic slots (I-slots). The statistics are only collected among the centre cells interfered by at least a full first tier of co-channel cells from about 3000 to 5000 calls. The cluster size is equal to seven or as specified.

- The propagation model has a pathloss of 35 dB per decade and the standard deviation of shadowing effects is 6.6 dB. The decorrelation length is 20 metres and consecutive shadowing samples are uncorrelated.

- All users use speech service and move in semi-directed trajectories with an average speed of 50 km/h.

- The simulation time duration has been 1200 s including a warm-up time of 200 s.

- The offered traffic is varied or in the range of about 1500 Erlangs.

- A speech detector termed as a slow detector characterised by a mean talkspurt duration of 1.41 and mean silence period of 1.74 s, resulting in a speech activity of approximately 45 % is used.

- Both up- and downlink are considered but results are often presented for the uplink only since the downlink is often very similar.

After agreed definitions of the issues above, it is possible to endeavour into the comparison of various systems on a "fair" technical basis. To facilitate this, the standardisation bodies are working on a complete set of selection criteria together with preselection criteria. The final stage for the selection is then the scoring process.

III. The ATDMA System Concept

Being based on the well-established TDMA access technique, a major innovative part of the ATDMA system concept are several novel radio control techniques described in more detail in this Section.

The management of radio resource allocation and handover between different adaptive radio links and the need to align the radio access system to an IN based UMTS network architecture causes a number of difficulties within the "traditional" radio access OSI-based layering structure.

To overcome these difficulties, the ATDMA project has developed a unified model as an alternative architecture, based on the division between Transport and Control functionalities and the use of a set of non-hierarchical, distributed functional groups. In the model, transport of both traffic and signalling are described in terms of a layered protocol model based on OSI methodologies, while all control aspects can be defined in terms of a non-hierarchical functional model, see **Fig. 2**. Research and therefore testing concentrates on the two lower parts, i.e. the Physical Transmission (Transport[1]) and Radio Access Control System (Control[2]).

---

1 **Transport** includes the complete transmission and reception chains including digital base band processing as well as analogue functionalities in the RF front-end and RF control devices.

2 The **Control** group holds all of the relevant processes required to control directly the radio access transmission system. The Control functions contain elements associated with each mobile as well as common elements that are responsible for location management and sharing of resources between services, mobiles and base stations. Most of these Control functions are of a peer relationship nature in the mobile and the UMTS infrastructure.
In general, it is assumed that all of these control processes follow a "master-slave" concept with the "master" components residing in the base station so that they can be changed or modified without affecting existing terminals.

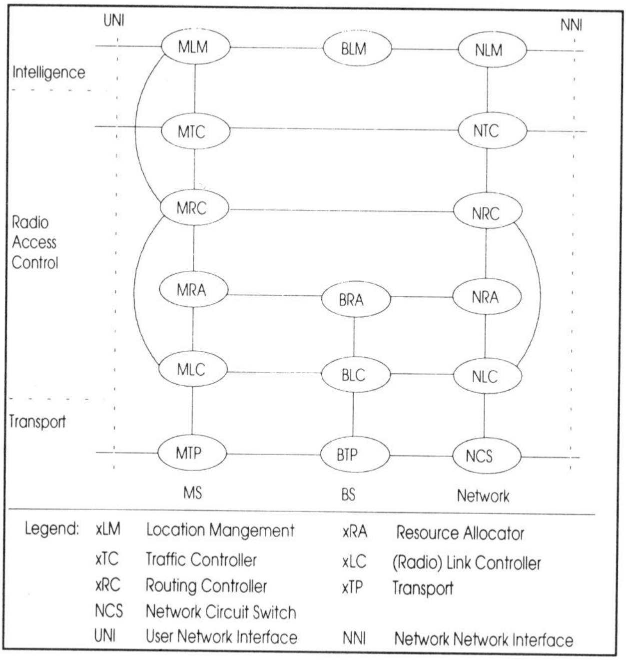

Figure 2: The Functional Model of the ATDMA system [5]

Let us consider that the adaptive radio access system normally is in a steady-state condition. The Control functions will then monitor the performance of the Transport functions and whenever an event is detected which indicates a disturbance of this steady-state operation adaptation functions will be activated to bring the radio link back into a steady-state with suitable performance. In other words: the Control functions can be thought of as moving the system between different states [5].

Concerning the bearer requirements of a UMTS radio access system the list given in Table 1 is considered to be a minimum set of bearer services such that the capabilities of a candidate UMTS can be supported. Based on these service requirements and a detailed study of the radio channel, the transport parameters of the physical transmission chain were designed. The resulting burst and frame structures - representing a static adaptation of the air interface - is based on a common frame duration of 5 ms. An example for a short macro cell is given in Figure 3.

Studies performed on control mechanisms comprised the optimisation and improvement of techniques known from GSM, like **Power Control** (with a fast open loop and a slower closed loop mechanism), **ARQ**, **Handover**, and **Macro Diversity** and the design and investigation of novel techniques like:

Link Adaptation: A technique used for delay-constrained services whereby the transmission chain (e.g. coding rate, modulation scheme) will be adapted dynamically or continuously to the radio channel characteristics.

Packet Access: PRMA++ (Packet Reservation Multiple Access), a technique well described in [6], allows for a very flexible slot allocation, according to the needs of the service.

Dynamic Channel Allocation: Based on channel segregation and a specially up-dated priority list, the radio resources can be changed dynamically. Thus the frequency planning process will be facilitated and the capacity, especially for hot spot situations, will be enhanced.

Table 1: Parameters for assumed list of ATDMA Bearer Service Types

Service	**Design constraint**	**Performance targets**
Speech	Delay < 30 ms	MOS > 4,0 Decoded BER < 10^{-3} Frame Error Rate < 2%
Data with low delay	Delay < 30 ms	BER < 10^{-6} Errored seconds < 10s/h
Data with high delay	Delay < 300 ms	BER < 10^{-6} Errored seconds < 10s/h
Data with unconstrained delay	Packet loss < 10^{-6}	Average Delay < 50 ms 90 % Delay < 100 ms

TDD and **FDD** can be used in Pico- and Macro Cells, using the available spectrum more flexibly.

A **control channel structure** similar as for GSM, with ACCH, DCCH, CCCH, and BCCH has been adopted. Noteworthy is the introduction of the Leash Control Channel (LCCH), designed to control each connection permanently, even when no traffic is being carried.

IV. Evaluation of the ATDMA System Concept

One of the key objectives of the System Performance Evaluation work was to evaluate the potential of ATDMA in terms of achievable capacity gains compared to existing second generation systems like GSM. This evaluation has been performed both in analytical studies and by simulation on the basis of a so-called "Reference System". This Reference System has been defined in ATDMA in order to give a basis for comparison with the full ATDMA system. The Reference System is a generic, circuit-switched TDMA system using the ATDMA transport chain (Figure 3) but without any of the advanced control functions. The key features of the Reference System are briefly summarised in the following list:

- ATDMA burst and frame structure and signalling channels.
- Always in circuit switched configuration.
- ATDMA handover.
- Slow frequency hopping.
- No power control, no link adaptation.

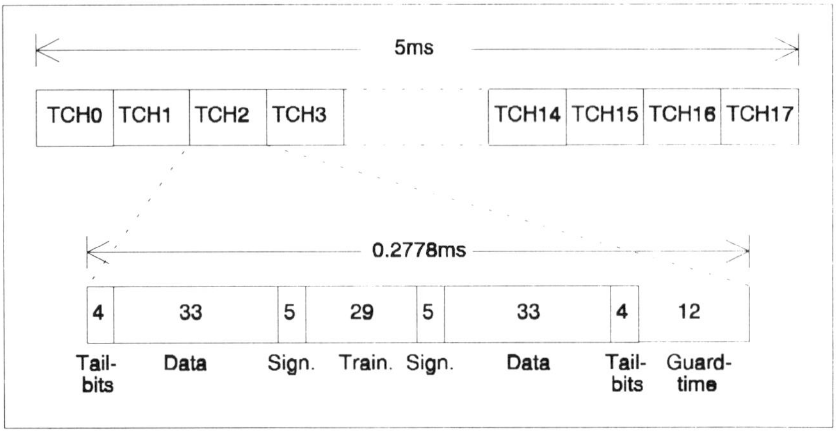

Figure 3: Burst and frame structure in ATDMA (short macro cell)

Figure 4 shows the capacity of the ATDMA system and the gain compared to the Reference System and second generation systems (100% level) for the speech service in macrocellular environment. The following basic assumptions have been taken into consideration for this comparison: macro-cellular environment described by a path loss slope of $\gamma = 3.6$ and a log-normal fading with standard deviation $\sigma = 6.6dB$.

The relative capacity figures given in Figure 4 assume a blocking probability of 1% and a coverage probability equal to 99% according to the UMTS requirement. Signalling has been neglected and all systems operate without power control. The ATDMA air-interface is thus in the position to give capacity gains of up to 3 and more over GSM. A part of the capacity gain is achieved by the ATDMA burst and frame structure which offers up to 18 traffic slots per TDMA frame (Figure 3) instead of 8 like in GSM. This gain is reflected by the beam for the Reference System. Further capacity gains are mainly initiated by ATDMA control techniques, namely link adaptation and packet access (PRMA++).

Figure 4: Relative capacity

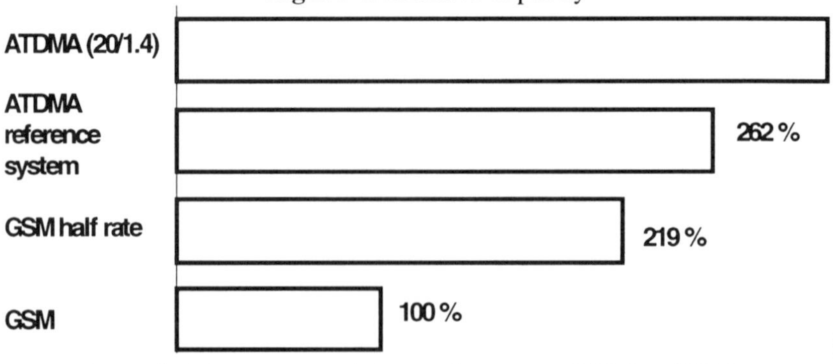

(speech service) in macro cellular environment (100% = GSM system; based on Erlang/cell/MHz figures)

For link adaptation 20% 2-slot speech transport mode usage and for packet access a statistical multiplexing gain of 1.4 (beam for ATDMA (20/1.4)) have been considered based on simulation results. That gives a slot occupancy of 74% and based on that an interference diversity gain due to frequency hopping of about 0.5 dB can be considered (a frequency hopping gain of 2 dB has been assumed for GSM and the Reference System). It has to be noted that the results are quite sensitive to the assumed C/I operating points, which are 9 dB for GSM and 5.9 dB for ATDMA

Capacity is allocated on demand using a technique called PRMA++ [6]. Like the original scheme proposed by Goodmann [7], the PRMA++ protocol avoids wasting capacity during breaks in traffic source activity (during silence period in speech or for highly bursty data services), or the need to permanently allocate capacity for the maximum bit rate a call may need. The configuration of this will be flexible to allow customisation by the operator.

Packet access has been mainly investigated concerning the evaluation of the statistical multiplexing gain supported by the PRMA++ mechanism. The speech detector [6] used in the simulations was characterised by a mean talkspurt duration of 1.41 seconds and mean silence period of 1.74 seconds, giving a speech activity ratio of approximately 45%.

From the mean number of dropped packets as a function of the number of mobiles the statistical multiplexing gain for less than 1% packet dropping gain can be derived. This is shown in Table 2 for the mean number of dropped packets with a constant loading per cell (mobiles are not allowed to leave their cell, i.e. no handover), along with values for the mean and three percentiles for a scenario allowing mobility between cells. For a constant cell loading a statistical multiplexing gain up to 1.7 with less than 1% packet dropping can be achieved. When mobility is allowed between cells, the achievable multiplexing gain goes down to 1.4. This is because calls will experience more than 1% packet dropping when there are many mobiles in a cell due to load fluctuations. For higher percentiles this decreases further. Statistical multiplexing gains up to 1.4, thus improving system capacity in the case of speech or highly bursty data services, is seen to be the maximum that can be achieved in a cellular context (see also [8]).

Table 2: Statistical multiplexing gains

Mean (no handover)	**Mean (with handover)**	**90 percentile (with handover)**	**95 percentile (with handover)**	**99 percentile (with handover)**
1.7	1.4	1.2	1.1	0.9

Although packet access is only marginally useful in terms of system capacity gain for speech only systems it is very useful as a generic access scheme to support required flexibility in terms of resource assignment and rapid channel assignment.

- Flexible resource assignment and mixed traffic types can be easily supported through the use of a common packet access scheme.
- The packet access scheme should be centrally controlled to support multiple priority levels. A possible candidate is the ATDMA PRMA++ algorithm which is based on separate access control and traffic channel resource allocations.

The main advantage of PRMA++ (capacity improvement due to statistical multiplexing of voice channels) compared to circuit-switched transmission can only be fully achieved after a reduction or compensation of co-channel interference through the use of advanced control techniques, either power control or link adaptation. This is mainly necessary because of the high UMTS QoS and coverage targets. This QoS advantage of packet access can also be seen quantitatively in Figure 5.

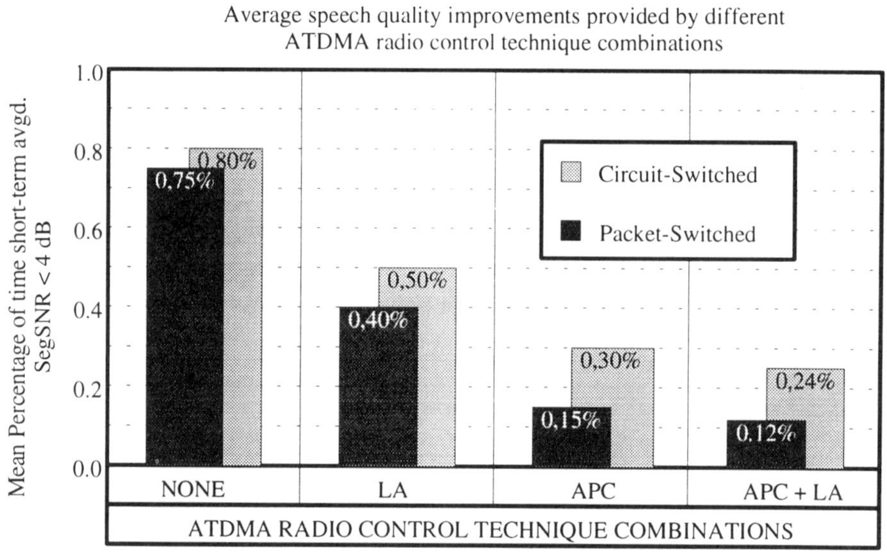

Figure 5: QoS for packet and circuit switched situations in combination with different control

Table 3 gives the exact capacity figures for these situations showing a maximum gain of up to 20% for Packet Switched Operation.

Table 3: Capacity figures for circuit and packet switched situations

Cluster SIZE	**Cell Capacity [User/ cell/ MHz] Circuit - Switched**				**Packet - Switched**			
	Ref. Sys.	**LA**	**APC**	**LA, APC**	**Ref. Sys. PA**	**LA**	**APC**	**LA, APC**
9	**2.8**	**2.8**	**4.4**	**4.4**	**2.8**	**3.6**	**5.2**	**5.2**

V. Conclusions and Summary

First, a brief description of three adaptive air interface projects was given. These projects are ATDMA (being based on Advanced TDMA techniques), CODIT (using DS-CDMA), and MBS (Mobile Broadband System, operating in the 60 GHz spectrum).

Next, an important prerequisite for comparing cellular system concepts, like the above, was described in some detail: the Quality of Service and the scenario definition. For the former, two major approaches can be distinguished, the "user perspective" and the "network perspective".

The major focus of this paper was on a description of the ATDMA system concept and its evaluation. This concept is based on TDMA and consists of techniques like link adaptation, packet access or dynamic channel allocation.

A detailed performance evaluation of packet access showed that besides a statistical multiplexing gain of up to 1.4, flexibility and improved QoS are important merits of this technique. Together with other advanced control techniques a potentially significant capacity increase, combined with new services (larger bit rates) and improved QoS, could be possible. Compared to current GSM a capacity increase of up to a factor of three was found.

For the final choice of the third generation air-interface, other important topics are the migration from current systems and the further development and the impact on the backbone network. These topics are, among others, studied in the ACTS (Advanced Communication Technologies and Services) project RAINBOW (Radio Access Independent Broadband on Wireless).

VI. ACKNOWLEDGEMENTS

This work has been partially funded by the Commission of the European Communities in the RACE II program under the RACE 2084 ATDMA Project. The author would like to thank DeTeMobil for the permission to publish this work and he would like to acknowledge the contributions of T. Kassing, J. Irvine, P. Blanc and of his colleagues from Siemens AG, Roke Manor Research, ALCATEL Mobile Communication France, ALCATEL Standard Eléctrica S.A., ALCATEL Italia S.P.A., Nokia Research Centre, Nokia Mobile Phones, Elektronik System GmbH, DeTeMobil GmbH, France Télécom CNET, Fondazione Ugo Bordoni, Télécom Paris, Universidad Politécnica Cataluña, and the University of Strathclyde, although the views expressed are those of the author and do not necessarily represent those of the project nor his organisation as a whole.

References

[1] A. Urie, M. Streeton, and C. Mourot, "An Advanced TDMA Mobile Access System for UMTS", IEEE Personal Communications, Vol. 2 No.1, pp. 38 - 47, Feb 95.

[2] P.G. Andermo and L.M. Ewerbring, "A CDMA-Based Radio Access Design for UMTS", ", IEEE Personal Communications, Vol. 2 No.1, pp. 48 - 53, Feb 95.

[3] L. Fernandes, "Developing a System Concept and Technologies for Mobile Broadband Communications", IEEE Personal Communications, Vol. 2 No.1, pp. 54 - 59, Feb 95.

[4] K. David, P. Blanc, and T. Kassing, "Performance Evaluation of the RACE/ATDMA System Concept for a 3rd Generation Air Interface by Simulation", accepted for 46th IEEE VTC, Atlanta, 1996

[5] K. David, N. Metzner, and T. Wierlemann, "An Adaptive Air Interface for 3rd Generation Mobile Access - Evaluation via realtime and simulation testbeds -", Tagungsband der 2. ITG Fachtagung Mobile Kommunikation '95, Neu-Ulm, pp. 125-132, 26-28 Sep 1995

[6] J. De Vile, "An Access Mechanism for ATDMA", *Proceedings of 7th European Conference on Mobile Radio and Personal Communications*, Brighton, UK, 13-15 December 1993.

[7] D. J. Goodman, "Cellular Packet Communications", IEEE Transactions on Communications, Vol. 38, August 1990.

[8] K. David, P. Blanc, J. Irvine, T. Kassing, M. Lopez, R. Maddalena, W. Mohr, and P. Ranta, "First Results of ATDMA´s Test Campaigns", Proceedings RACE Mobile Telecommunications Summit, pp. 468-472, Portugal, 22-24 Nov. 1995

Feasibility Study for a GSM Private Cordless Base Station Based on Total Frequency Hopping

Maija Kuusela, Marko I. Silventoinen and Mika Raitola

Nokia Research Center, P.O. Box 45
FIN-00211 Helsinki, Finland

Abstract - This paper continues addressing the problem of allocating frequencies in a GSM-based Private Cordless Base Station (PCBS) system introduced in [1] by the term of a GSM Home Base Station. Total Frequency Hopping (TFH) is presented in order to overcome the frequency allocation and the inter-PCBS synchronisation problems. The PCBS system will use the same frequency band as the overlaying GSM network and thus create more interference to it. Earlier simulations have revealed that the negative impact of the underlying PCBS system to the GSM network performance is minimal if the PCBS transmission power is low. In this paper we investigate if the PCBS system with low transmission power will actually itself be able to function despite the interference from the overlaying system. Furthermore, a simple way to synchronise to the frequency hopping BCCH is presented with time analysis.

Introduction

In the contemporary and near future mobile communications systems the subscribers will be provided with the possibility of total mobility, i.e., of using their handsets in different environments including homes, offices, highways and shopping centres. One technique is to use different systems in different environments, for example Digital Enhanced Cordless Telecommunications (DECT) at home and in the office and Global System for Mobile Communications (GSM)[1] outside the DECT coverage area or when GSM features are needed. However, in this way the subscriber needs either two handsets or an expensive dual-mode handset. An alternative approach is to provide GSM users with a possibility of enjoying the benefits of cordless phone at home using their GSM handsets. This can be achieved by PCBS, a low cost consumer device that anybody can purchase, connect to the Public Switched Telephone Network (PSTN) and have low rate calls when using it.

The PCBS devices are currently available e.g. in Personal Handyphone System (PHS) and in Advanced Mobile Phone Service (AMPS), but not yet in GSM.

In a GSM based PCBS system an essential feature is its coexistence in the same frequency band as the GSM network.

In this paper we will first introduce shortly the GSM-based PCBS concept and describe what limitations it has with respect to the real GSM base station. There are some inherent problems in the PCBS concept that must be solved somehow before the real system can be launched. One potential solution for these problems is the use of TFH.

If the PCBS system uses TFH it will increase the interference in the overlaying network. In [1] it is shown that the negative impact is minimal if low power PCBS is used. In [1] as well as in this paper the simulation parameters are chosen carefully in order to avoid too optimistic conclusions. Despite this the results in [1] clearly demonstrate the negligible effect of the PCBS system to the overlaying GSM network. In this paper we address the problem of interference in the opposite direction, i.e., we study whether the low power PCBS system will be able to function in a satisfactory way in spite of the interference produced by the overlaying GSM network using the same frequency band. The study is carried out by means of simulation models: the acceptable signal to interference (C/I) ratio for a hopping channel is determined by a link level simulation with COSSAP tool and the C/I distribution in the network is simulated in MATLAB. The simulators and their assumptions are described followed by a presentation of the results. The results can be applied in other contexts, too, where randomly placed low power transmitters are using the same frequency as another independent

---

[1]In this presentation GSM refers to any GSM based system, e.g. DCS1800, PCS 1900, etc.

system. Finally, a simple yet effective method of synchronising to the frequency hopping BCCH is presented.

Private cordless base station

A PCBS is a device a GSM subscriber can buy and connect to his PSTN line in order to achieve two advantages: to make calls at the rate of a fixed network when at home and to have a better indoor coverage. The GSM operator benefits from the PCBSs since the infrastructure cost of achieving a good indoor coverage will be lower. This is especially important in the early phases of the network build-up. Besides, customary users of a PCBS will eventually be likely to exploit their Mobile Stations (MS) more also in the GSM network and hence create more revenues to the GSM operator. The PCBS can be installed in the domestic environment or possibly in small offices. It can be connected to any PSTN where it uses the number allocated to that line by the PSTN operator. From the PSTN's point of view the PCBS is a normal terminal.

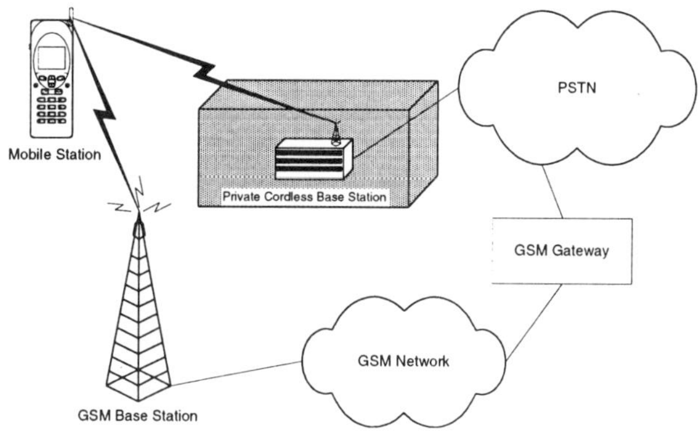

Fig. 1. PCBS System

The MS uses GSM Base Stations (BS) when it is not in the coverage area of a PCBS it is allowed to use. When a PCBS is available, the MS will camp to it. Then, using an IN function called Follow Me Diversion (FMD) the calls to the MS number are forwarded to the PCBS via PSTN.

The key idea in PCBS HW implementation is to derive PCBS from the existing MS by adding needed functionalities rather than trying to miniaturise the current BTS. This will lead to a low-cost, low-power device where only one or two Time Slots (TS) out of eight available in a carrier are used. In this study we have assumed a two-slot PCBS where the first slot is reserved for signalling channels (BCCH, CCCH, FCCH and SCH, see [1] for a description of the channels) that are transmitted constantly to ensure that MS finds PCBS. The second timeslot is reserved for Traffic Channel (TCH) when one is needed. It can be seen easily that in a normal case, where only one fixed line connection is coming to the home, only one TCH is really needed and it is unnecessary to provide more capacity on the air interface.

In the PCBS system a limited set of GSM features is applied. The most important limitations from this presentation's point of view are that there is no possibility for handover (HO) to or from PCBS and that the number of possible PCBSs that a MS can use is limited.

Compared to a Public Land Mobile Network (PLMN) the traffic load on each PCBS TCH is moderate. The PCBS is a personal base station and its use is tightly limited to those MSs whose bill the owner of the PCBS is ready to pay. In practice this means that only the MSs of the family members are allowed to use the PCBS. Consequently, subscribers will not share PCBSs nor TCHs. It is assumed that the load per PCBS subscriber at peak hour is 100 mErl. Hence, the utilisation rate of a TCH is only 10% at the maximum.

Problems In the PCBS System

The PCBS environment has some inherent and unique problems:

1) There is no possibility of centralised frequency planning because the exact location of a PCBS is not known and it may change. Even if the location were known, the high number of PCBSs in use would effectively prohibit their exact control by a centralised system.

2) Inter-PCBS synchronisation is difficult to obtain, because PCBSs do not have a common reference clock. The PSTN does not provide PCBSs with a synchronisation signal and the methods that can be used in the GSM BS, such as high accuracy master frequency generators or exploiting Global Positioning System (GPS), are far too expensive for a low-cost product. Because of this it is not possible to use more than one PCBS per carrier, even though only 1 or 2 slots per PCBS are used. If the PCBSs shared the same carrier the Time Slots (TS) would inevitably overlap sooner or later due to drifting PCBS reference clocks and the lack of synchronisation. Hence, the spectral efficiency will be low.

3) The Dynamic Frequency Allocation (DCA) or any other autonomous channel allocation scheme is difficult to implement in a cost effective way. DCA requires the BS to be able to select a free channel and the selection is based on the measurement of neighbour BS transmission. In GSM a Frequency Division Duplex (FDD) is adopted with link separation of 45 MHz . If a PCBS is required to monitor a transmission band of another PCBS, the bandwidth of the frequency synthesiser must be widened a lot and thus non-trivial extra cost is included. Naturally, in Time Division Duplex (TDD) systems this problem does not occur and DCA is more easily implementable. Without the possibility of monitoring the radio environment the PCBS has no information about the channels at its disposal and hence it is unable to select a proper channel for communication.

Total Frequency Hopping

With the Total Frequency Hopping (TFH) we mean a feature where all the channels, i.e., TCH and the signalling channels transmit each burst on a different carrier frequency and where the hopping takes place over all the available carriers the operator has. The hopping sequences for individual PCBSs are independent of each other.

In the current GSM it is forbidden to hop on BCCH since this would introduce problems in the neighbour BS measurements. In the PCBS environment this restriction is not valid because there is no HO between PCBSs and thus no need to measure neighbours. Also, in GSM the maximum number of frequencies used in hopping sequences and the number of available sequences is limited [2]. Naturally, due to the frequency hopping BCCH the TFH system is not compatible with the current MS and calls for changes in specifications.

The frequency hopping itself will increase the performance of the system because FH will help in fighting against the frequency selective fading [3]. In the next section the results of simulations applicable to the PCBS context are presented.

The TFH will eliminate the need for frequency planning since all the frequencies will be used in a pseudo-random manner and the additional interference to the overlaying system is evenly distributed over the whole spectrum in use.

There are no more problems with overlapping TS due to the lack of inter-PCBS synchronisation. If the TSs overlap, a marginal average interference is generated but due to hopping the interference lasts only a short period of time.

Finally, the moderate TCH usage can be exploited by the system. If there is no traffic on the TCH, the interference is reduced and all the users will benefit from it. With a more rigid channel allocation system, an unused channel will still occupy bandwidth.

Link Level Simulations

This section describes a simulation model used to estimate the C/I limit leading to acceptable performance in a frequency hopping channel. This is done by means of bit level simulations of the GSM full rate speech channel. The simulation model includes blocks for the transmitter, channel, interference and receiver. The simulation principle is the Monte Carlo method and the COSSAP simulation tool is used. The following basic simplifications are done.

1. Complex envelope presentation for the radio channel and RF-parts of the receiver and the transmitter.
2. Analogue radio signals are presented in digital domain.
3. The time and frequency synchronisation is assumed to be ideal.
4. One dominant interferer per frequency is assumed and other interferers are negligible.

Total frequency hopping is aimed to perform over 25 or 75 frequencies. In this case we can assume even in the urban and indoor environments that adjacent bursts are almost uncorrelated. Now we can model hopping over large number of frequencies by using ideal frequency hopping, i.e., the power of each burst is randomly selected. In the simulation both the signal and the dominant interferer are Rayleigh fading and ideally hopping. The following figure illustrates the simulation principle.

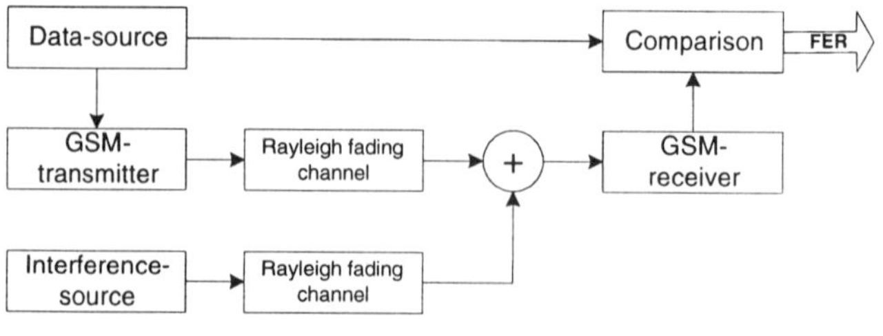

Fig. 2. The simulation environment

The channel model used is Typical Urban with mobile speed 3 km/h, specified in [6], and the carrier frequency is 1900 MHz.

The following figure shows the frame erasure rate (FER) as a function of C/I. The non-hopping case is included for reference.

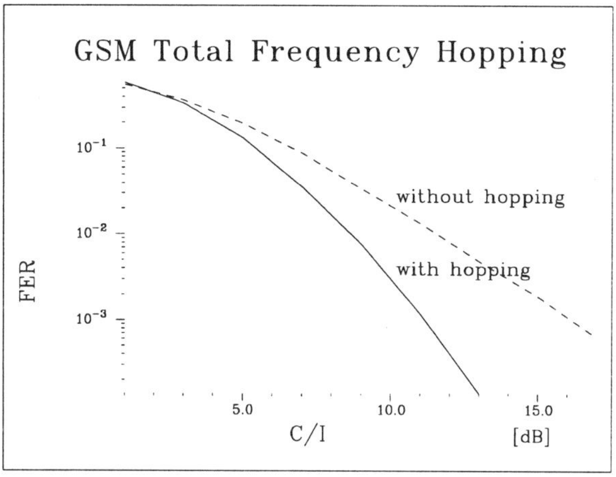

Fig 3. Frame erasure rate as a function of C/I

The FER value of 3% assures a reasonable speech quality. One may conclude that this leads to the C/I limit of at most 7 dB which will be applied in the next section. However, the results presented in this paper are not particularly sensitive to the limit used.

Our simulation ignores the possible load dependent interference diversity gain [3] due to the averaging effect of frequency hopping. This approximation emphasise the pessimistic nature of the results.

C/I Simulations

This section describes a MATLAB simulation model used to estimate the impact of the interference originating from an overlaying GSM network to a PCBS system. One of the design features of a PCBS system is its low power level. However, since the power levels in the overlaying GSM network are much higher one may expect that PCBS suffers from serious interference from the overlaying network. The objective of the simulation study was to find an upper limit for the ratio of the transmission powers of the overlaying network and the PCBSs so that the performance of the PCBS system does not deteriorate.

As a measure of the transmission quality one may use the outage percentage. By an outage we mean a snapshot where the C/I value drops below a predefined quality threshold. The outage percentage is related to the call dropping rate, but the call dropping rate depends also on the distribution of the C/I values in time, not only its mean value. Hence, the outage probabilities presented here can not be converted directly to call drop rates. The limit for the C/I outage was chosen to be 7 dB in accordance with the COSSAP simulations.

Simulations were done in the downlink direction, i.e., the receiving end was a MS in the PCBS network, since this was estimated to be more sensitive to the GSM interference.

The attenuation due to walls plays a significant role in the study; accordingly, two cases were studied separately, the MS being indoors and outdoors.

In the study we used US PCS characteristics, i.e., 1900 MHz band and band allocations of 10 and 30 MHz representing the band for D-F blocks and A-C blocks for PCS, respectively.

Simulation Model

The simulation uses a snapshot technique where each snapshot consists of a MS connected to a randomly chosen PCBS. The downlink simulation model calculates the carrier to interference ratio (C/I) for the MS using a traffic channel. The interference is made up of two parts, one due to the other private cordless base stations in the area and the other originating from the GSM BSs transmitting in the same frequency as the PCBS under study. We assume that

- No power control is used, i.e., all BSs and PCBSs transmit on a constant maximum power Txp and Txp_H respectively. There is no frequency hopping in the overlaying network. However, the results also apply in the case where the overlaying network uses frequency hopping since in both cases the frequency of the interference experienced by the PCBS varies from burst to burst.

- The overlaying network base stations have three carriers. The antenna radiation pattern is omnidirectional.

- The overlaying network has a reuse factor 7. Two tiers of interfering BSs are taken into account, i.e., there are 13 BS interferers.

- The distance of the BSs is $d = \sqrt{3}$ km. The PCBSs are spread randomly with a uniform distribution and average density $\rho_H = 150$ PCBSs/km^2.

- The overlaying network has a 80 % load whereas the traffic channel in the PCBS system is in use with probability $p_H = 10\ \%$.

- The multiframes of the PCBSs are unsyncronised. Because of that and the fact that the constantly transmitted signalling channels (BCCH, SCH and FCCH) require only 14 slots in a 51-slot multiframe, the probability that another PCBS transmits on the same frequency and in the same slot as the MS under study receives is $\frac{1}{8}(\frac{14}{51} + p_H)\frac{1}{n_{hop}}$, where n_{hop} is the number of channels used in frequency hopping.

The simulation parameters are the number of hopping channels n_{hop}, for which we use values 25 and 75 corresponding to the D-F blocks and A-C blocks respectively in the US market, and the transmission power ratio Txp/Txp_H , for which we use 60 dB, 40 dB and 20 dB. These values correspond to possible transmission powers, for example, Txp = 10 W and Txp_H = 0.01 mW, 1 mW or 100 mW.

The assumptions on the signal attenuation are

- The path loss model in the pico cell for the signal from the PCBS to its MS is given by $L = 38 + 20 \log(d) + L_{wall}$, where L is the pathloss in decibels, d is the distance in meters and L_{wall} is total attenuation due to inner walls ranging from 0 to 12 dB. No shadowing is used in the pico cell.

- The pathloss for the interference was modelled by two possible models describing different radio environments: $L = 125 + 35 \log(d) + L_n$ or $L = 143 + 38 \log(d) + L_n$. Here d is the distance in kilometers and L_n is the shadowing modelled by a lognormal distribution with $\sigma = 6$ dB. In the case where the receiving MS is assumed to be indoors, an additional attenuation of the average value 10 dB is added to the interference from the GSM network and 15 dB to the interference from the other PCBSs.

- A private cordless base station is usually inside a building but the MS may well be inside or outside. The attenuation due to an outer wall varies typically between 5 dB and 15 dB [4]. Since a private cordless base station might also be outside (in the garden, for example) or surrounded by several layers of walls, the wall attenuation is modelled by an additional attenuation factor equally distributed between 0 and 20 dB.

Simulation Results

The figure below shows the outage percentages in simulations with 15000 snapshots using the less severe attenuation model $L = 125 + 35\log(d) + L_n$ for macrocell pathloss. Outage values approximately from 0 to 10 % may be considered small enough for the PCBS service quality to be acceptable.

Table 1. Outage percentages for a MS communicating with a PCBS. In a) the MS is indoors at the distance of 1 - 10 m from the sending PCBS. In b) the receiving MS is outdoors, 1-50 m away from the sending PCBS.

a) MS indoors:

Txp / Txp_H:	n_{hop} =25	n_{hop} =75
60 dB	7	3
40 dB	1	0
20 dB	0	0

b) MS outdoors:

Txp / Txp_H:	n_{hop} =25	n_{hop} =75
60 dB	77	25
40 dB	35	11
20 dB	3	1

In all the cases above, the outage is due to the interference from the overlaying GSM network rather than the other PCBSs. This is seen clearly from the cumulative C/I distributions in Figure 4.

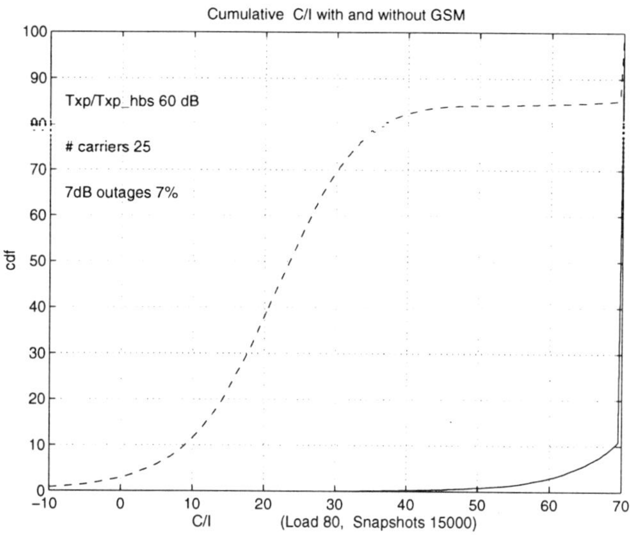

Fig. 4. The cumulative C/I distribution due to the total interference from both the BSs in the overlaying network and the other PCBSs is shown with the dotted line. The role of the PCBS interference is negligible as is seen from the C/I distribution using only PCBS interference (represented by the solid line).

The notable effect of walls can be seen clearly from these figures, outdoor results being drastically worse than in the indoor case. Even in the indoor case the outage drops considerably if we assume that the PCBS and the MS are in the same room. The outage percentage drops then to 3 % (instead of 7 %) in the case of 25 carriers and 60 dB power ratio.

Another significant factor is the macro cell attenuation model used. The coefficients of the attenuation model vary a great deal depending on the local conditions such as geographic features, building and antenna heights etc. For comparison, we include here the corresponding outage percentages for un outdoor mobile for the more severe attenuation model mentioned $L = 143 + 38 \log(d) + L_n$, which is in accordance with the ETSI recommendation [5]. In this case, the outage is virtually zero for an indoor MS.

Table 2. Outage percentages for a MS outdoors for a more severe attenuation model.

$Txp\,/\,Txp_H$:	n_{hop} =25	n_{hop} =75
60 dB	38	12
40 dB	5	1
20 dB	0	0

The reliability of the PCBS service increases with the number of channels used for frequency hopping and with the increasing transmission power for the PCBSs. On the other hand, if the transmission power is increased too much, the PCBS system produces too much interference to the overlaying GSM system, a situation not acceptable for the network operator. However, results in [1] show that for the power range cited above the increased interference is small enough not to pose any problems.

Synchronisation

In the TFH scheme one problem is how the MS can find the frequency hopping BCCH, i.e., how to synchronise to the hopping sequence. The synchronisation has to take place when the MS camps to the PCBS cell. This procedure is not a frequent one, it is assumed to happen only a limited number of times a day. The process is essentially equivalent to the new network selection, since the PCBS network is not part of GSM but another independent system. Thus the time requirements are not so tight as they would be if the synchronisation would take place more often, for example as part of a HO process which would be the case if THF were applied in GSM.

When a PCBS is taken into use it has to be informed which MSs are allowed to use it; correspondingly, the MS has to be told which PCBSs it may use. This introduction phase is needed to ensure only authorised MSs can use PCBS. At the same time, the frequency hopping sequence of the PCBS can be given to the MS. Hence, the MS will have a priori information about the TFH sequences of the PCBSs it is allowed to use. The TFH sequences will be unique for each PCBS and they can be defined as a list of carriers in the sequence.

It is beneficial to transmit the two bursts (traffic channel and signalling channel) in the same frame in the same carrier. Otherwise PCBS would not have enough time to switch the transmission frequency.

To reach synchronisation MS has to be able to decode two bursts: first burst from FCCH (Frequency Correction Channel) and then from SCH (Synchronisation Channel). The SCH burst will contain enough information for MS to determine the Frame Number (FN). If the starting point of the TDMA frames that the PCBS sends is the same as the starting point of the hopping sequence, i.e., the first frame that the PCBS sends is sent in the first frequency in the hopping sequence, MS can from that on calculate the carrier of any frame.

If the length of the FH sequence is not a multiple of 51 (the length of the multiframe FCCHs are sent in), the FCCH will be sent in every carrier. Now the MS can start to listen to any carrier and be sure that the FCCH will be sent on that sooner or later if a PCBS is nearby.

Let's assume first that every carrier is used only once in the FH sequence. When MS detects FCCH it will hop to the next carrier in the sequence to find and decode SCH and that is enough for synchronisation.

If some carriers can be used more than once in the sequence, the previous method does not work. In that case we propose to break the pseudo randomness in a case of FCCH and SCH, i.e., when FCCH is sent the following SCH will be sent either on the same carrier ($C_{n+1} = C_n$) or on another carrier with certain offset ($C_{n+1} = (C_n+X) \bmod N$, where X is the offset and N is the number of carriers).

The required time for synchronisation can be calculated as follows: a 51-frame multiframe lasts for T = 255.38 ms and contains 5 FCCH-SCH pairs. Hence, if the hopping sequence has length N, the maximum time needed for reaching synchronisation is N x T / 5 and the average time is half of that. In practice the times are far less than 5s (in our case 1.3 s for 25 carriers and 3.8 s for 75 carriers) and hence the time for achieving synchronisation is small enough.

Summary

In this study we have investigated if the low power PCBS system using the same frequency band as the overlaying GSM system can function. A simulation model was designed with MATLAB tool to investigate the downlink which was assumed to be more prone to the deterioration due to large GSM BS transmission power. The various parameters were set in a pessimistic way to represent a worst case scenario.

The conclusion was that for realistic power ratios (20 - 60 dB) of the GSM and PCBS networks the PCBS system is more sensitive to the interference generated by the other network. In particular, the system parameters, such as wall attenuation and coefficients of the macro cell attenuation mode, affect the system's feasibility. However, this study indicates that there are in general no problems in the service quality indoors. The possible problems arise if the PCBS is indoors but the MS e.g. in the garden. In that case the PCBS service quality can not be guaranteed under all circumstances. The subscriber has then two alternatives: either move indoors to make his call at the reduced rate through the PCBS or use the GSM network. The main inconvenience is the reduced freedom of moving to and from the house during an ongoing call, since the present PCBS concept does not allow handover to or from the GSM network. In places where the GSM coverage is not adequate there should be no problems in using the PCBS even from on outside MS.

Acknowledgements

The authors thank colleagues at Nokia Research Center, in particular Pekka Ranta and Riku Pirhonen, for useful discussions.

References

[1] Silventoinen, Kuusela, Ranta: Total Frequency Hopping for Home Base Stations, Proceedings of the IEEE International Conference on Personal Wireless Communications, New Delhi, India, 1996

[2] GSM 05.02, Multiplexing and multiple access on the radio path, ETSI, Sept. 1994

[3] Olofsson H., Näslund J. & Sköld J.: Interference Diversity Gain in Frequency Hopping GSM, Proceedings of the IEEE International Conference on Communications, 1995

[4] Paulsen et al, Radiowave Propagation Model Document, RACE RMTB/CC/R157 Issue 1.1, London 18.12 1991

[5] GSM 03.30, Radio network planning aspects, ETSI, Jan. 1994

[6] GSM 5.05, Radio Transmission and Reception, ETSI, March 1995

On Satellite Path Diversity for GSM Rate Compatible TDMA Big LEOs

Tor E. Wisløff

Norwegian University of Science and Technology, Department of Computer Systems and Telematics, N-7034 Trondheim, Norway [1)]

Abstract

The mobile link of the proposed non-geostationary satellite systems is subject to multipath fading and shadowing of the required line-of-sight path, resulting in link-outage. This paper explains how satellite path diversity can be employed to overcome this link-outage in a TDMA based non-geostationary satellite communications system. The GSM halfrate TDMA format is used as an example. Timing and implementation issues related to dynamic satellite path diversity are considered. Two TDMA structures are derived. Both accommodate diversity, but differ in the assumed usage of diversity. Neither of the structures necessitate a duplicated receiver chain nor a duplexer in the mobile terminal. The diversity schemes divide the 16 timeslot halfrate GSM TDMA frame in different slots for communicating with different satellites. The slots represent how much the timing adjustment at the mobile terminal can be changed before timing constraints are violated. The paper also considers mechanisms to determine whether a call should be set-up with diversity, and whether the diversity state should be allowed to toggle during a connection's lifetime. It is demonstrated how diversity toggling can be used to achieve a seamless handover, and message sequence charts are shown for both handover and diversity toggling.

1.0 Introduction

The advantage of using satellite path diversity has been reported on before, e.g. [CA95b], [VOJ94], [DE93], [WER95], [NEE93] and [HAR95]. Most have assumed CDMA, dual receivers at the mobile, or not considered implementation details for a TDMA system. This paper will show that satellite path diversity can be employed in a TDMA based Non-Geostationary (NGEO) Big LEO satellite system with only one receiver / transmitter pair and no duplexer at the mobile. The GSM [MOU92] halfrate TDMA structure is used as an example in the discussion, and GSM terminology is mirrored throughout the paper.

Section 2.0 reviews the ICO, Globalstar and Iridium Big LEOs and details a reference system. Section 3.0 discusses diversity, and factors influencing the decision as to whether diversity should be allocated at call set-up, and whether diversity could be switched on or off (toggled) during a call's lifetime. Section 4.0 considers timing

[1)] *Now with Nera AS, P.O. Box 91, N-1361 Billingstad, Norway. E-mail: twi@nera.no*

Table 1: A brief overview of three Big LEOs [WIS95]

	ICO	**Globalstar**	**Iridium**
Orbit class	MEO, 10355 km	LEO, 1400 km	LEO, 780 km
Satellites	10 + 2 spare	48 + 8 spare	66 +6 spare
Bit rates (kbps)	4.8 voice, 2.4 data	Adaptive 2.4 / 4.8 / 9.6 voice, 7.2 data	2.4 / 4.8 voice, 2.4 data
System cost (M US$)	2600	2000	3700
Call rates, US$/ min.	1-2	0.35 - 0.55	3
Operations scheduled	2000	1998	1998
Planes / inclination	2 / 45°	8 / 52°	6 / 86.4°
Orbital period (minutes)	358.9	114	100.13
Propagation delay (ms)	34.5 - 48.0	4.63 - 11.5	2.60 - 8.22
Multiple access method	TDMA	CDMA	TDMA / TDD
Beams pr. satellite	163	16	48
Dual satellite visibility	Usually ≥ 2	"Substantial"	At poles
Diversity exploited?	Yes	Yes	No
Link margins, dB	8 - 12	11 - 16 equivalent	16 voice, 35 page

implications, and proposes two timing structures both accommodating diversity within the GSM half-rate TDMA frame. Section 5.0 discusses different handover types, and shows how diversity establishment and release can be used to achieve a seamless handover. It is also shown examples of signalling sequences required for a connection experiencing both handovers and diversity toggling.

2.0 System examples

2.1 Three Big LEOs

The Big LEO systems are satellite communications systems with sufficient power and bandwidth to provide high quality voice services to hand-held mobile terminals (MTs). Other typical services are short message services (SMS) familiar from GSM, data transmission at various bit-rates, paging, facsimile, and position determination and reporting. Several non-geostationary mobile satellite communications systems are proposed by different organisations. Table 1 (extracted from [WIS95]) summarises for comparison the ICO, Globalstar and Iridium systems. Note that figures for e.g. system cost may entail different elements for the different systems, and are intended to serve as an indication only.

2.2 Reference system

The remainder of this paper will assume a TDMA based transparent transponder spot-beam S/L-band MEO circular orbit reference system at an altitude of 10355 km. The reference system is illustrated in Figure 1. The proposed Big LEOs are designed to augment and interoperate with GSM (though they will not be integrated), and strive to utilise existing software and resources with as few modifications as possible. The GSM terminology and structure is therefore assumed for the reference system, as this provides a well know fundament on which to base the discussion. The reference TDMA format is the half-rate GSM, as this matches the proposed service profiles of

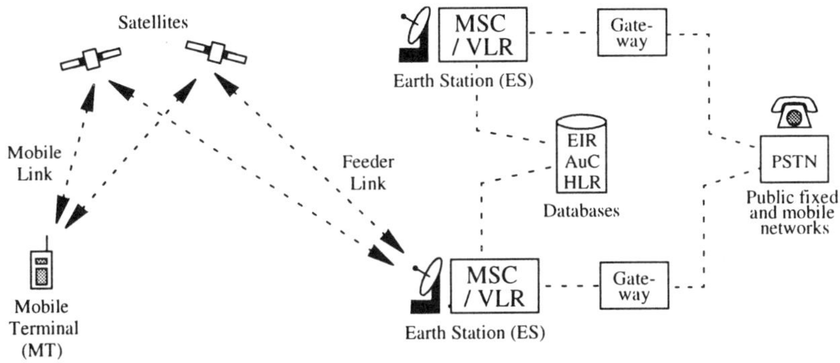

Figure 1: Reference satellite system architecture.

the Big LEOs better than the full-rate format. The motivation for using the GSM TDMA format is to enable a discussion on how a TDMA structure can be exploited to offer satellite path diversity without needing to define a new TDMA format.

In Figure 1, the Earth Station (ES) performs the equivalent functions of the GSM BSS. The ES comprises the radio transmission and reception devices, and also manages the radio interface, mainly through the allocation, release and handover of radio channels. An MSC/VLR is assumed integral to each ES, performing the same functions as in GSM. Gateways are needed to provide an interface to the terrestrial fixed and mobile networks. In addition to what is shown in Figure 1, functions will be needed for Telemetry, Tracking and Control functions (TT&C). The number and interconnection of MSC/VLRs, ESs, databases are not discussed, as these topics are not the interest of this paper. During both idle and active mode it is important for the ES to know roughly where the MT is located. The required position can determined by measuring the delay and Doppler characteristics of the burst received from the MT at the ES [ZHA95]. The accuracy of the estimation can be increased when diversity is employed.

3.0 Diversity

3.1 Introduction

The mechanism of diversity is to combine two or more correlated information signals having uncorrelated noise and / or fading characteristics. The purpose is to improve the signal quality through the combination of the two diversity signals. There are several different forms of diversity, e.g. antenna, time, frequency, field, code or satellite path diversity. The latter consists of transmitting the same signal to the same destination via (usually two) different satellites. The underlying assumption is that the Line of Sight (LOS) to each satellite will not be obstructed both at the same time. Unless explicitly stated, "diversity" will mean "satellite path diversity" throughout this paper.

Note that satellite path diversity will be a blend of frequency, time, path and even antenna diversity in a TDMA system when no duplexer or dual transmitter/receiver

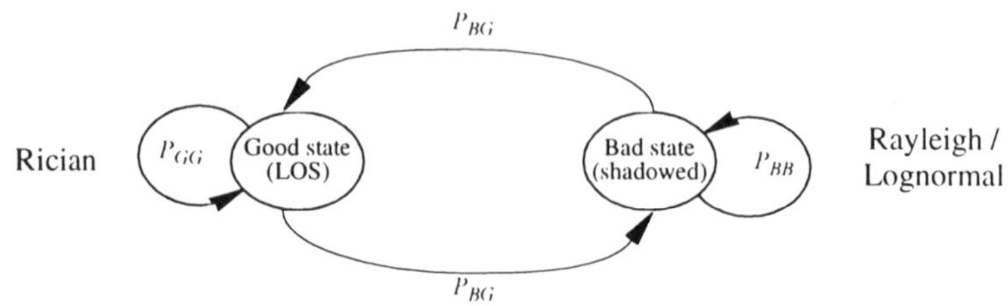

Figure 2: Two-state Markov channel model. The an MT in the good state experiences Rician fading, whereas a mobile in the bad or shadowed state experiences Rayleigh / Lognormal fading.

pair is assumed at the MT. In a CDMA system, path diversity would also provide code diversity on the mobile downlink.

3.1.1 Diversity combination

At reception, the signals are combined such that the probability of correct detection is increased. The combination can be of three different forms, either selection, maximal ratio, or equal gain combination. In terms of performance, [PAR89] reports that maximal ratio is marginally better than equal gain whilst selection has the poorest performance. The difference between maximal ratio and selection diversity is on the order of 1 ~ 1,5 dB in average when only two diversity branches are employed.

3.1.2 Channel model

Several reports have been made on a digital channel model suitable for the Big LEOs, e.g. [LUT91], [BUT95], [CA95a], [CA95b], [COR94]. They seem to agree that the channel can be modelled as being in one of two states, as Figure 2 illustrates.

The *good* state when there is no shadowing can typically be characterised by a Rician distribution. The *bad* state corresponds to where the MT is experiencing an obstruction of LOS to the satellite, which may happen as the MT and satellite have moved relative to each other. The bad state can be characterised by Rayleigh / Lognormal distributed fades [LUT91].

3.2 Satellite Path Diversity

The advantage offered by satellite path diversity in a satellite communications system is that communication can be maintained even when the LOS path to the active satellite becomes temporarily obstructed due to the relative movement of the satellite and MT. Figure 3 illustrates the principle. Shadowing is more likely at low elevation angles than at high elevation angles. The low elevation angles correspond to the outer beams in a satellite footprint. For equal-size apertures on the satellite antenna, these beams cover the largest geographical area, and therefore might be required to carry the largest amount of traffic. Shadowing may also occur at high elevation angles, e.g. if the user were to walk behind a building or a tree.

Obviously, the cost of employing diversity is that potentially twice the bandwidth is used. It is therefore of interest to use diversity as little as possible, whilst using it frequently enough to minimise link outages and the number of lost connections.

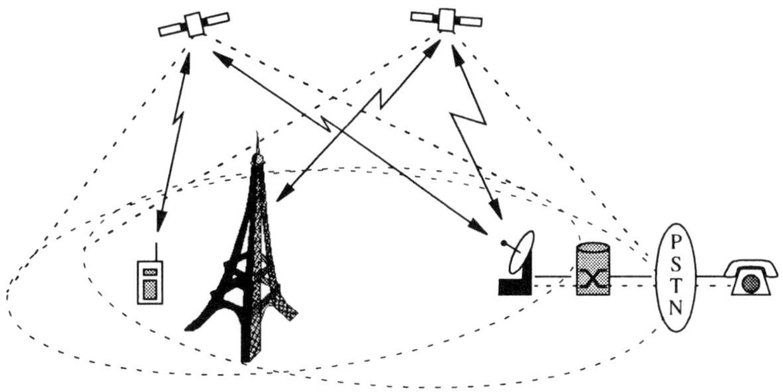

Figure 3: Satellite path diversity. If one LOS path from the MT to the satellite is shadowed, communications can still take place over the diversity path to the other satellite.

3.2.1 Prerequisites

Some prerequisites must be met before diversity can be used on a particular connection:

- The MT must support diversity. Diversity may be restricted to a class of MTs only, and the use of diversity can be offered as a subscribed service at an additional subscription cost. The operation may also be controlled by the user, e.g. if the use of diversity results in a higher charge for the air time.
- There must be dual satellite visibility. If the constellation does not provide ubiquitous dual satellite coverage, the availability of a second satellite is an obvious additional prerequisite.
- There must be an available channel. Channels may be temporarily unavailable due to local congestion, reservation schemes, or because none of the available channels match the timing constraints (see Section 4.0).

3.2.2 Initiation

Diversity establishment can be initiated by either the ES or the MT. If the ES initiates diversity, the "old" channel must be available for communication (not obstructed). Otherwise, if the old channel is shadowed, the MT must locate and scan un-shadowed paging channels for call announcements indicating when and over which satellite to establish the diversity connection. At the same time, the MT should monitor the old channel in case the connection was only momentarily broken. Further, even though the ES has knowledge of the MTs position, it will not be able to know whether an un-shadowed path is available to page MT, and might waste unnecessary bandwidth by paging the MT using a satellite the MT cannot see. Thus, this page mechanism is not practical.

As for the ES, the MT could easily request the establishment of the diversity connection if the old channel is available. If it is not, the diversity link could be established using an approach similar to the emergency handover procedure. The MT would then

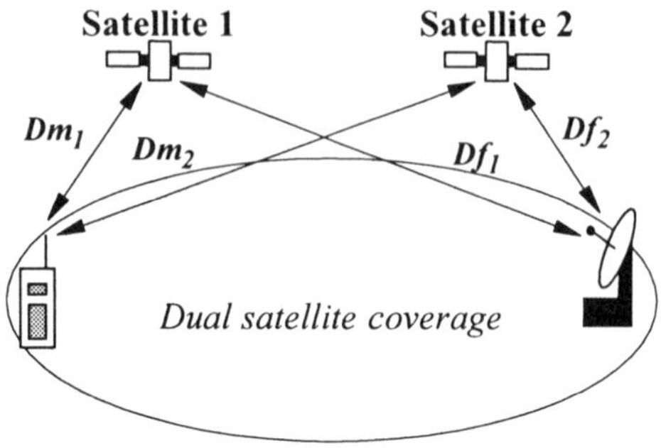

Figure 4: Mobile link delay (*Dm*) and feeder link delay (*Df*)

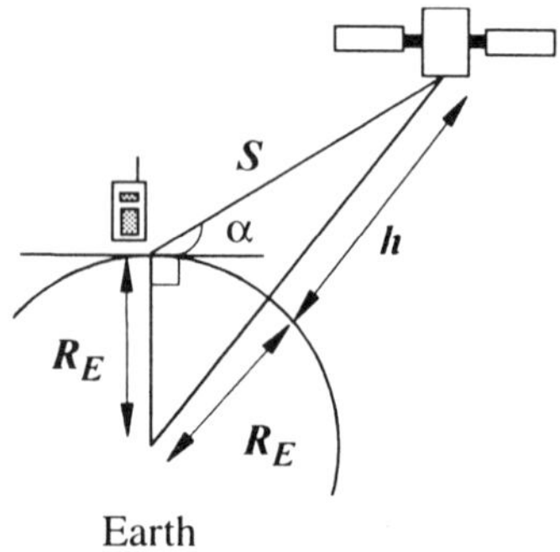

Figure 5: Satellite geometry

issue a channel request message over the RACH to visible satellites, using "diversity request" as the establishment cause.

3.3 Dynamic diversity

The diversity allocation mechanisms is said to be dynamic when the system allows the diversity to be switched on when needed and off when it is not. This switching of states is referred to as diversity toggling.

3.3.1 Allocation

There are several options for deciding when to allocate a satellite path diversity channel (assuming that the prerequisites of Section 3.2.1 are met):

1. The shadowing problem decreases for higher elevation angles. Diversity could therefore be established when the elevation angle is less than some threshold, either at call set-up, or if this elevation angle threshold is crossed due to the MT and satellite's relative movement during a connection. This will imply that the connection is lost when shadowing occurs at an elevation angle higher than the threshold, unless an emergency handover is successfully performed.
2. Diversity could be established as the connection to the current satellite is lost for some given time. This criteria can be average signal level (in dB), quality (in BER), and / or fade duration, or the GSM SACCH counter approach. The threshold could either be fixed, or a function of a number of parameters (elevation angle, available capacity, anticipated future demand, current mobility pattern of the MT, etc.).
3. Diversity can be applied as a "stand-by channel" option: a channel can be reserved for satellite path diversity, and used whenever the MT detects that a threshold is crossed. The stand-by channel would only be used when the primary channel is assumed obstructed; at all other times, it would be not be in use. Assuming that diversity establishment is relatively rare event, the stand-by channel can be shared by several MTs, provided a) there is some means of resolving

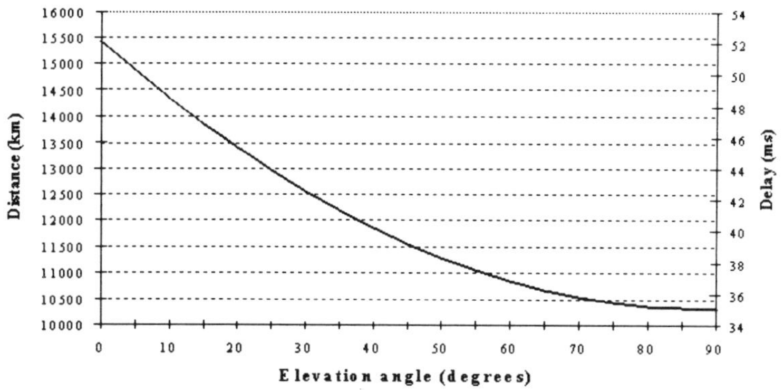

Figure 6: Plot of the variable one-way distance (left y-axis) and propagation path delay D (right y-axis) between mobile terminal and satellite as a function of elevation angle α (x axis) for orbit altitude h = 10355 km.

collisions, b) the MT's ID can be discerned on the stand-by-channel, and c) the other MTs sharing the stand-by channel can be allocated a new stand-by channel. This option implies that communications is momentarily lost whenever the primary channel becomes obstructed, but offers the potential advantage of a higher capacity (more simultaneous connections).

4. Diversity can also be "allocated" as an emergency handover: once the connection over the active satellite is lost, the MT attempts an emergency handover. This is similar to the case above, but here contention is resolved prior to the subsequent establishment of the diversity channel. This will (normally) result in a larger time-period when no communication is available and traffic is lost. Contrasting the GSM emergency handover structure, the old connection is subject to additional criteria before being released.

3.3.2 Release mechanisms

In the event that only one (or neither) of the allocated channels are no longer employed by the MT, the ES should detect this, and release the channel. The detection mechanisms could be similar to the SACCH counter approach employed for link management in GSM. Further, if the threshold levels described in Section 3.3.1 are no longer crossed, the diversity connection should be released.

3.3.3 Stability

If the operation environment of the MT is such that the threshold levels for diversity allocations and release are frequently crossed, the connection could become unstable. Unless precautions are made, the diversity channel could end up being constantly toggled on and off. Two mechanisms are readily identifiable to further stability: a hysteresis loop, and a hangover state after a toggle during which diversity toggling is not permitted.

3.4 Static diversity

Static diversity refers to the case where the allocation and release of a diversity channel is unlikely to happen during a connection. Diversity would be established at call set-up and normally not allowed to toggle, irrespective of the elevation angle or other criteria. This means that shadowing of a single channel will not result in even a temporarily broken connection for MTs employing diversity, but at the expense of capacity and reduced flexibility.

4.0 System timing

4.1 Variable delay components

Figure 4 illustrates the largest variable delay components; the feeder and mobile link path delays *Df* and *Dm* respectively, each subscripted with the index number of the satellite they are communicating through. (EQ 1) expresses the variable delay component D between an observer and the satellite in terms of elevation angle α, earth radius R_E (6 370 km), orbital altitude h and speed of light c (see Figure 5):

$$D = \frac{1}{c}\left[\sqrt{(R_E + h)^2 - R_E^2 \cos^2\alpha} - R_E \sin\alpha\right] \quad \text{(EQ 1)}$$

(EQ 1) is plotted in Figure 6. The minimum round trip delay (MT - satellite - ES or vice versa) will occur when the MT and ES are co-located at the satellites subsatellite point ($\alpha = 90°$), and the maximum when the MT and ES are located both at the minimum elevation angle (typical $\alpha = 10°$). The round-trip delays are then 96 ms maximum and 69 ms minimum for an altitude of 10355 km.

4.2 Timing

The timing requirements are listed for the satellite, ES, and MT.

4.2.1 Satellite.

Bursts from MTs sharing the same frequency (but obviously utilising different timeslots) must not overlap at the satellite, irrespective of the difference in path delay to the different MTs. This calls for timing adjustment at the MT, synchronised to the satellite. For the diversity case, the MT must then maintain possibly different timing adjustments for each satellite. A similar argument can be carried out for controlling the Doppler shift towards each satellite. As for other TDMA systems, a guard interval will be required in the burst structure to allow small inaccuracies in the timing control. This guard interval should be as small as possible for minimum overhead.

4.2.2 Earth Station.

The ES requires that bursts emanating from the same MT arrives at a single end point for combination to take place. This raises a question of routing rather than timing, in that both satellites employed for the diversity should be controlled by the same ES to avoid routing through a land network line. The ES must be able to discern which bursts are to be combined to successfully use diversity, and a terrestrial line could introduce uncertainties depending on the chosen implementation. The time delays via each satellite is already known, as the ES must have detailed knowledge of each satellites movement for e.g. tracking purposes.

4.2.3 Mobile Terminal

The MT should for cost reasons be implemented with only one receiver and transmitter and no duplexer, as is the case for e.g. the GSM mobile station. In addition to the usual constraint that the MT cannot receive and transmit simultaneously, the MT must be able to receive from and transmit to both satellites with no overlap in time at the MT within one TDMA frame. The MT must also be able to maintain synchronisation with each satellite. The relative movement of the satellite is known, and a timing adjustment function for each satellite could be communicated to the MT at call set-up. The timing advancement need then only be updated according to the MTs movement and for possible corrections to the movement function at e.g. handovers, and the GSM timing advance (carried in the layer one header of the SACCH burst once every 480 ms) rate should be sufficient.

4.2.4 Diversity combination

If the synchronisation reference is taken at the satellite or at the ground, the ES can compensate for the variable path delay to the satellite, and only the mobile link delay must be compensated for. Choosing the synchronisation reference at the ES requires that the diversity slots must compensate for the variable delay on both the feeder link and mobile link. A synchronisation reference at the ES will thus require more frequent synchronisation handovers as the propagation path delay changes twice as rapidly as if the synchronisation reference is taken at the satellite or ground.

Both the ES and MT will delay the first received burst until the second burst is received for diversity combination. Figure 6 plots the variance in path delay and distance between a satellite and an observer on the earth as a function of elevation angle (EQ 1). As the satellite rises over the horizon, the elevation angle from the user will increase from the minimum (typically 10°) to the maximum (90°) when the satellite is immediately above the observer, before the delay reduces again as the satellite sets over the horizon. The maximum difference in arrival times is on the order of 27 ms if the synchronisation reference is taken at the ES, and 13.5 ms if the reference is taken at the ground or satellite.

4.3 Implementation

4.3.1 Principle of operation

The timing implementation relies on dividing the TDMA frame in different slots for communicating with different satellites. These slots represent how much the timing at the MT can be adjusted before the timing requirements are violated. As the propagation delay is changed due to the relative movement of the satellite and MT, the start of burst transmission time is advanced (for a rising satellite) or retarded (for a setting satellite) to accommodate the change in path delay as seen by the MT. The start of burst reception time behaves inversely to the start of transmission time: as the satellite is rising it is retarded, and increased as the satellite sets. As discussed in Section 4.2.3, the timing adjustment is performed according to a pre-defined function for each connection modified by conventional timing advancement. When the timing can

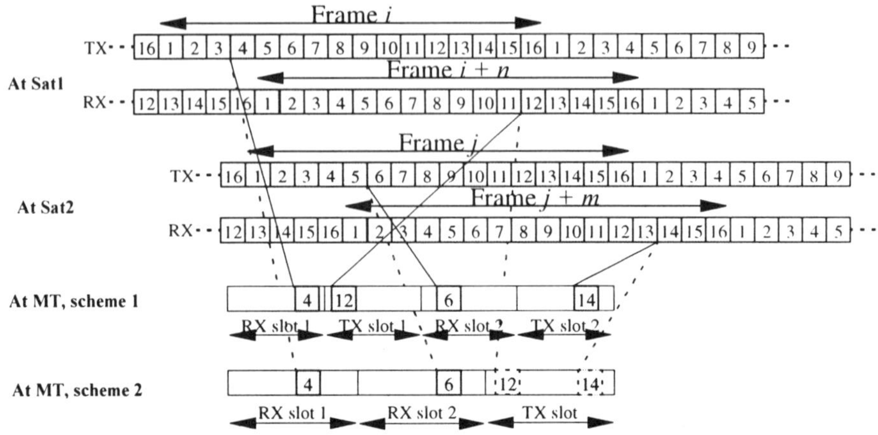

Figure 7: System timing. At the satellites, bursts emanating from different MTs must not overlap, and the MT employs adaptive time alignment to prevent burst collisions. In scheme 1, one half of the 16 burst half-rate TDMA frame is used for transmissions with satellite 1, and the second half of the TDMA frame for transmissions with satellite 2. In scheme 2, where diversity is employed most if not all of the time, the TDMA frame is split in three: the first two thirds are used for receptions from satellite 1 and 2, and the last third is used for communications with that satellite which had the best signal, using the timing adjustment required for that satellite.

no longer be adjusted without requiring that the MT transmits or receives simultaneously, the MT has no more room within the slots for timing adjustment, and a new channel must be allocated. This changing of timeslots and possibly carriers is effectively a handover, and is henceforth called a *synchronisation handover*. Figure 7 shows the principles of the timing system, and indicates that two different schemes can be thought of. These are described in more detail in the next sections.

4.3.2 Scheme 1: Dynamic diversity.

This scheme assumes that several active mode MTs will employ diversity at any given time, but also that a significant number of MTs will not. The TDMA frame is divided into four equal size slots; one transmit and one receive slot for each satellite. The MTs employing diversity will transmit in both TX slots, and receive in both RX slots. A MT not employing diversity will use only one RX and one TX slot over the same satellite. Figure 8 shows how the burst transmission and reception times slides within the diversity slots at the MT as one satellite is rising and the other is setting.

As the satellite's footprint passes over the MT due to their relative movement, handovers will become necessary between beams. In addition, synchronisation handovers are required when the timing adjustment can no longer compensate for the changing delay. The restriction that the TCH timeslots matches the delay required at the MT is an additional requirement when choosing a channel for handover. This reduced flexi-

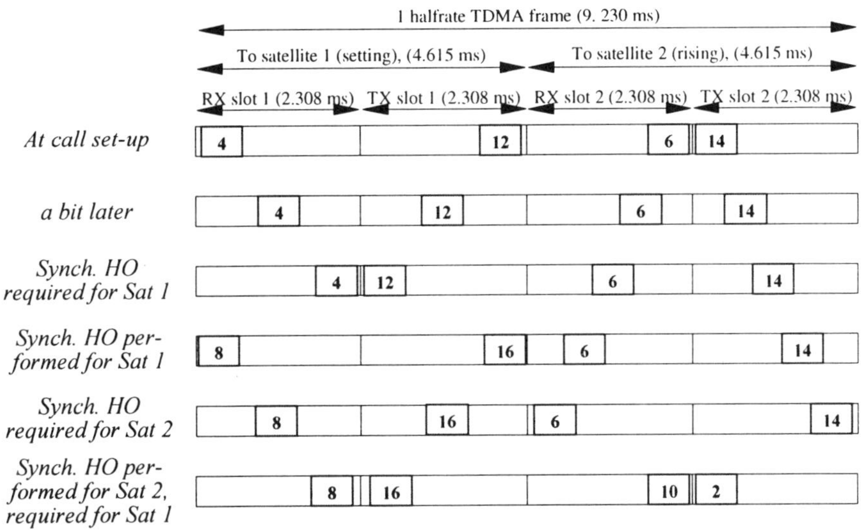

Figure 8: Transmit and receive times for scheme 1 at the MT for different time instants. Satellite 1 is setting over the horizon (increasing path delay), and satellite 2 is rising (path delay is decreasing). Notice that the rate of change of path delay can vary at different rates for the two satellites.

bility increases the probability of a losing a diversity connection when a handover is attempted.

4.3.3 Scheme 2: Static diversity.

For a satellite system where diversity is normally allocated and rarely toggled, the transmit and receive slots at the MT can be adapted. Noting that the selection combination is only slightly worse than equal gain or maximal ratio combination, an asymmetric TDMA structure where there are two mobile downlink receive slots for every mobile transmit slot can be constructed, as the bottom part of Figure 7 shows.

When the MT has received a burst from each satellite within the appropriate receive slot, the MT uses selection diversity and determines which received burst had the best quality based on e.g. received power. During the transmit slot, the MT will transmit a burst to the satellite which had the best quality on the downlink, having switched to the frequency and timing adjustment required for that satellite.

The result is an improved spectrum utilisation on the mobile uplink, which might be traded for improved link margins, battery conservation and / or reduced radiation at the MT (using coding mechanisms trading larger bandwidth for reduced transmit power whilst maintaining the BER) or offering new, asymmetric services. To aid the ES in maintaining channel synchronisation, the MT should when possible alternate between the two channels, such that time and frequency synchronisation can be maintained.

The underlying assumption is that the fading characteristics are correlated within one TDMA frame duration on the forward and return link frequencies. The structure exploits the fact that on the uplink there are two receivers (one in each satellite) and hence it is possible to receive the burst emanating from a MT at each receiver simultaneously. This is not possible in the reverse direction, i.e. having one receive slot for two transmit slots at the mobile, as this would require two receivers at the MT.

5.0 Handover and diversity

5.1 Handover

The most common cause for a handover is the relative movement of the satellite(s) and the MT. Other potential causes for a handover are regulatory issues, frequency re-arrangements due to traffic balancing with neighbouring beams, interference with other systems, or dynamic channel assignments. The normal classes of handover are intra-beam (typically synchronisation handovers), inter-beam, inter-satellite, inter-ES, and inter-system handovers.

5.1.1 Initiation

The decision to handover would normally be taken by the ES, which has detailed knowledge of the MT's location relative to the beams. This knowledge is used to direct the MT at the appropriate time to measure potential beams to handover to. The actual initiation and handover execution is triggered by these measurements. The MT is required to measure the downlink signal quality continuously to determine whether the link becomes obstructed, but measure candidate handover beams only as the MT is ordered to do so or has become shadowed. Only as an emergency handover caused by shadowing would the decision to handover be taken by the MT.

For the satellite path diversity to be of use, both diversity channels must necessarily convey identical information. This adds the requirement for the *measurement report* messages to identify which channel the measurements pertain.

5.1.2 Planning ahead

The expected traffic demand in a beam is short-term predictable, as the ES is able to register the currently served traffic in the areas that a particular beam is approaching. This enables reservation schemes for handover, as the location of the MT in the short term future is known. Dynamic rearrangement of capacity between beams in the satellite to meet the expected demand will be required to serve the expected high variation in offered traffic.

5.1.3 Seamless handover

Inter-satellite seamless handover is similar to diversity toggling in terms of signalling over the air interface. The handover is said to be seamless if the communication path between the communicating entities (the MT and ES) is not broken during the handover procedure. Both TDMA schemes permit seamless handovers. If diversity is employed, the handover is initiated by first releasing the channel to be handed over, and, if required, initiating measurements of candidate channels whilst continuing the connection via the remaining diversity channel. When the new channel has been suc-

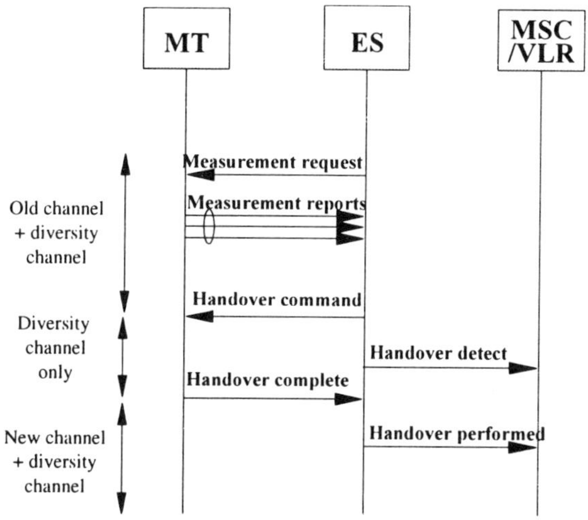

Figure 9: Intra-ES handover, assuming diversity is employed. The available channels are indicated on the left.

cessfully determined the handover can be completed. During the entire operation, the diversity channel has been available for communications and signalling. If diversity is not established when the intersatellite handover is to be initiated, the channel to be handed over to is first identified, and the connection is established (diversity toggle on). Only then is the old channel released (diversity toggle off), and a seamless handover has taken place exploiting the diversity capabilities.

5.2 Handover signalling

Intra-ES handover is appropriate for synchronisation handovers and changing of beams on the same or over different satellites. Figure 9 shows that the ES initiate the handover by issuing the *measurement request* message. This message will be different from GSM, as it to indicates which channel is no longer needed and will be handed over. There is no time-compatible window at the MT for monitoring neighbouring beams and reporting measurements back to the ES when diversity is employed. The MT must therefore refrain from using the channel to be handed over while it is measuring candidate channels. The MT should therefore be commanded to measure candidate channels, and also be told which channel it will handover from.

Once an appropriate channel has been selected, the actual handover is initiated by issuing the *Handover command* message to the MT, and the ES notifies the MSC / VLR that a handover has been initiated (mainly for audit purposes). The MT confirms the handover with the *Handover complete message*, acknowledging that the previous channel is released. The message exchanges can in most cases be carried by the SACCH since the handover time is know in advance, and only rarely would the FACCH be needed for time critical signalling.

5.2.1 Intra-satellite handover

If the location of the mobile is known such that the choice of beam to handover to is known unambiguously, the new channel could be engaged for communications by the MT without prior measurements. This could for instance be the case where a synchronisation handover results in the allocation of a new timeslot on the same carrier. Measurements might still be required prior to the allocation to discern undue co-channel or neighbouring channel interference. Alternatively, the new channel could be engaged while these measurements are performed, thus speeding up the handover process if the interference was not adverse. Measurements are in some instances still required to resolve ambiguities in determining the appropriate handover channel based on the MT's position, e.g. when the MT is located at an intersection between two beams where this intersection is parallel to the satellites movement.

5.2.2 Inter-satellite handover

When the connection is handed over to another satellite, special precautions must be taken to make sure that the path delay to the new satellite is determined accurately enough. If the MT positioning scheme is not accurate enough to determine the exact path delay to the new satellite, this must be determined by another mechanism. This can be achieved if the MT acquires synchronisation on the downlink BCCH carrier of the candidate handover satellite and beam, and reports the time delay measured back to the ES. If diversity is not established, the "old" link to the ES would be used to carry the measurement. The ES would in turn notify the MT of the correct timing adjustment for the new channel. The reporting can be part of the indicated *measurement request* and *measurement report* exchanges shown in Figure 9, and the required timing adjustment could be carried as a field in a suitably modified *handover command* message.

5.2.3 Failures

In GSM, the channel which is handed over from is not released at the BSS before the *handover complete* message is received. This is because GSM specifies that if the handover fails, the old channel should be attempted before the emergency handover procedure is initiated. This is not necessary in the satellite case. As described in Section 5.1.3, one communication path is always maintained in the satellite case. Further, as the satellite's movement relative to the MT is the normal case for handover, the channel to which the MT was connected will typically not be available for long. There is therefore little need in the satellite case to maintain the handover channel after the *handover command* message is received by the ES.

5.3 Diversity toggling

Changing from the two channel diversity case to the single channel case (or vice versa) during a connection is called diversity toggling, and is illustrated in Figure 10 for ES initiated diversity toggling.

5.3.1 Diversity toggle on

When the ES initiates diversity, the first message it will send the MT is a *measurement request* in order to identify the new channel on the new satellite. Once the *meas-*

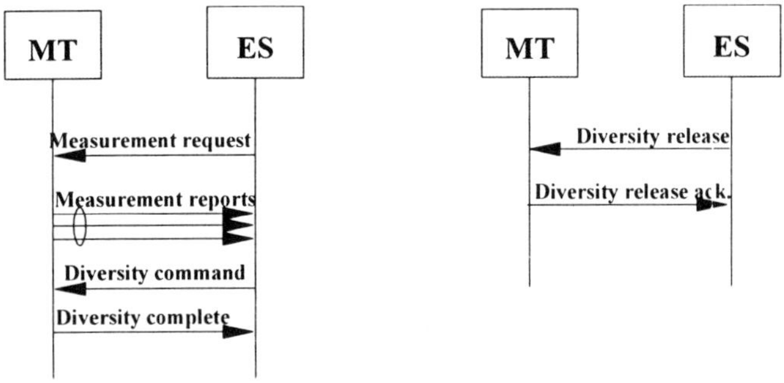

Figure 10: Example of ES initiated diversity toggling. Establishment during an existing connection (left) and release (right). The MSC/VLR is not involved.

urement reports are received, a *diversity command* message can be used to establish the diversity connection. An approach similar to the inter-satellite handover timing explained in Section 5.2.2 would be required to discern the timing adjustment for the new channel. The diversity establishment is acknowledged in the *diversity complete* message.

If a connection is currently without diversity, the in-band signalling over the SACCH messages could be used to update the MT as to which satellite, beam and channels to attempt re-registration if the connection is lost. This will speed up the diversity toggle on procedure when initiated by the MT.

5.3.2 Diversity toggle off

This is the simplest of the exchanges shown. When initiated by the ES, the message *diversity release* is sent to the MT, identifying which channel is to be released. The MT acknowledges the release by the *diversity release acknowledge* message, which can be carried by the SACCH over the remaining channel. The MT can initiate the release either by sending a *diversity release* message, or simply stop using a channel. The ES will detect this as discussed in Section 3.3.2, and clear the connection.

6.0 Conclusion and further work

This paper has discussed how satellite path diversity can be operated in a non-geostationary satellite communication systems without the need for a costly duplicated transmitter/receiver pair or duplexer at the MT, using the GSM halfrate TDMA format as an example. Mechanisms for controlling the diversity operation were discussed, and two TDMA structures were derived, both shown to support diversity. Different handover scenarios were discussed, and diversity was shown to enable seamless handover in a TDMA system.

Future work will focus on showing how diversity channels can be regarded as a stand-by resource pool, using the diversity channels to improve performance when the network load is low. At high network loads approaching congestion, the quality experienced by individual links can be traded for capacity, making the TDMA system's upper capacity bounds less hard limited.

Acknowledgements

This work is carried out as part of an ongoing Ph.D. (Dr.Ing) program at the Norwegian University of Science and Technology, Department of Computer Systems and Telematics.

Special thanks are due to Sudeep Kumar Palat for invaluable objections, fruitful comments and enduring patience.

References

[BUT95] **Butt, G., Evans, B.G. Parks, M.** *Modelling the Satellite Channel for Communication System Design*, ICAP'95, Antennas & Propagation, IEE conference publication no. 407, ISBN 0 85296 637 7, p. 387 - 394.

[CA95a] **Carter, P. and Beach, M.A.** *Evaluation of Handover Mechanisms in Shadowed Low Earth Orbit Land Mobile Satellite Systems*, International Journal of Satellite Communications, Vol. 13, 1995, pp. 177-190.

[CA95b] **Carter, P. and Beach, M.A.** *Diversity in low and Medium Earth Orbit Satellite Systems*, ICAP'95, Antennas & Propagation, IEE conference publication no. 407, ISBN 0 85296 637 7, p. 225 - 230.

[COR94] **Corazza, G. E., Jahn, A., Lutz, E., Vatalaro, F.** *Channel Characterisation for Mobile Satellite Communications, EMPS* 94.

[DE93] **De Ridder, Jean-Jacques and Dosiere, Frederic.** *Statistical analysis of mobile channel fading of MEO Communication Systems*, European Conference on Satellite Communications ECSC 93, pp. 367 - 373.

[HAR95] **Hart, Nicholas, Goerke, Thomas and Jahn, Axel, A** *Discussion on Mobile Satellite Systems and the Myths of CDMA and Diversity Revealed*, 4th International Mobile Satellite Conference 1995, Ottawa, Canada, June 6-8, 1995, pp 469-473.

[LUT91] **Lutz, E., Cygan, D., Dippold, M. Dolainsky, F. Papke, W.** *The Land Mobile Satellite Communications Channel: Recording, Statistics, and Channel Model*, IEEE Transactions on Vehicular Technology, Vol. 40 #2 1991.

[MOU92] **M. Mouly, M. Pautet**, *The GSM System for Mobile Communications*, Book Published by authors, 49, rue Louise Bruneau, F-91120 Palaiseau, France, 1992. ISBN 2-9507190-0-7

[NEE93] **van Nee, Richard D. J. and Prasad, Ramjee** *Spread-Spectrum Path Diversity in a shadowed Rician Fading Land-Mobile Satellite Channel*, IEEE Transactions on Vehicular Technology, Vol. 42, No 2, May 1993.

[PAR89] **Parsons, J. D. and Gardiner, J. G.** *Mobile Communications Systems*, Chapter 6. Published by Blackie of Glasgow and London. ISBN 0-216-92261-5

[VOJ94] **Vojcic, B. R., Milstein, B. L., Pickholtz, R. L.** *The Effect of Dual Satellite Diversity on the Total Capacity of a Multiple Band-Shared CDMA LEOS Systems*, Supercomm / ICC '94, ISBN 0-7803-1825-0/94 pp 1141 - 1143.

[WER95] **Werner, M., Bischl, H., Lutz, E.** *Mobile User Environment and Satellite Diversity for NGSO S-PCNs*, 4th International Mobile Satellite Conference 1995, Ottawa, Canada, June 6-8, 1995, pp 476 - 481.

[WIS95] **Wisløff, Tor E.** *Tor E. Wisløff's Big LEO Tables*, document retrieved from *www.idt.unit.no/~torwi/synopsis.html*, document dated 28 December 1995

[ZHA95] **Zhao, W., Tafazolli, R. Evans, B. G.**, *A UT Positioning Approach for Dynamic Satellite Constellations*, 4th International Mobile Satellite Conference 1995, Ottawa, Canada, June 6-8, 1995, pp 35 - 41

AN OPTIMAL CHANNEL ALLOCATION ALGORITHM IN A LINEAR CELLULAR REUSE PARTITIONING NETWORK

Partha P. Bhattacharya* and Leandros Tassiulas**

* *IBM T. J. Watson Research Center, P.O. Box 704, Yorktown Heights, NY 10598, USA*
fax: +1 914 784-6205; e-mail: partha@watson.ibm.com
** *Electrical Engg. Dept, University of Maryland, College Park, MD 20742, USA*
fax: +1 301 314-9920; e-mail: leandros@eng.umd.edu

ABSTRACT

In this paper we consider a model of a linear network with reuse partitioning. An instance of the network is considered where a number of call requests are present at the different zones of each cell. Interference constraints prevent the use of the same channel in certain combination of zones at different cells. We obtain an algorithm that allocates a channel to each call request without violating the interference constraints and uses the minimum number of channels.

1 INTRODUCTION

One technique for increasing the capacity of a cellular network is to use multiple frequency reuse factors in order to pack more calls using the limited number of available channels. This is called reuse partitioning. The main objective of reuse partitioning is to provide an increase in system capacity over that which can be achieved with a single factor, without relaxing SIR performance requirements. The underlying principle behind reuse partitioning is to degrade SIR performance for these mobiles that already have more than adequate transmission quality while offering greater protection to those mobiles that require it. The available channels are split among several reuse patterns with different reuse factors N_i. Channel allocation within the i-th group is then determined by the reuse factor N_i. Mobile units with the best received signal quality will be preferentially assigned to the group of channels having the smallest reuse factor value, while those with the poorest received signal quality will be only assigned to the group of channels having the largest reuse factor value. Typically mobile units close to a cell-site are served by channels from a group having a small value for N_i since they receive high signal level and therefore are more resistant to interference than a mobile close to the perimeter of the cell. Mobile units that are far from the cell-site are served by channels from a group having a large value for N_i. One may visualize this structure as if each cell is divided into concentric zones, where a mobile in each zone-i achieves a sufficient SIR level by being assigned a channel from reuse pattern-i or some pattern associated with a zone closer to the cell perimeter.

The use of multiple reuse patterns in a cellular network for capacity increase has been studied extensivelly in the literature [4, 2, 8, 7, 6]. In this paper we focus on a linear network with reuse partitioning and we consider a scenario that may arise in a dynamic channel allocation framework. An instance of the network is considered where a number of call requests are present at the different levels of each cell. Interference constraints prevent the use of the same channel in certain combinations of zones at different cells. We obtain an algorithm that allocates a channel to each call request without violating the interference constraints and uses the minimum number of channels. The above problem may arise in several stages of a dynamic channel allocation algorithm, where all the channels are available in every cell and the objective of the algorithm is to accomodate all the call requests with the limited number of available channels.

2 LINEAR ARRAY

Consider a linear cellular network with cells that are numbered sequentially from 1 to N. Each cell is partitioned in K concentric zones as shown in Figure 1; with zone K being the outermost and zone 1 the innermost.

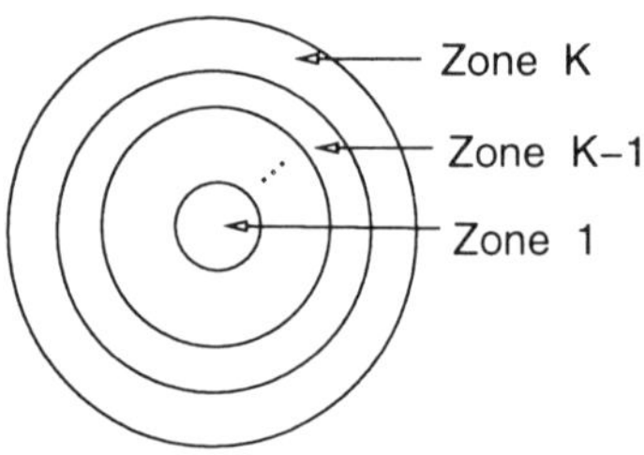

Figure 1 A cell with several zones

Calls in different cells may reuse the same channel subject to constraints on the spatial proximity of the calls. Let (i, k) denote the kth zone of the ith cell. Two calls in zones (i, k) and (j, l) may use the same channel if and only if $|i - j| \geq \max\{k, l\}$. This implies for example, that two calls in innermost zones of any two cells may reuse the same channel; also two calls in zone 2 of two cells can reuse the same frequency only if the cells are not adjacent etc. Figure 2 illustrates the collection of zones which cannot use the same channels as those used in zone (i, k).

We consider the following channel allocation problem. A number of call requests C_{ik} are present at each zone (i, k). Channels have to be assigned to calls so that each call gets one channel, there are no conflicts and the minimum number of channels is used.

The channel allocation problem specified above can be formulated as a node coloring problem for an appropriately defined *interference graph*. This graph contains one node for each zone and two nodes in the graph are adjacent if a channel can not be simultaneously assigned to two calls in the corresponding zones. With each node in the interference graph associate a weight equal to the number of call requests in the corresponding zone. Then the above problem is identical to finding the chromatic number of the weighted interference graph.

Node coloring problems are NP-hard for general graphs but efficient algorithms exist for some special classes of graphs, e.g. perfect graphs. A graph $G = (V, E)$ is a perfect graph [3] if its independent set polytope can be completely characterized by the clique

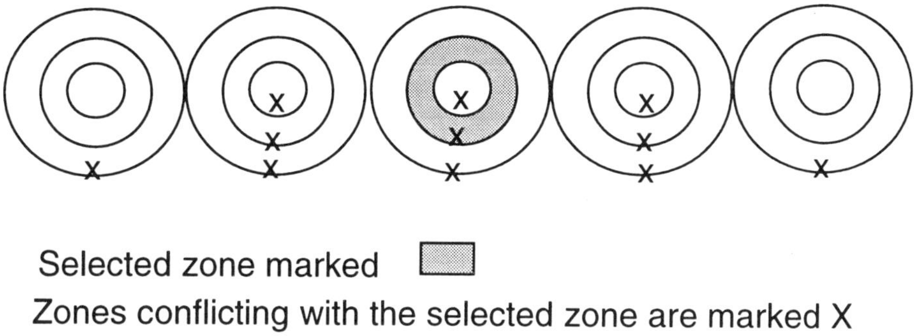

Figure 2 Illustration of channel reuse: $K = 3$

constraints. The independent set polytope of the graph G is the convex hull of vectors of the form $1(S)$, $S \subset V$, S an independent set; where $1_i(S)$, the ith component of $1(S)$, equals 1 if node $i \in S$ and 0 otherwise. A simple example of a perfect graph is an interval graph, a graph whose nodes can be represented by intervals on the real line such that two nodes are adjacent if and only if the corresponding intervals do not overlap.

If each cell has only two zones, then the associated interference graph is an interval graph and hence, a perfect graph for which efficient coloring methods exist. To show that the interference graph is an interval graph, associate interval $(2i, 2i+1)$ on the real line to node $(i, 1)$, and interval $(2(i-1)+3/4, 2(i+1)+1/2)$ to node $(i, 2)$, $i = 0, 1, 2, \cdots$.

Interestingly however, if there are more than 2 zones per cell then the interference graph may not be an interval graph (see Appendix). Our main result in this paper is that the interference graph in the general case is still a perfect graph and hence the number of channels can be completely described by clique constraints. These constraints can be easily enumerated and hence the minimum number of channels can be computed as well. Our proof is constructive and instead of relying on general coloring methods for perfect graphs [3], we develop a simple coloring/channel allocation scheme for establishing the perfectness of the graph.

We now describe an optimal channel allocation scheme. The channels are allocated to the cells sequentially starting from cell 1 to cell N. In each cell the channels are allocated to the zones sequentially as well, starting from the calls in zone K and moving up to the calls in zone 1. All the calls in a certain zone are allocated before the algorithm moves on with the allocation of the calls in the next zone.

Assume that channels have been allocated to all calls in cells 1 through $i-1$ as well as to all calls in zones K through $k-1$ of cell i. The calls in zone (i, k) are then allocated by the following steps:

1. Consider all the channels that have been used already in one of the cells 1 through i. Reuse first those channels for the calls in zone (i, k) as follows.

   Start from cell $i-1$ and go up to cell 1 doing the following for each cell j. Consider all the channels used in zones of cell j which are not conflicting with zone (i, k), if any. From these channels identify those that are not already allocated to calls of zone (i, k) and allocate as many as necessary to calls of (i, k). The order of selection of zones of cell j is not important. If there are remaining calls without a channel allocated in zone (i, k) then proceed to cell $j - 1$ and do the same.

2. If all the channels that are already used are exhausted and there are calls remaining in (i, k) without an allocated channel, then use new channels arbitrarily.

We claim that the number of channels required by the above algorithm is equal to the size of the maximum weighted clique of the interference graph. Note that the number of required channels must be no less than the size of the maximum weighted clique of the interference graph. Hence it suffices to show that the algorithm requires a number of channels equal to the size of some clique in the interference graph. In fact, the following stronger property holds for every zone (i, k).

P1. For any cell m, $m \leq i$, consider all the calls in the cells m through $i - 1$ as well as the calls in the zones (i, K) through (i, k). Consider the weighted interference subgraph that corresponds to just those calls. The total number of distinct channels used for the allocation of calls in this group of cells is equal to the size of some clique in the interference subgraph.

Note that Property $P1$ is sufficient for establishing the fact that the interference graph is a perfect graph, since the chromatic number is determined only by the clique constraints.

Theorem 1 *Property P1 holds for any zone* (i, k), $i = 1, \cdots, N$, $k = 1, \cdots, K$.

Proof. The proof is by induction. $P1$ clearly holds for zone $(1, K)$. We show that it will hold for zone (i, k) when it holds for zone $(i, k + 1)$, $k = K - 1, \cdots, 1$; and that it will hold for zone (i, K) when it holds for zone $(i - 1, 1)$.

Let m be an arbitrary cell, $m \leq i$. We consider two cases in verifying P1.

Case A: All calls in zone (i, k) are allocated channels that are used already either by calls in cells m through $i - 1$ or by calls in zones (i, K) through $(i, k + 1)$.

In this case the total number of distinct channels, say C, used in cells m through $i-1$ and zones (i, K) through $(i, k + 1)$ is equal to the total number of distinct channels used in cells m through $i-1$ and zones (i, K) through (i, k). The induction hypothesis ensures that there is a clique in the interference graph corresponding to the cells m through $i - 1$ and zones (i, K) through $(i, k + 1)$, with size equal to C. That clique is the one for which property P1 is satisfied for zone (i, k) as well.

Case B: The calls in zone (i, k) need additional channels than those used by calls in cells m through $i - 1$ and in zones (i, K) through $(i, k + 1)$.

Consider the weighted interference graph corresponding to calls in cells $m, \cdots, i-1$ and zones $(i, K), (i, K - 1), \cdots, (i, k)$. We specify the maximal weighted clique U in this graph. The set U consists of three types of nodes.

(a) All the nodes corresponding to zones $(i, K), (i, K - 1), \cdots, (i, k)$.

(b) All the nodes corresponding to any zone of cells $m, \cdots, i-k$ that are conflicting with zone (i, k).

(c) From the induction hypothesis, there exists a clique Q in the weighted interference subgraph consisting of nodes corresponding to calls in the set of cells $i-k+1, \cdots, i-1$ and the size of this clique is equal to the total number of distinct channels needed by the algorithm to allocate channels to calls to this set of cells. Consider the set of nodes in Q.

We claim first that U is a clique. This follows from the following arguments.

- All the nodes of type (a) are conflicting among themselves; they are also conflicting with nodes of type (b) by definition and finally they are conflicting with nodes of type (c) since they are conflicting with any zone in the cells $i-k+1, \cdots, i-1$.
- A zone (j,l) of type (b) has the property that $i - j < \max\{k,l\}$, since it conflicts with zone (i,k). Furthermore, $\max\{k,l\} = l$, otherwise $i - j < k$, a contradiction. Hence a zone (j,l) of type (b) conflicts with a zone (j',l') of type (c) since $j'-j < i-j < l \leq \max\{l,l'\}$. Furthermore, any two zones of type (b) are conflicting among themselves since they each conflict with (i,k). This can be seen as follows. Take two nodes (j,l) and (j',l') of type (b). Suppose $j < j'$. Since they both conflict with (i,k) arguing as before, we have $\min\{l,l'\} > k$. Therefore $j' - j < i - j < \max\{k,l\} = l \leq \max\{l,l'\}$ which shows that (j,l) and (j',l') do indeed conflict.
- All nodes of type (c) are conflicting among themselves by definition.

This establishes that U is a clique.

We now show that the size of U equals the total number of distinct channels used by the algorithm to allocate channels to calls in cells $m, \cdots, i-1$ and zones $(i,K), (i,K-1), \cdots, (i,k)$. The channels allocated to calls in zones $(i,K), (i,K-1), \cdots, (i,k)$ are certainly included in U by definition. Also, the channels allocated to calls in cells $i-k+1, \cdots, i-1$ is equal to the size of Q and $Q \subset U$ by construction. Finally, consider channels allocated by the algorithm to cells $m, \cdots, i-k$. If a zone in this set of cells is conflicting with (i,k) then the channels allocated to that zone are included in U as well by construction. If on the other hand, a zone in this set of cells does not conflict with (i,k), then all these channels are allocated by the algorithm to (i,k), and hence these channels are also included in U. The proof the theorem is complete. □

In summary, we have provided an algorithm which provides a conflict free allocation of channels to calls in a linear array. The number of channels used by this algorithm is equal to the maximal weighted clique of the associated interference graph and is the minimum possible. This also establishes the fact that the interference graph is a perfect graph since its chromatic number is only determined by clique constraints.

One way to determine the miminum number of required channels is of course, to run the algorithm. But it is also of interest to see if there is a way to characterize all the cliques which would then lead to an expression for the minimum number of required channels.

We first show that there is a smaller set of cliques which determine the minimum number of channels. Consider cliques U of the following type: for each level k, $i \leq k \leq K$, U contains k consecutive zones, with the property that the k cells for level k consitute a subset of the $k+1$ cells of level $k+1$. As an example, for $K = 3$, two cliques of the above form are given by (a) $(i,3)$, $(i+1,3)$, $(i+2,3)$, $(i,2)$, $(i+1,2)$, $(i,1)$ and (b) $(i,3)$, $(i+1,3)$, $(i+2,3)$, $(i+1,2)$, $(i+2,2)$, $(i+2,1)$. We claim that all the maximal cliques that need to be considered for the evaluation of the bound are of this type. Consider an arbitrary clique Q. Let m_k, M_k be the minimum and maximum cells respectivelly where a zone of level less than or equal to k that belongs to Q may arise. Consider the set of zones S that includes for each level k all the zones $(m_k,k), (m_k+1,k), .., (M_k,k)$. It can be easily verified that this set of zones is a clique and clearly it is a superset of Q. Furthermore it can be easily verified that S is a subset of a maximal clique of the type specified earlier.

One can derive an recursive expression to determine the number of required channels. Recall that C_{ik} denotes the number of required channels in zone (i, k). Let $\{N_i^{(k)}\}_{1 \le i \le N, 1 \le k \le K}$ denote some intermediate quantities that are related by the recursion

$$N_i^{(k+1)} = \sum_{l=i}^{i+k} C_{l,k+1} + \max\left\{N_i^{(k)}, N_{i+1}^{(k)}\right\}.$$

The initial condition in the recursion is given by

$$N_i^{(2)} = C_{i,2} + C_{i+1,2} + \max\{C_{i,1}, C_{i+1,1}\}$$

and the total number of required channels is given by

$$\max_{0 \le i \le N-K} N_i^{(K)}.$$

The proof follows from the above discussion on the special class of determining cliques and is omitted for brevity.

3 OTHER CELLULAR TOPOLOGIES

In this section, we present the case when the cells are arranged ina circular array. The two dimensional topology will be the subject of a future paper.

N cells are arranged in a circular array, that is, cell $N-1$ is adjacent to cell 0. Each cell is partitioned into K concentric zones. Two cells (i, k) and (j, l), $i < j$ can reuse the same frequency if $i + \max\{k, l\} \le j$ holds and either $j + \max\{k, l\} < N$ holds or $(j + \max\{k, l\}) \pmod N \le i$ holds. Let C_{ik} denote the number of call requests in zone (i, k). We wish to determine a channel allocation that assigns one channel to a call without conflicts and uses the minimum number of channels.

It is well known [1] that the special case of the above problem with 1 zone per cell is a hard problem because of the presence of complex combinatorial structures like holes, antiholes etc. in the associated interference graph. Therefore, we concentrate on suboptimal allocations.

Consider opening up the circular array after cell $N-1$. Then it becomes an linear array for which we know the number of required channels to meet the call requirements. Let N^* denote this quantity. Since a circular array presents more constraints, we know that at least N^* channels would be required in this case as well. We adapt an argument presented in [1] to construct an algorithm that needs $\frac{N}{N-K}N^*$ channels. Note that the algorithm becomes optimal for large arrays as $N/K \to \infty$.

The algorithm works in N cycles. In cycle n, $0 \le n \le N-1$, cells n, $(n+1) \pmod N$, $\cdots n + N - K \pmod N$ are considered for allocation. The channel requirements in each cycle are reduced to $C_{ik}/(N-K)$. We make the following observations.

- The cells considered in each cycle form a linear array since the additional constraints from the endpoints are absent. Hence the algorithm of the previous section can be used to assign, in each cycle, $C_{ik}/(N-K)$ frequencies to zone (i, k) using a total of $N^*/(N-K)$ channels.

- Each zone (i, k) is allocated channels in $N-K$ cycles out of the N cycles. Hence after N cycles, zone (i, k) gets its required C_{ik} channels.

- After N cycles, a total of $\frac{N}{N-K}N^*$ channels are needed

The fractional frequencies can be handled naturally. If zone (i,k) requires $x.y$ frequencies in a cycle, it is assigned x frequencies in all but the last cycle and $x+1$ frequencies in the last cycle.

Appendix: Linear reuse partitioning networks may not correspond to interval graphs

In this appendix, we show that the interference graph corresponding to the linear array with 3 zones per cell is not an interval graph.

Suppose the interference graph is indeed an interval graph. Therefore, there exists an interval graph representation. Let I_{ik} denote the interval on the real line associated with the node of the interference graph corresponding to zone (i,k). Also let $L(I_{ij})$ (respectively $R(I_{ij})$) denote the left (respectively right) end-point of the interval I_{ij}.

We first consider the case where interval I_{11} is on the left of interval I_{21}, that is $R(I_{11}) < L(I_{21})$. Since I_{12} intersects both I_{11} and I_{21}, we have $L(I_{12}) \leq R(I_{11}) \leq L(I_{21}) \leq R(I_{12})$ as shown in Figure 3(a). Since I_{31} does not intersect any one of I_{11}, I_{21} and I_{12}, it must be that interval I_{31} is either on the right or left of I_{11}, I_{21} and I_{31}. In other words, either $L(I_{31}) > \max\{R(I_{21}), R(I_{12})\}$ as shown in Figure 3(a) or $R(I_{31}) < \min\{L(I_{12}), L(I_{11})\}$ as shown in Figure 3(b). Accordingly consider the following two subcases.

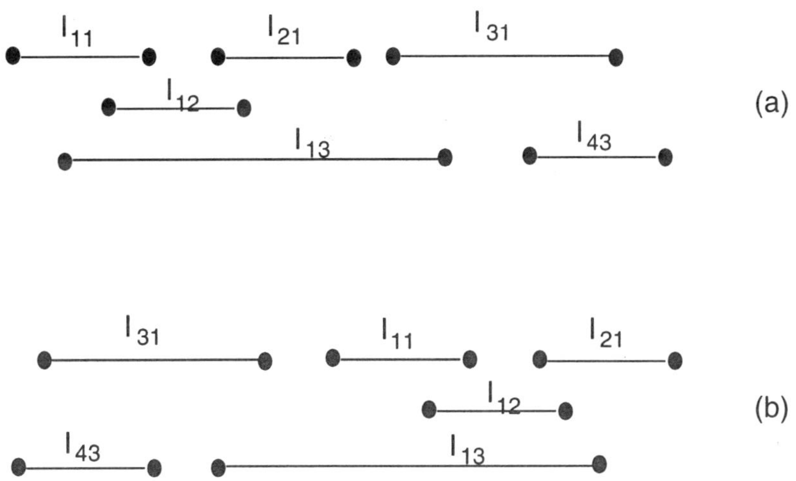

Figure 3 Illustration of the two cases

Case 1: I_{31} is on the right of I_{11}, I_{21} and I_{31}. We will show that I_{21} and I_{43} cannot intersect which is a contradiction. Since I_{31} is on the right of I_{21}, $R(I_{21}) < L(I_{31})$. Since I_{13} and I_{31} intersect, $L(I_{31}) \leq R(I_{13})$. Also, since I_{13} and I_{43} do not intersect and I_{31} and I_{43} intersect, we have $R(I_{13}) \leq L(I_{43})$. Combining the above three inequalities, we derive $R(I_{21}) < L(I_{43})$ which implies that I_{21} and I_{43} do not intersect, a contradiction. This case is illustrated in Figure 3(a).

Case 2: I_{31} is on the left of I_{11}, I_{21} and I_{31}. Arguments similar to that given in Case 1 can be given to show that I_{21} and I_{43} cannot intersect, a contradiction.

The discussion of the case where interval I_{11} is on the right of interval I_{21} is very similar and is skipped for brevity.

REFERENCES

[1] P. P. Bhattacharya, L. Georgiadis and A. Krishna, "Distributed channel allocation for PCN with bursty traffic", preprint. Also, *Proc. INFOCOM*, pp.113-120, 1995.

[2] M. Frodigh, " Reuse partitioning combined with traffic adaptive channel assignment for highway microcellular systems", *Proc. GLOBECOM*, pp.1414–1418, 1992.

[3] M. Grotschel, L. Lovasz and A. Schrijver, *Geometric Algorithms and Combinatorial Optimization*, Springer Verlag, 1985.

[4] S. W. Halpern, "Reuse partitioning in cellular systems", *Proc. Veh. Tech. Conf.* , pp. 322–327, 1983.

[5] S. Papavassiliou, L. Tassiulas, and P. Tandon, "Meeting QOS requirements in a cellular network with reuse partitioning", *IEEE Journal on Selected Areas in Communications*, 12(8):1389–1400, October 1994.

[6] K. Sallberg, B. Stavenow, and B. Eklundh, " Hybrid channel assignment and reuse partitioning in a cellular mobile telephone system", *Proc. Veh. Tech. Conf.* , pp.405–411, 1987.

[7] J. Zander, "Generalized reuse partitioning in cellular mobile radio", *Proc. Veh. Tech. Conf.*, pp. 181–184, 1993.

[8] J. Zander and M. Frodigh, "Capacity allocation and channel assignment in cellular radio systems using reuse partitioning", *Electronics Letters*, 28(5):438–440, February 1992.

δ-ADJUSTED MTH ORDER MULTIUSER DETECTOR

Lin-Lang Yang and Robert A. Scholtz*

Qualcomm
6455 Lusk Boulevard
San Diego, CA 92121-2779
USA

**University of Southern California*
Los Angeles, CA 90089-2565
USA

ABSTRACT

The δ-adjusted M-th order multiuser detector employs a new approximation algorithm to decorrelate the received signals from different users. This proposed algorithm can completely detect the received signal from each user at asymptotically high signal-to-interference ratios without error if the signals' cross-correlation matrix is available at the receiver site.

1 INTRODUCTION

Code Division Multiple Access(CDMA) has emerged as a key technology for communication services. Its application has been demonstrated in a number of systems in use. For example, the CDMA cellular system [1], its PCS derivative, and the Globalstar LEO satellite system. The advantages of CDMA communication systems include high capacity, high resolution time-of-arrival for positioning measurement, low probability of interception, and independent interference rejection. It is one of the most promising candidates for multiuser communications.

The problem of conventional multiuser detectors in CDMA systems is that they ignore all the cross-correlation information among users in signal detection, which results in worse error performance and reduced capacity. The capacity loss is most significant when the processing gain of the system is high. To overcome the above problem, a new multiuser detector which employs the cross-correlation information to achieve much higher system capacity is proposed here. A practical implementation of the new multiuser detector is also proposed.

Verdú initiated the idea of the optimal multiuser detector [2]. Several sub-optimal multiuser detectors have been introduced ever since [3] [4], aiming at reducing the complexity of Verdú's algorithm. Some other studies of multiuser detectors were presented in [5], [6]. The focus of [5] is on decentralized sequence detection as compared to Verdú's centralized detection algorithms. On the other hand, [6] is based on a multiuser detection strategy of successive multiple-access interference rejection.

The scheme proposed here is an attempt to solve the complexity problem encountered in Verdú's multiuser detection scheme. We will proceed with the presentation of this topic as follows: (i) present the new sampling algorithm which is different from the conventional ones; (ii) propose a new alternative multiuser detector which provides much higher capacity than the conventional ones and has a lower complexity requirement than Verdú's algorithm; and (iii) present simulation results.

2 PROPOSED SAMPLE SCHEME

For an asynchronous direct sequence spread-spectrum multiple access system with K users operating in an additive white Gaussian noise (AWGN) environment, the received signal, $\mathbf{r}(t)$, at the receiver is the sum of K spread-spectrum signals $s_k(t-\tau_k,\, \mathbf{b})$, $1 \leq k \leq K$, plus an AWGN random process, $\mathbf{n}(t)$, with two-sided power spectrum density $\frac{N_0}{2}$:

$$\mathbf{r}(t) = \mathbf{s}(t,\, \mathbf{b}) + \mathbf{n}(t), \qquad t \in \mathbf{R} \tag{1.1}$$

where

$$\mathbf{s}(t,\, \mathbf{b}) = \sum_{k=1}^{K} s_k(t-\tau_k,\, \mathbf{b}) = \sum_{i=0}^{M}\sum_{k=1}^{K} b_k(i)\, s_k(t-iT-\tau_k) \tag{1.2}$$

and

$$s_k(t) = \sqrt{2w_k}\, a_k(t)\, \cos(\omega_c t + \theta_k). \tag{1.3}$$

In the above equations, T denotes the symbol length which is the same for each user, $s_k(t)$ is the waveform transmitted by the kth user during a T-second

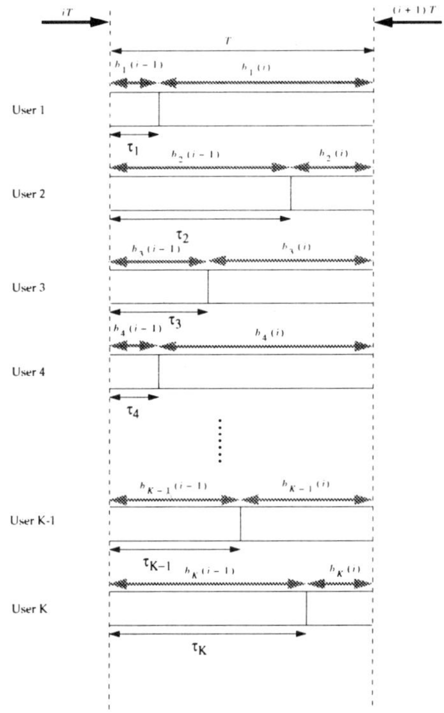

Figure 1 The delay of each user and the corresponding transmitted data symbols during observation time iT to $(i+1)T$.

interval, $a_k(t)$ is the kth user's spectral-spreading signal (normalized to have unit power), τ_k is the time delay of the kth user, θ_k is the carrier phase of the kth user, ω_c is the common carrier frequency, w_k is the power of the kth user which is assumed constant for each user during the observation, $b_k(i)$ is the ith transmitted data symbol of the kth user with $b_k(i) \in \{-1, 1\}$ for all k and i, and $(M+1)T$ is the message length for each user. Figure 1 illustrates the different delay, τ_k, and the corresponding transmitted data symbols during the observation time iT to $(i+1)T$ for each user.

When the carrier frequency, the carrier phase, the PN sequence epoch and the symbol time for each user separately are synchronized, the received signal $\mathbf{r}(t)$ is despread and demodulated by the local PN sequence and the local oscillator

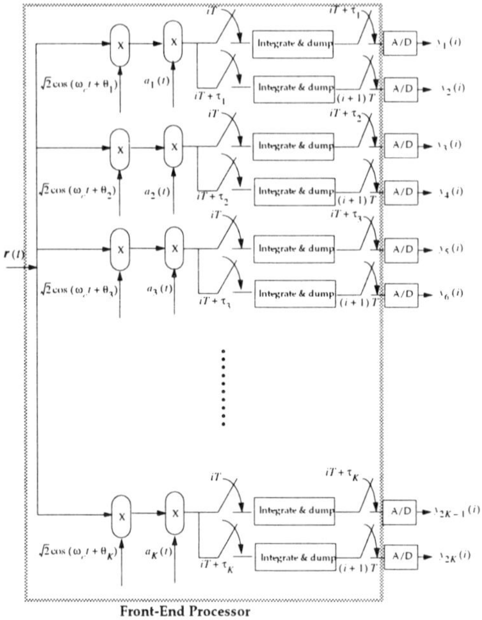

Figure 2 The proposed sampling scheme.

for each user. The K demodulated outputs are then integrated, sampled, and converted to digital signals for post-end signal processing.

Here, a new sampling method is proposed. We propose that the receiver for each user's signal consist of two integrators. The first integrator for user k integrates from time iT to time $iT + \tau_k$ and outputs value $y_{2k-1}(i)$. The second integrator for user k integrates from time $iT + \tau_k$ to time $(i+1)T$ and outputs value $y_{2k}(i)$. Figure 2 illustrates the proposed scheme.

Define vector $\mathbf{y}(i)$ as

$$\mathbf{y}(i) \triangleq \begin{bmatrix} y_1(i) \\ y_2(i) \\ y_3(i) \\ y_4(i) \\ \vdots \\ y_{2K-1}(i) \\ y_{2K}(i) \end{bmatrix}. \tag{1.4}$$

Let vector $\mathbf{z}(i)$ represent the received data corresponding to vector $\mathbf{y}(i)$ with

$$\mathbf{z}(i) \triangleq \begin{bmatrix} \sqrt{w_1}b_1(i-1) \\ \sqrt{w_1}b_1(i) \\ \sqrt{w_2}b_2(i-1) \\ \sqrt{w_2}b_2(i) \\ \vdots \\ \sqrt{w_K}b_K(i-1) \\ \sqrt{w_K}b_K(i) \end{bmatrix}. \tag{1.5}$$

Similarly, let vector $\mathbf{a}(t,i)$ represent the waveform of the received PN sequence corresponding to $\mathbf{y}(i)$ with

$$\mathbf{a}(t,i) \triangleq \begin{bmatrix} a_1(t,i) \\ a_2(t,i) \\ \vdots \\ a_{2K-1}(t,i) \\ a_{2K}(t,i) \end{bmatrix}, \tag{1.6}$$

where

$$\begin{aligned} a_{2k-1}(t,i) &= \begin{cases} a_k(t) & \text{if } iT \le t < iT + \tau_k, \\ 0 & \text{otherwise;} \end{cases} \\ a_{2k}(t,i) &= \begin{cases} a_k(t) & \text{if } iT + \tau_k \le t < (i+1)T, \\ 0 & \text{otherwise.} \end{cases} \end{aligned} \tag{1.7}$$

Denote the correlation between $a_k(t,i)$ and $a_l(t,i)$ as

$$a_k(t,i) \bullet a_l(t,i) \triangleq \int_{iT}^{(i+1)T} a_k(t,i) a_l(t,i)\, dt. \tag{1.8}$$

In addition, let matrix $\mathbf{C}(i) = [c_{kl}(i)]$ represent the correlation matrix of the received PN sequence waveform as given by

$$\mathbf{C}(i) = \mathbf{a}(t,i) \bullet \mathbf{a}'(t,i) \tag{1.9}$$

where $\mathbf{a}'$ denotes the transpose of $\mathbf{a}$.

Ignoring the additive white Gaussian noise term, $\mathbf{n}(t)$, the sampled digital output $y_{2k-1}(i)$ in Figure 2 can be expressed by

$$\begin{aligned}
&y_{2k-1}(i) \\
= \ & \sqrt{w_k} b_k(i-1)\frac{\tau_k}{T} + \sum_{l=1, l\neq k}^{K} \sqrt{w_l} b_l(i-1)\cos(\theta_k - \theta_l) c_{(2k-1)(2l-1)}(i) \\
&+ \sum_{l=1, l\neq k}^{K} \sqrt{w_l} b_l(i)\cos(\theta_k - \theta_l) c_{(2k-1)(2l)}(i).
\end{aligned} \tag{1.10}$$

Similarly, the sampled output $y_{2k}(i)$ can be represented as

$$\begin{aligned}
y_{2k}(i) &= \sqrt{w_k} b_k(i)\left(1 - \frac{\tau_k}{T}\right) \\
&+ \sum_{l=1, l\neq k}^{K} \sqrt{w_l} b_l(i-1)\cos(\theta_k - \theta_l) c_{(2k)(2l-1)}(i) \\
&+ \sum_{l=1. l\neq k}^{K} b_l(i)\cos(\theta_k - \theta_l) c_{(2k)(2l)}(i).
\end{aligned} \tag{1.11}$$

Therefore, the sampled digital output vector $\mathbf{y}(i)$ relates to the received data vector $\mathbf{z}(i)$ by

$$\mathbf{R}(i)\mathbf{z}(i) = \mathbf{y}(i), \tag{1.12}$$

with $\mathbf{R}(i) = [r_{kl}(i)]$ given by

$$\begin{aligned}
r_{(2k-1)(2k-1)}(i) &= \frac{\tau_k}{T}, \\
r_{(2k)(2k)}(i) &= 1 - \frac{\tau_k}{T}, \\
r_{(2k-1)(2k)}(i) &= 0, \\
r_{(2k)(2k-1)}(i) &= 0, \\
r_{(2k-1)(2l-1)}(i) &= \cos(\theta_k - \theta_l) c_{(2k-1)(2l-1)}(i), \\
r_{(2k-1)(2l)}(i) &= \cos(\theta_k - \theta_l) c_{(2k-1)(2l)}(i), \\
r_{(2k)(2l-1)}(i) &= \cos(\theta_k - \theta_l) c_{(2k)(2l-1)}(i), \\
r_{(2k)(2l)}(i) &= \cos(\theta_k - \theta_l) c_{(2k)(2l)}(i), \\
r_{kl}(i) &= r_{kl}(i).
\end{aligned} \tag{1.13}$$

Note that $r_{ll}(i)$ is independent of the data symbol index i, that is, $r_{ll}(i)$ is time invariant. The derivations of $\mathbf{C}(i)$ and $\cos(\theta_k - \theta_l)$ at receiver are shown in [10]. In addition, the mathematical background is also illustrated in [10].

3 PROPOSED MULTIUSER DETECTOR

Firstly, let the matrix $\mathbf{R}$ in Equation 1.12 be decomposed into two parts: a matrix $\mathbf{D} = [d_{ll}]$ containing only the diagonal components of the matrix $\mathbf{R}$, and a matrix $\mathbf{O}_f$ containing only the off-diagonal components of the matrix $\mathbf{R}$. That is,

$$\mathbf{R} = \mathbf{D} + \mathbf{O}_f. \tag{1.14}$$

The result in Equation 1.13 shows that

$$\begin{aligned} d_{(2k-1)(2k-1)} &= r_{(2k-1)(2k-1)} = \frac{\tau_k}{T} \\ d_{(2k)(2k)} &= r_{(2k)(2k)} = 1 - \frac{\tau_k}{T}. \end{aligned} \tag{1.15}$$

Also, define $\mathbf{D}_\delta$, $\mathbf{O}_\delta$ and $\mathbf{F}_\delta$ as

$$\begin{aligned} \mathbf{D}_\delta &= (1+\delta)\mathbf{D}, \\ \mathbf{O}_\delta &= \mathbf{O}_f - \delta\mathbf{D}, \\ \mathbf{F}_\delta &= \mathbf{D}_\delta^{-1}\mathbf{O}_\delta, \end{aligned} \tag{1.16}$$

where δ is an arbitrary positive number. Then the matrix $\mathbf{R}$ can be decomposed in terms of the matrix $\mathbf{D}_\delta$ and the matrix $\mathbf{F}_\delta$ as

$$\begin{aligned} \mathbf{R} &= \mathbf{D} + \mathbf{O}_f \\ &= (1+\delta)\mathbf{D}\left(\mathbf{I} + \frac{1}{1+\delta}\mathbf{D}^{-1}\mathbf{O}_f\right) - \delta\mathbf{D} \\ &= (1+\delta)\mathbf{D}\left(\mathbf{I} - \frac{\delta}{1+\delta}\mathbf{I} + \frac{1}{1+\delta}\mathbf{D}^{-1}\mathbf{O}_f\right) \\ &= (1+\delta)\mathbf{D}\left(\mathbf{I} + \frac{1}{1+\delta}\mathbf{D}^{-1}(\mathbf{O}_f - \delta\mathbf{D})\right) \\ &= \mathbf{D}_\delta(\mathbf{I} + \mathbf{D}_\delta^{-1}\mathbf{O}_\delta) \\ &= \mathbf{D}_\delta(\mathbf{I} + \mathbf{F}_\delta), \end{aligned} \tag{1.17}$$

where $\mathbf{I}$ is the identity matrix. The reasons that a non-zero δ being used are: i) the matrix $\mathbf{F}_0$ is not necessarily a convergent matrix; and ii) with proper choice of δ will make the convergent speed faster.

Secondly, Equation 1.12 can be rewritten as

$$\mathbf{D}_\delta(\mathbf{I}+\mathbf{F}_\delta)\mathbf{z} \;=\; \mathbf{y}. \tag{1.18}$$

Define a matrix $\mathbf{H}^{(M)}$ and a vector $\hat{\mathbf{z}}^{(M)} = [\hat{z}_l^{(M)}]$ as

$$\begin{aligned}\mathbf{H}^{(M)} &\triangleq \left(\sum_{m=0}^{M}(-1)^m\mathbf{F}_\delta^m\right)\mathbf{D}_\delta^{-1},\\ \hat{\mathbf{z}}^{(M)} &\triangleq \mathbf{H}^{(M)}\mathbf{y}.\end{aligned} \tag{1.19}$$

Multiplying both sides of Equation 1.18 by $\mathbf{H}^{(M)}$, we get the relation between the estimate, $\hat{\mathbf{z}}^{(M)}$, and the vector, $\mathbf{z}$, as follows.

$$\begin{aligned}\hat{\mathbf{z}}^{(M)} &= \mathbf{H}^{(M)}\mathbf{y}\\ &= \left(\sum_{m=0}^{M}(-1)^m\mathbf{F}_\delta^m\right)\mathbf{D}_\delta^{-1}\mathbf{D}_\delta(\mathbf{I}+\mathbf{F}_\delta)\mathbf{z}\\ &= \left(\mathbf{I}-(-1)^{M+1}\mathbf{F}_\delta^{M+1}\right)\mathbf{z}\\ &= \mathbf{z}-(-1)^{M+1}\mathbf{F}_\delta^{M+1}\mathbf{z}.\end{aligned} \tag{1.20}$$

Using Equation 1.5, the estimate $\hat{b}_k^{(M)}(i)$ of the received data $b_k(i)$ can be determined by either $\hat{z}_{2k}^{(M)}(i)$ or $\hat{z}_{2k-1}^{(M)}(i+1)$. In the case of $w_k = w$ for all k, we propose that the estimate $\hat{z}_{2k}^{(M)}(i)$ be weighted by $\frac{T-\tau_k}{T}$, while the estimate $\hat{z}_{2k-1}^{(M)}(i+1)$ be weighted by $\frac{\tau_k}{T}$. Therefore, the received data estimate, $b_k(i)$, for user k of the ith symbol, can be expressed as

$$\begin{aligned}\hat{b}_k^{(M)}(i) &= \frac{T-\tau_k}{T}\frac{1}{\sqrt{w}}\hat{z}_{2k}^{(M)}(i)+\frac{\tau_k}{T}\frac{1}{\sqrt{w}}\hat{z}_{2k-1}^{(M)}(i+1)\\ &\triangleq b_k(i)+\mathbf{n}_{b_k}^{(M)}(i),\end{aligned} \tag{1.21}$$

where the estimation noise $\mathbf{n}_{b_k}^{(M)}$ is defined as a function of $\mathbf{F}_\delta^m(i) = [f_{lj}^{(m)}(i)]$ by

$$\begin{aligned}\mathbf{n}_{b_k}^{(M)}(i) \triangleq -(-1)^{M+1}&\left(\frac{T-\tau_k}{T}\sum_{j=1}^{K} f_{(2k)(2j-1)}^{(M+1)}(i)b_j(i-1)\right.\\ &+\sum_{j=1}^{K}\left(\frac{T-\tau_k}{T}f_{(2k)(2j)}^{(M+1)}(i)+\frac{\tau_k}{T}f_{(2k-1)(2j-1)}^{(M+1)}(i+1)\right)b_j(i)\\ &\left.+\frac{\tau_k}{T}\sum_{j=1}^{K} f_{(2k-1)(2j)}^{(M+1)}(i+1)b_j(i+1)\right).\end{aligned} \tag{1.22}$$

Thirdly, with the relation shown in Equation 1.21, the goal is to find proper choice of δ such that

$$\mathbf{E}\{(\mathbf{n}_{b_k}^{(M+1)}(i))^2\} < \mathbf{E}\{(\mathbf{n}_{b_k}^{(M)}(i))^2\} \qquad \forall k, i, M, \tag{1.23}$$

with fast convergent speed. In the above Equation, the $\mathbf{E}\{\}$ denotes the expected value operator. Equation 1.23 implies that the estimation noise in Equation 1.22 contributes less variance as M increases. The details of choosing the optimal δ are shown in [10], and the result is as follows.

$$\delta_{opt} \approx \frac{K}{N}\left(1 + \frac{1}{3}\bar{g}\right) \tag{1.24}$$

where $\bar{g}$ is function of symbol delay τ_i, and is ranged through 6 to 8. The details of deriving $\bar{g}$ are given in [10].

Finally, because Equation 1.21 can also be rewritten as

$$\begin{aligned}
\hat{b}_k^{(M)}(i) &= \frac{T-\tau_k}{T}\frac{1}{\sqrt{w}}\hat{z}_{2k}^{(M)}(i) + \frac{\tau_k}{T}\frac{1}{\sqrt{w}}\hat{z}_{2k-1}^{(M)}(i+1) \\
&= \frac{(1+\delta)(T-\tau_k)}{(1+\delta)T}\frac{1}{\sqrt{w}}\hat{z}_{2k}^{(M)}(i) + \frac{(1+\delta)\tau_k}{(1+\delta)T}\frac{1}{\sqrt{w}}\hat{z}_{2k-1}^{(M)}(i+1) \\
&= \frac{1}{\sqrt{w}(1+\delta)}\left(\hat{z}_{2k}^{(M),wt}(i) + \hat{z}_{2k-1}^{(M),wt}(i+1)\right),
\end{aligned} \tag{1.25}$$

where $\hat{\mathbf{z}}^{(M),wt}(i) = [\hat{z}_l^{(M),wt}(i)]$ is given as

$$\hat{\mathbf{z}}^{(M),wt}(i) = \mathbf{D}_\delta(i)\mathbf{H}^{(M)}(i)\mathbf{y}(i). \tag{1.26}$$

Since only binary modulation is considered, Equation 1.25 indicates that the proposed system can be implemented as follows.

1. Compute the vector $\hat{\mathbf{z}}^{(M),wt}(i) = [z_l^{(M),wt}(i)]$ as

$$\begin{aligned}
\hat{\mathbf{z}}^{(M),wt}(i) &= \mathbf{D}_\delta(i)\mathbf{H}^{(M)}(i)\mathbf{y}(i) \\
&= \left(\sum_{m=0}^{M}(-1)^m(\mathbf{O}_\delta(i)\mathbf{D}_\delta^{-1}(i))^m\right)\mathbf{y}(i) \\
&\triangleq \sum_{m=0}^{M}(-1)^m\mathbf{x}^{(m)} \\
&\triangleq \mathbf{s}^{(M)},
\end{aligned} \tag{1.27}$$

where $\mathbf{x}^{(M)}$ and $\mathbf{s}^{(M)}$ are defined to be

$$\begin{aligned} \mathbf{x}^{(M)} &= \left(\mathbf{O}_\delta \mathbf{D}_\delta^{-1}\right)^M \mathbf{y}, \\ \mathbf{s}^{(M)} &= \sum_{m=0}^{M} (-1)^m \mathbf{x}^{(m)}. \end{aligned} \tag{1.28}$$

2. If $\hat{z}_{2k}^{(M),wt}(i) + \hat{z}_{2k-1}^{(M),wt}(i+1) > 0$, decide $\hat{b}_k^{(M)}(i)$ to be +1; otherwise, decide $\hat{b}_k^{(M)}(i)$ to be -1.

Note the special structure that Equation 1.27 show. The recursion of $\mathbf{x}^{(M)}$ and $\mathbf{s}^{(M)}$ can be easily established as

$$\begin{aligned} \mathbf{x}^{(M+1)} &= \mathbf{O}_\delta \mathbf{D}_\delta^{-1} \mathbf{x}^{(M)}, \\ \mathbf{s}^{(M+1)} &= \mathbf{s}^{(M)} + (-1)^{M+1} \mathbf{x}^{(M+1)}, \end{aligned} \tag{1.29}$$

with $\mathbf{x}^{(0)} = \mathbf{y}$ and $\mathbf{s}^{(0)} = \mathbf{y}$. The estimate, $\hat{\mathbf{z}}^{(M),wt}$, can be extracted from the last stage of $\mathbf{s}^{(M)}$. This shows that the proposed system can be implemented in a neatly recursive manner.

To summarize, the proposed multiuser detector is illustrated in Figure 3 where DB, APB and CCB stand for decision block, approximation processor block and correlation computation block. The function of each block is demonstrated in Figure 4, Figure 5 and Figure 6, respectively.

4 SIMULATION RESULTS

Figure 7 shows the symbol error rate of each user for the proposed δ-adjusted Mth order multiuser detector with M equal to 0, 2, and 4. In this simulation, the number of users in the system, K, is set to 40, and the symbol processing gain, N, is 42. From Equation 1.24, with $\bar{g}$ equals 6.9, the optimal value of δ is given by $\delta_{opt} = \frac{K}{N}(1 + \frac{1}{3}\bar{g}) = 3.143$.

Each users' PN sequence is generated by using the following generator function,

$$\begin{aligned} g(x) = {} & x^{42} + x^{35} + x^{33} + x^{31} + x^{27} + x^{26} + x^{25} + x^{22} + x^{21} + x^{19} + x^{18} \\ & + x^{17} + x^{16} + x^{10} + x^{7} + x^{6} + x^{5} + x^{3} + x^{2} + x + 1. \end{aligned} \tag{1.30}$$

The symbol delay, τ_k, the received carrier phase, θ_k, and the transmitted data, $b_k(i)$, for all users are randomly generated. We obtain the input $\mathbf{y}$ to the

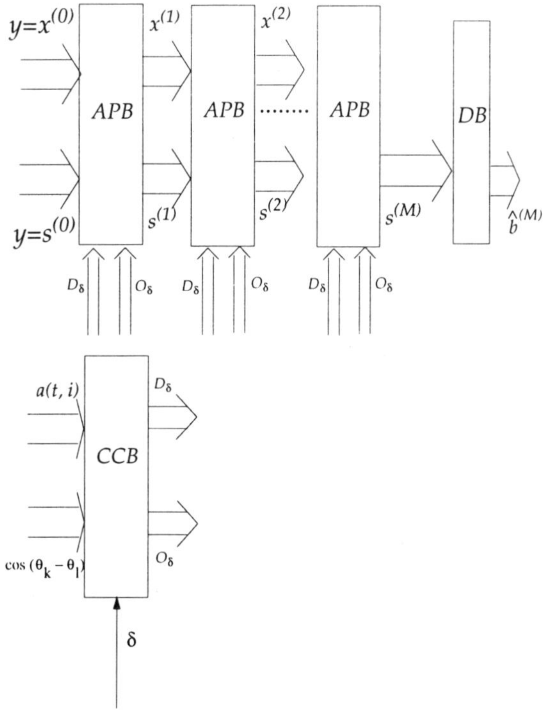

Figure 3 The δ-adjusted Mth order multiuser detector.

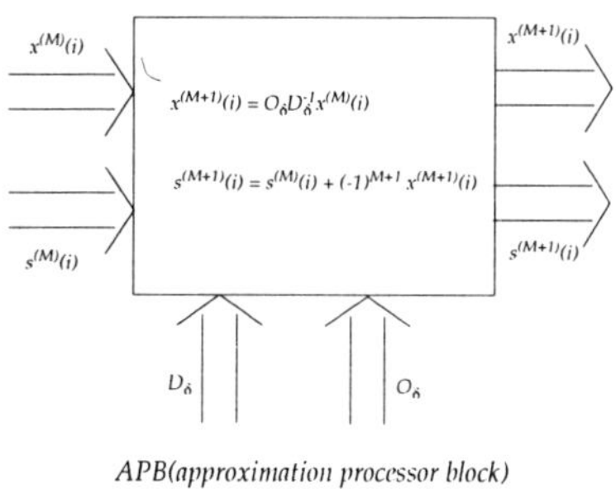

Figure 4 The approximation processor block.

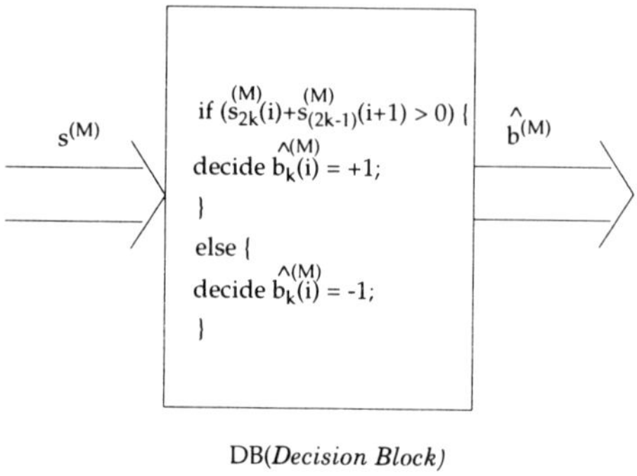

Figure 5 The decision block.

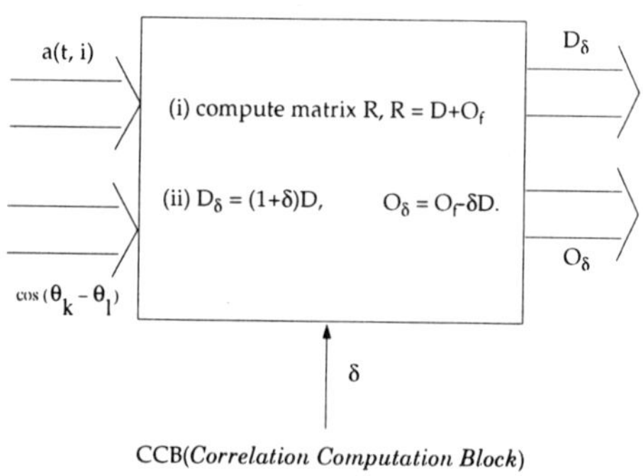

Figure 6 The correlation computation block.

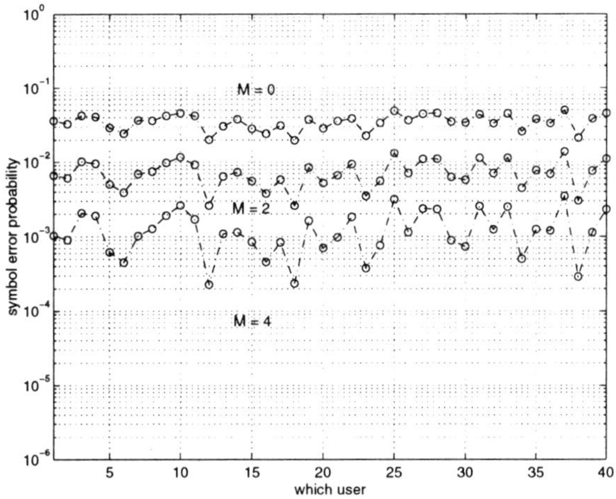

Figure 7 Symbol error probability for each user.

proposed system in Figure 3 using Equation 1.12, then compare the estimated data, $\hat{b}_k(i)$, and the transmitted data, $b_k(i)$, to derive the symbol error rate. Each point in Figure 7 is generated with 10^6 simulated symbols.

Figure 8 illustrates the average symbol error rate of all users against $\frac{3N}{K}$ in dB for $M = 0, 2, 4$, and 6. The parameters, except K and δ_{opt}, are set to the same as in the first simulation. Again, each point is generated with 10^6 simulated symbols.

Note from Figure 8 that the system capacity increases as M increases. The system capacity is defined as the number of users that can be accommodated in a system with the average symbol error rate below some threshold value. For example, if the performance target is set to keep the average symbol error rate below 10^{-2}, then Figure 8 shows that there is a capacity gain of about 2.5 dB for $M = 2$ as compared to $M = 0$.

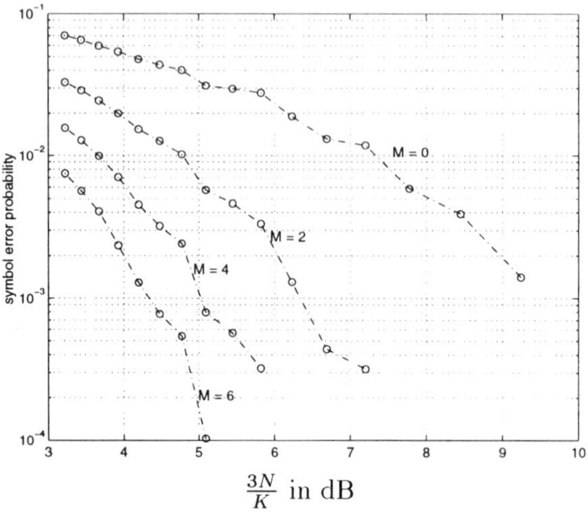

Figure 8 Average symbol error probability of all users versus $\frac{3N}{K}$ in dB.

5 CONCLUSIONS

In this paper, we proposed a system which has less complexity than Verdú's algorithm while achieving the same system performance asymptotically as M increases. In addition, we carefully investigated the noise variance generated by each user to determine the adjusting parameter δ such that the noise variance of the system is guaranteed to converge as M increases. Finally, by making use of the special structure of the estimation matrix $H^{(M)}$, a recursive implementation of the proposed system was presented.

The main contribution of this work lies in the feasible implementation of the proposed real-time optimal multiuser detector with today's digital technology. Furthermore, the proposed systems can be used to mitigate the interference-limited problem that is encountered in the multiuser detection. By mitigating the problem, it implies that the system capacity can be increased to support more subscribers. Finally, for the further research, more real environment needs to be considered. For example, Rayleigh fading and multipath mitigation have not been studied in this paper.

REFERENCES

[1] K. S. Gilhousen, I. M. Jacobs, R. Padovani, A. J. Viterbi, L. A. Weaver, Jr. and C. E. Wheatley III, "On the capacity of a cellular CDMA system," *IEEE Trans. Veh. Technol.*, vol. VT-40, No. 2, pp. 303-312, May 1991.

[2] S. Verdú, "Minimum probability of error for asynchronous Gaussian multiple access channels," *IEEE Trans. Inform. Theory*, vol. IT-32, No. 1, pp. 85-96, January 1986.

[3] Z. Xie, C. K. Rushforth and R. T. Short, "Multiuser signal detection using sequential decoding," *IEEE Trans. Comm.*, vol. 38, No. 5, pp. 578-583, May 1990.

[4] R. Lupas and S. Verdú, "Near-far resistance of multiuser detectors in asynchronous channels," *IEEE Trans. Comm.*, vol. 38, No. 4, April 1990.

[5] H. V. Poor and S. Verdú, "Single-user detectors for multiuser channels," *IEEE Trans. Comm.*, vol. COM-36, pp.50-60, January 1988.

[6] M. K. Varanasi and B. Aazhang, "Multistage detection in asynchronous code-division multiple access communications," *IEEE Trans. Comm.*, vol. COM-38, No. 4, pp. 509-519, April 1990.

[7] G. Strang, *Linear Algebra and Its Application*, third edition, Harcourt Brace Jovanovich, 1988.

[8] A. Papoulis, *Probability, Random Variables, and Stochastic Processes*, second edition, McGraw-Hill, 1984.

[9] S. W. Golomb, R. E. Peile, and R. A. Scholtz, *Basic Concepts in Information Theory and Coding*, Plenum Press, 1994.

[10] L. L. Yang, *Digital Algorithms for Multiuser Detector, Decision-Directed Phase Tracking Loop, and A/D & D/A Converter*, Dissertation, University of Southern California, 1995.

NON-COHERENT DIVERSITY RECEIVERS FOR MOBILE AND PERSONAL SATELLITE COMMUNICATIONS*

Dimitrios Makrakis †
Department of Electrical Engineering
Lakehead University
955 Oliver Road
Thunder Bay, Ont., P7B 5E1
CANADA

Dimitrios P. Bouras
Panayotis T. Mathiopoulos
Department of Electrical Engineering
University of British Columbia
2356 Main Mall
Vancouver, British Columbia, V6T 1Z4
CANADA

Abstract

Improved non-coherent receivers for mobile and personal satellite communications are presented and evaluated. The channel is assumed as being corrupted by shadowing and multipath fading. The results demonstrate that the receivers are capable of reducing the error floors considerably and provide improvements higher than 12 dB. The technology is promising and a good candidate for use in the development of future wireless multimedia personal communication terminals.

*This research was supported by a Research Infrastructure Grant from Lakehead University, by the Natural Sciences and Engineering Research Council of Canada (NSERC) under Research Grant (Individual) OGP0184108 and by the Advanced Systems Institute of BC.

†Also Adjunct Professor with the Dept. of Electrical Eng. of the Technical University of Nova Scotia (TUNS), Halifax, Nova Scotia.

1 Introduction

The last few years are characterized by a spectacular interest in Wireless Personal Digital Communication Systems (WPDCS) [1], [2]. Today it is evident that wireless personal communications emerge as a key wide-sweeping technology with dramatic impact on our society.

Although the first generation of modern PWDCS systems offers improvements in the capacity of the network and the quality of service, it is unquestionable that in the future, higher capacities and power efficiencies will be needed to accommodate the increasing number of users, to support new applications such as video (teleconferencing, videophone) as well as to be able to enter new promising markets such as distance education, telemedicine etc. VHF/UHF and L-band (which are already congested) are not capable and adequate to achieve the new targets of the wireless personal communications industry. Presently, the Ka-band is considered to servicing personal communications. In addition to bandwidth availability, the small wavelength of the Ka-band frequencies (it is in the range of 1-1.5 cm) will allow the use of powerful micro-diversity techniques. Also, miniaturized electronically steerable phase array antennas (based on microstrip technology) can be used to design relatively inexpensive briefcase type and hand-held terminals. Use of micro-diversity will be considerably useful, considering the strong impairments of the Ka-band channels [3], [16], [17].

The relatively high implementational complexity of the conventional coherent receiver and its poor performance in the presence of fading and shadowing, discourage its use in the performance and cost sensitive new personal communications services. This increases the popularity of the non-coherent detection techniques which have already been used extensively in the first generation of WPDCS (i.e. the North American digital cellular radio, the spectrally efficient digital coded modems for mobile-satellite communications [5] etc.). Objective of the present work is to propose more powerful non-coherent systems with superior performance, able to combat the high degree of impairments associated with the Ka-band channel. We shall achieve it by expanding the non-coherent detection technology we proposed in the past [6] - [14] (and as early as 1986 [6]) so that it can be applicable with space diversity techniques.

The remainder of this paper is sectioned as follows. In section 2, the block diagram of the transmitter is presented. Section 3 discusses the model of the channel simulator. Section 4 describes the structures of the non-coherent receivers with diversity. Section 5 presents the performance evaluation results of these receivers in the Ka-band communications channel. Finally, section 6 concludes the present work.

2 Transmitter

The block diagram of the transmitter is presented in Fig. 1. It consists of a convolutional encoder (CON. ENC.), a signal mapper (SM), a differential encoder (DE) and a quadrature modulator. Input to the convolutional encoder is the sequence $\bar{A} = [\bar{a}_1^p, \bar{a}_2^p, \ldots, \bar{a}_{Z-1}^p]$ of the p-bit information words $\bar{a}_k^p = [a_k^1, a_k^2, \ldots\ldots, a_k^p]$. a_k^i ($a_k^i \in \{0,1\}, 1 \leq i \leq p$) are independent and equally probable. Z is an integer. The number of possible combinations $\bar{a}_k^p$ can take is equal to $M_p = 2^p$. Output of the convolutional encoder is the sequence $\bar{B} = [\bar{b}_1^q, \ldots, \bar{b}_{Z-1}^q]$ with $\bar{b}_k^q$ being the q-bit word $\bar{b}_k^q = [b_k^1, b_k^2, \ldots, b_k^q]$. b_k^ν are binary ($b_k^\nu \in \{0,1\}$, $1 \leq \nu \leq q$). The rate of the code equals p/q. $\bar{B}$ is the input to the signal mapper. For uncoded systems, $p = q$ and $\bar{a}_k^p = \bar{b}_k^p$.

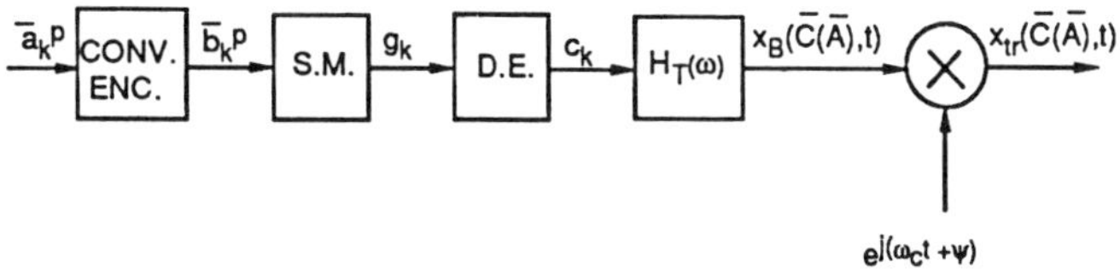

Figure 1: Block diagram of the transmitter.

The signal mapper converts the information word $\bar{b}_k^q$, to an M-ary PSK symbol $g_k = e^{j\phi_k}$, with

$$\phi_k = \frac{2\pi}{M} \sum_{i=1}^{q} (2b_k^i - 1)2^{i-1} \tag{1}$$

($M = 2^q$). After being differentially encoded, g_k produce the symbols c_k as follows:

$$c_k = e^{j\theta_k} = g_k c_{k-1} = e^{j(\phi_k \oplus \theta_{k-1})} \tag{2}$$

where θ_k represents the phase of the transmitted symbol c_k[1]. The relation between θ_k and ϕ_k is

$$\theta_k = \phi_k \oplus \theta_{k-1} \tag{3}$$

At this point, it is convenient to define the parameter $\Delta\Theta_l(k)$, it being equal to:

$$\Delta\Theta_l(k) = Arg[c_k c_{k-l}^*] = Arg[e^{j(\theta_k \ominus \theta_{k-l})}] = \theta_k \ominus \theta_{k-l} \tag{4}$$

[1]From now on, in order to distinguish between g_k and c_k, we shall be refering to them as information and transmitted symbols respectively.

The $\oplus$ and $\ominus$ represent summation and subtraction in the modulo 2π (mod (2π)) sense respectively (unless if otherwise indicated). Also, $Arg[\cdot]$ represents *argument of* $\{\cdot\}$. Eqs.(2), (3) give:

$$c_k c_{k-l}^* = e^{j(\Delta\Theta_l(k))} = \Pi_{i=0}^{l-1} g_{k-i} = e^{j(\phi_k \oplus \phi_{k-1} \oplus \ldots\ldots \oplus \phi_{k-l+1})} \tag{5}$$

and

$$\Delta\Theta_l(k) = \phi_k \oplus \phi_{k-1} \oplus \ldots. \oplus \phi_{k-l+1} \tag{6}$$

The transmitted signal is given by:

$$x_{tr}(\bar{C}(\bar{A}), t) = x_B(\bar{C}(\bar{A}), t) e^{j(\omega_c t + \psi)} = \left(\sum_{k=0}^{Z-1} c_k h_T(t - kT)\right) e^{j(\omega_c t + \psi)} \tag{7}$$

where $x_B(\bar{C}(\bar{A}), t)$ is the baseband signal at the output of the premodulation low-pass filter $H_T(\omega)$. $h_T(t)$ represents the impulse response of $H_T(\omega)$. $\bar{C}(\bar{A}) = [c_0, c_1, \ldots, c_{Z-1}]$ is the sequence of transmitted symbols, which are generated by the information sequence $\bar{A}$, ω_c is the radian frequency of the carrier and ψ is the initial phase of the modulator. Since non-coherent detection is considered, ψ is unknown to the receiver and it is treated as random variable, uniformly distributed in the $(0, 2\pi)$ interval, i.e., $p(\psi) = \frac{1}{2\pi}$ for $0 \le \psi \le 2\pi$ and $p(\psi) = 0$ elsewhere.

3 Channel Model

We assume that the transmitted signal is received through L different channels (L-branch diversity).

The disturbances appearing in each diversity channel are additive white Gaussian noise (AWGN) multipath fading and shadowing. We assume that the multipath faded channel behaves as non-frequency selective (flat fading).

Depending on the nature of the application, the receiver might or might not have a direct line of sight (LOS) with the transmitter. In most applications (i.e. mobile-satellite) the LOS signal experiences shadowing. C. Loo [15] has found that the shadowing appearing in L-band mobile-satellite communication channels follows a log-normal distribution. He distinguishes three categories of shadowing, light, average and heavy. These three cases are reflected as different values in the average and standard deviation describing the log-normal probability density function (pdf).

The model of the software simulator used in the computer simulations is presented in Fig.2. The shadowing and fading are generated by white Gaussian random processes ($n_w^{sh}(t)$, $n_w^{f,I}(t)$, $n_w^{f,Q}(t)$) passing through lowpass filters ($H_{sh}(\omega)$ for the shadowing, $H_{f,I}(\omega)$, $H_{f,Q}(\omega)$ for the multipath fading). $f_{sh}(t)$ simulates the shadowing. $f_{mul}(t) = f_I(t) + jf_Q(t)$ simulates the multipath fading ($f_I(t)$, $f_Q(t)$ are the outputs of the $H_{f,I}(\omega)$ and $H_{f,Q}(\omega)$ respectively).

$f^l(t) = f_{sh}(t) + f_{mul}(t) = \rho^l_f(t)e^{j\phi^l_f(t)}$ is the overall fading (shadowing and multipath) which corrupts the transmitted signal traveling through the lth diversity channel ($1 \leq l \leq L$). $\rho^l_f(t)$, $\psi^l_F(t)$ are the envelope and phase fluctuations introduced (to the signal) by the fading process. The μ parameter appearing in Fig. 2 represents the mean of the log-normally distributed shadowing.

In order to guarantee that the fading processes applied on the individual diversity channels are independent from each other, we used different seeds in our random noise generators (that were producing the $n^{sh}_w(t)$, $n^{f,I}_w(t)$, $n^{f,Q}_w(t)$) for the different fading processes.

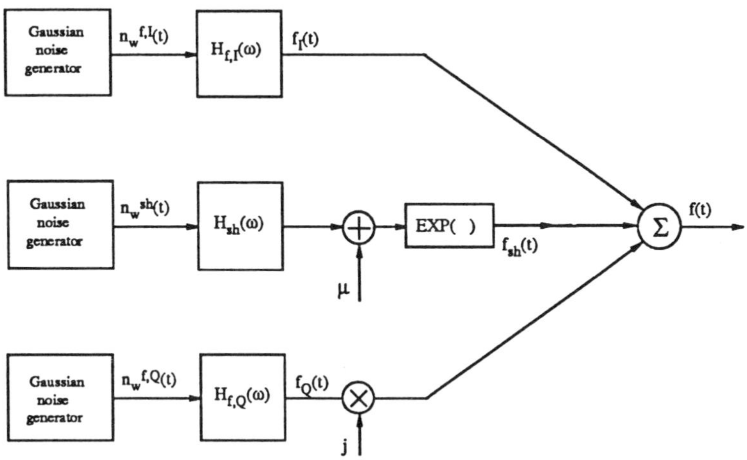

Figure 2: Model of the fading channel used in the simulations.

As mentioned in the introduction, our main interest lies in the case of Ka-band. In our simulations, we apply the effect of shadowing in the Ka-band according to a model we developed in the past [16], [17]. The model is based on measurements and the use of scaling techniques based on the Modified Exponential Decay (MED) model [19].

The signal reaching the receiver through the lth diversity channel ($1 \leq l \leq L$) can be expressed as

$$\begin{aligned} x^l_r(t) &= f^l(t)x_{tr}(\bar{C}(\bar{A}), t) + n^l_w(t) = f^l(t)x_B(\bar{C}(\bar{A}), t)e^{j(\omega_c t + \psi)} + n^l_w(t) \\ &= \rho^l_f(t)e^{j\psi^l_F(t)}x_B(\bar{C}(\bar{A}), t)e^{j(\omega_c t + \psi)} + n^l_w(t) \end{aligned} \tag{8}$$

where $n^l_w(t)$ is the additive white Gaussian noise corrupting the lth diversity channel.

4 Receiver Configurations

In [18], the analysis leading to the derivation of improved non-coherent sequence estimators with diversity is given. Due to the space constraints, in the present submission, we shall limit ourselves to the description of these schemes.

Conceptually, the receiver can be described as follows. It first demodulates the incoming signal from the ith $(1 \leq i \leq L)$ channel by multiplying it with $e^{-j(\omega_c t)}$ (notice that the demodulating signal is locked in frequency but not in phase with the incoming carrier). Afterwards, it filters the demodulated signal with a post detection filter $H_R^i(\omega)$, matched to the transmit filter $H_T(\omega)$ (i.e. $H_R^i(\omega)$ is a square root raised cosine filter as well). The output of $H_R^i(\omega)$, $y^i(t)$ is sampled. The $y_k^i(= y^i(kT))$ samples are passed to the decoder, together with the outputs of the other diversity channels.

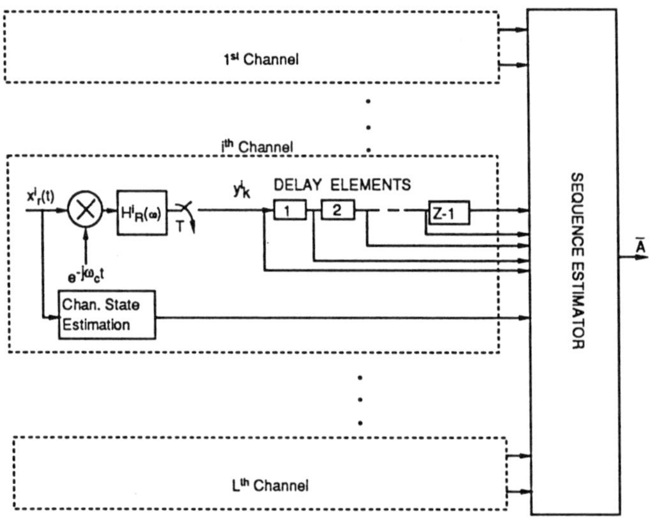

Figure 3: Generic Block Diagram of the Receiver.

The structure of the non-coherent decoder depends on the channel-state information available to the receiver. In the present work, three different configurations of non-coherent diversity receivers will be described.

4.1 Configuration 1

This receiver assumes knowledge of the the sampled amplitudes $\rho_F^i(kT)$ and phase $\psi_F^i(kT)$ of the fading processes (described by the $\rho_F^i(t)$ and $\psi_F^i(t)$ $(1 \leq i \leq L)$) which corrupt the L diversity channels. The optimal non-coherent sequence estimation metric corresponding to this case is the following:

$$\aleph^L_{OP}(\bar{y}^1,\bar{y}^2,\ldots,\bar{y}^L,\bar{C}(\bar{A})) = |\sum_{k=0}^{Z-1}(\sum_{i=1}^{L}\hat{\Re}^i_F(kT)y^i_k e^{-j\hat{\psi}^i_F(kT)})(c_k)^*| \tag{9}$$

where $\hat{\Re}^i_F(t)$, $\hat{\psi}^i_F(t)$ are the estimates of the fading gain $\rho^i_F(t)$ and phase $\hat{\psi}^i_F(t)$ respectively. Also, $\bar{y}^i = [y^i_0, y^i_1, \ldots, y^i_{Z-1}]$.

When the fading is slow enough to the point that it can be considered as time invariant in the time interval $\{0,(Z-1)T\}$ (during which the transmission takes place) the metric can be expressed as:

$$\aleph^L_{OP}(\bar{y}^1,\bar{y}^2,\ldots,\bar{y}^L,\bar{C}(\bar{A})) = |\sum_{k=0}^{Z-1}(\sum_{i=1}^{L}\hat{\Re}^i_F(0)e^{-j\hat{\psi}^i_F(0)}y^i_k)(c_k)^*| \tag{10}$$

Maximization of $\aleph^L_{OP}(\bar{y}^1,\bar{y}^2,\ldots,\bar{y}^L,\bar{C}(\bar{A}))$ is equivalent with maximization of its square. By carrying out this squaring process and by eliminating terms which do not contribute to the decision, we come to the following equivalent metric expression (to be maximized):

$$\begin{aligned}\aleph^L_{DDOP}(\bar{y}^1,\bar{y}^2,\ldots,\bar{y}^L,\bar{C}(\bar{A})) &= \sum_{k=0}^{Z-1}\sum_{l=1}^{k}[\sum_{i=1}^{L}\sum_{m=i}^{L}(\hat{\Re}^i_F(kT)\hat{\Re}^m_F((k-l)T))\\ [d^{i,m}_{l,I}(k)\cos(\hat{\Psi}^{i,m}_l(k) &+ d^{i,m}_{l,Q}(k)\sin(\hat{\Psi}^{i,m}_l(k))]\cos(\Delta\Theta_l(k))\\ +[d^{i,m}_{l,Q}(k)\cos(\hat{\Psi}^{i,m}_l(k)) &- d^{i,m}_{l,I}(k)\sin(\hat{\Psi}^{i,m}_l(k))]\sin(\Delta\Theta_l(k)))\end{aligned} \tag{11}$$

where $\hat{\Psi}^{i,m}_l(k) = \psi^i_F(kT) \ominus \psi^m_F((k-l)T)$. Also,

$$d^{i,m}_l(k) = y^i_k(y^m_{k-l})^*, \quad d^{i,m}_{l,I}(k) = Re\{d^{i,m}_l(k)\}, \quad d^{i,m}_{l,Q}(k) = Im\{d^{i,m}_l(k)\} \tag{12}$$

The term $d^{i,m}_l(k)$ is the product between the signal sample y^i_k received through the ith channel and the complex conjugate of the y^m_{k-l} sample, received lT seconds before through the mth diversity channel. This operation represents differential detection, with the differential detector having a delay element equal to lT. From Eq.(11) we realize that the non-coherent maximum likelihood sequence estimator processes the output of several differential detectors with delay elements integer multiples of the symbol period T. Notice that the structures we introduced in the past apply differential detection strictly on signals traveling through the same channel [6] - [14]. The receiver presented here applies differential detection between signals received through different channels as well. This represents a generalization of the multiple differential detection concept. We call this structure *Cross-Channel Multiple Differential Detection* (CC-MDD).

We mentioned earlier that according to the structure displayed in Fig. 3 the signals received through the L diversity channels have to be demodulated and low pass filtered prior to being processed by the decoder. In order to apply the demodulation processes, a replica of the carrier signal has to be generated locally at the receiver site, which adds complexity to the receiver. The use of the CC-MDD structure allows the development of a receiver which uses differential detectors implemented in IF form. By following an approach similar to the method described in [13], [10], the receiver can be built using only L squared envelope detectors and $(3L-2)$ differential detectors with delay elements equal to the symbol period T. The rest of the required signal parameters can be easily produced through some simple digital signal processing.

4.2 Configuration 2

For the second and third configurations, the receiver has knowledge of only the amplitudes $\rho^i_F(kT)$ $(1 \leq i \leq L)$ of the fading processes. In this case, two different sequence estimators will be presented. The sequence estimator examined in this subsection is asymptotically optimal for low E_b/N_o and its decoding metric is equal to

$$\aleph^L_{LAO}(\bar{y}^1, \bar{y}^2, \ldots, \bar{y}^L, \bar{C}(\bar{A})) = \sum_{i=1}^{L} | \sum_{k=0}^{Z-1} \hat{\Re}^i_F(kT) y^i_k c^*_k |^2 \tag{13}$$

When the channel is slow enough to be considered static over the period where the information is transmitted, the decision metric becomes:

$$\aleph^L_{LAO}(\bar{y}^1, \bar{y}^2, \ldots, \bar{y}^L, \bar{C}(\bar{A})) = \sum_{i=1}^{L} (\hat{\Re}^i_F(0))^2 | \sum_{k=0}^{Z-1} y^i_k c^*_k |^2 \tag{14}$$

Maximization of $\aleph^L_{LAO}(\bar{y}^1, \bar{y}^2, \ldots, \bar{y}^L, \bar{C}(\bar{A}))$ is equivalent with the maximization of its square value. After carrying out the squaring process and eliminating terms which are independent from the considered sequence $\bar{C}(\bar{A})$, we end up to the following metric expression (to be maximized):

$$\aleph^L_{DDLAO}(\bar{y}^1, \bar{y}^2, \ldots, \bar{y}^L, \bar{C}(\bar{A})) = \sum_{k=1}^{Z-1} \sum_{l=1}^{k} [(\sum_{i=1}^{L} (\hat{\Re}^i_F(kT) \hat{\Re}^i_F((k-l)T))$$

$$d^{i,i}_{l,I}(k)) \cos(\Delta\Theta_l(k)) + (\sum_{i=1}^{L} (\hat{\Re}^i_F(kT) \hat{\Re}^i_F((k-l)T)) d^{i,i}_{l,Q}(k)) \sin(\Delta\Theta_l(k))] \tag{15}$$

In this case, the multiple differential detection involves only signals received through the same diversity channel. The cross-channel differential detection terms have disappeared.

When channel information is absent (i.e. $\Re^i_F(kT)$ are unknown), all the $(\Re^i_F(kT))^2$ in Eqs. (13), (15) are set equal to 1.

4.3 Configuration 3

The third configuration is asymptotically optimal for high values of E_b/N_o. The metric expression of the sequence estimator is the following:

$$\aleph_{HAO}^{L}(\bar{y}^1, \bar{y}^2, \ldots, \bar{y}^L, \bar{C}(\bar{A})) = \sum_{i=1}^{L} |\sum_{k=0}^{Z-1} \hat{\Re}_F^i(kT) y_k^i c_k^*| \tag{16}$$

and for slow faded channels

$$\aleph_{HAO}^{L}(\bar{y}^1, \bar{y}^2, \ldots, \bar{y}^L, \bar{C}(\bar{A})) = \sum_{i=1}^{L} \hat{\Re}_F^i(0) |\sum_{k=0}^{Z-1} y_k^i c_k^*| \tag{17}$$

Again, when channel state information is not available, we set all the $\hat{\Re}_F^i(kT)$ equal to 1.

5 Performance Evaluation Results

We have evaluated uncoded and trellis coded $\frac{\pi}{4}$-shift QPSK [4] schemes over the Ka-band channel. We have assumed a carrier frequency around 21 GHz, vehicle speed of about 65 km/h and antenna pointing elevation of 20°[2]. We assume a symbol rate 24 KBaud. The resulting maximum Doppler frequency comes to about 1200 Hz (giving a $B_F T$ product for the diffused signal component equal to 0.05). The bandwidth of the time varying and lognormally distributed LOS component is approximately equal to 43 Hz.

Figs. 4, 5 present results for an uncoded $\frac{\pi}{4}$-shift QPSK signal and for the case of average shadowing. Fig. 4 displays BER curves corresponding to systems without diversity ($L = 1$). Figure 5 gives results for the diversity case ($L = 2$). The curves corresponding to a receiver without channel information are labeled as (Z=??, L=??, NCI). The curves corresponding to a system where both amplitude and phase channel information is available are identified as (Z=??, L=??, A&PCI). Curves corresponding to systems where amplitude information is available and the estimator uses the squared envelope metric (see Eq.(13)) are labeled as (Z=??, L=??, SACI). Finally, the curves corresponding to the sequence estimator based on Eq.(16) with fading amplitude information are labeled as (Z=??, L=??, ACI). Notice that the receiver with $Z = 2$ is identical to the differential detector.

From Fig. 4 we see that when diversity is not used and channel information is not available, the multiple differential receiver and the differential detector appear almost with the same performance. Also, both of them experience a high error floor. Even when amplitude channel information is available, the performance remains practically at the same level.

[2]An elevation angle of 20° makes the effective speed with respect to the satellite equal to 65km/h × cos(21°) which is approximately 61 km/h.

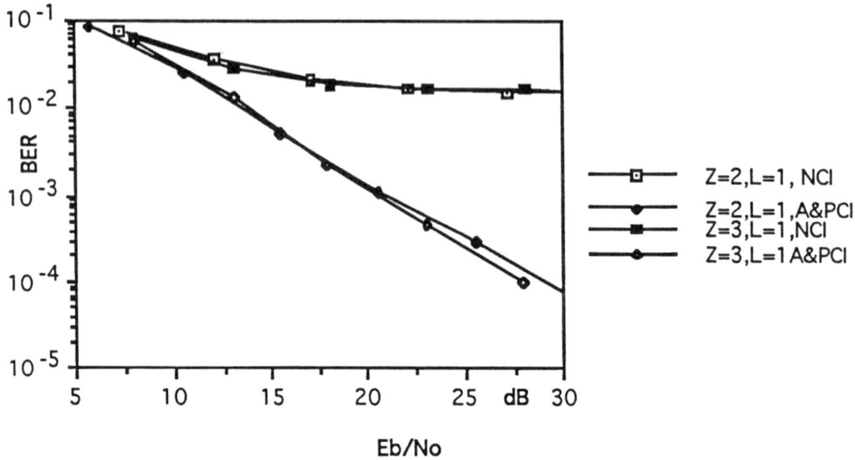

Figure 4: BER Performance curves of non-coherent receivers without Diversity for uncoded $\frac{\pi}{4}$-QPSK in a Ka-band channel with average shadowing.

From the same figure we see that when phase channel state information is added to the receiver, the performance of the non-coherent receivers improve dramatically. We also realize that the MDD receiver with $Z = 3$ provides in this case an improvement of 1 dB. For higher values of Z, the improvement increases furthermore.

From Fig. 5 we realize that when diversity is used, the performance of non-coherent receivers improves considerably. Even the structures without channel state information achieve a reduction in error floors more than one order of magnitude. Use of amplitude channel state information improves the performance furthermore. The best performance is achieved when the amplitude and phase information of the faded channels is available. Notice that with the use of diversity, the performance of the MDD receiver is always superior as compared to differential detector.

For the coded case (1/2 rate trellis coded $\frac{\pi}{4}$-shift QPSK), we consider only the configuration that does not make use of any channel state information at the receiver. There is no doubt that this is the simplest structure, since no special subsystem is needed for providing an estimate of the amplitude and/or the phase of the faded signal.

The results for light shadowing conditions, for $Z = 2, 3$ and 4, and diversity $L = 1$, 2 and 3, are illustrated in Fig. 6. The results demonstrate a large performance improvement by using diversity with $L = 2$ as compared to the no diversity case ($L = 1$). At a BER level of 10^{-3} and using sequence length $Z = 4$ this gain is approximately 10 dB, while at a BER level of 10^{-4} it increases

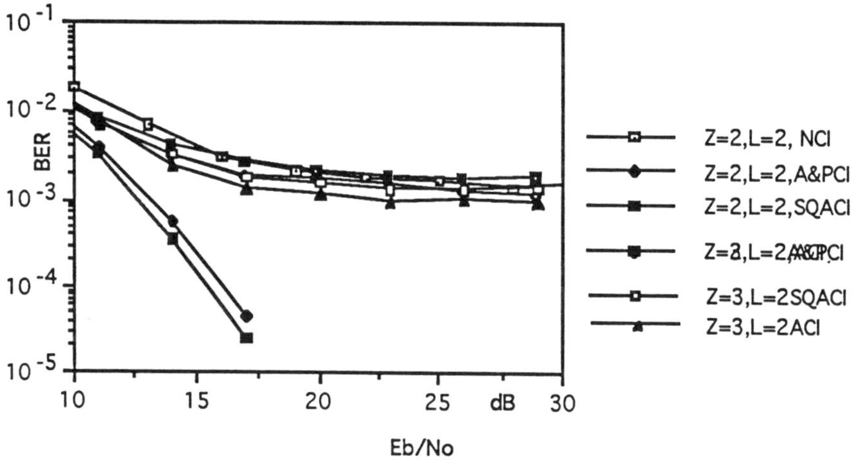

Figure 5: BER Performance curves of non-coherent receivers with Diversity for uncoded $\frac{\pi}{4}$-shift QPSK in a Ka-band channel with average shadowing.

to over 13 dB. When using diversity order $L = 3$ besides the additional gain available, we also note a greater incremental gain when going from using $Z = 2$ to $Z = 3$ and 4, as compared to the case with L = 2. This is attributed to the fact that most of the distortion imposed by fading has been offset by the high diversity order, which in turn decreases the penalty in performance inherent to employing this type of receivers in fast fading channels. It is interesting to note that the receiver performance with $Z = 4$ and $L = 3$ is only approximately 1 dB worse than the rate-1/2 trellis-coded $\frac{\pi}{4}$-shift QPSK performance in AWGN.

Figs. 7 and 8 illustrate the results for average and heavy fading conditions respectively. It is interesting to note that even for heavy shadowing, the receivers perform reasonably well, if sufficient order of diversity is employed. The inaccuracies due to fast fading are all the more visible as we go from light, to average and then to heavy shadowing. They are manifested as performance degradation for increasing values of Z, and are more pronounced in higher SNR values, where the randomization effect of AWGN is negligible.

6 Conclusions

This work has presented the structure and examined the performance of non-coherent diversity receivers for mobile and personal satellite communications operating at Ka-band. The channel is corrupted by severe shadowing and multipath fading. Performance evaluations have shown that the proposed schemes can provide considerable improvements compared to the conventional differ-

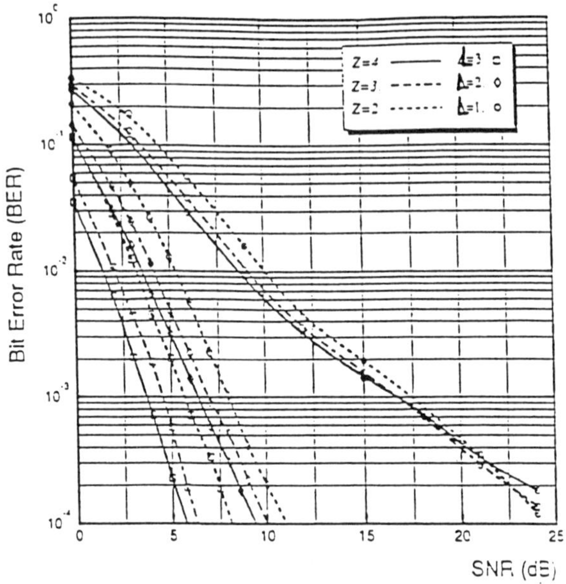

Figure 6: BER Performance curves of non-coherent receivers with and without Diversity for a 1/2 trellis coded $\frac{\pi}{4}$-shift QPSK in a Ka-band channel with light shadowing.

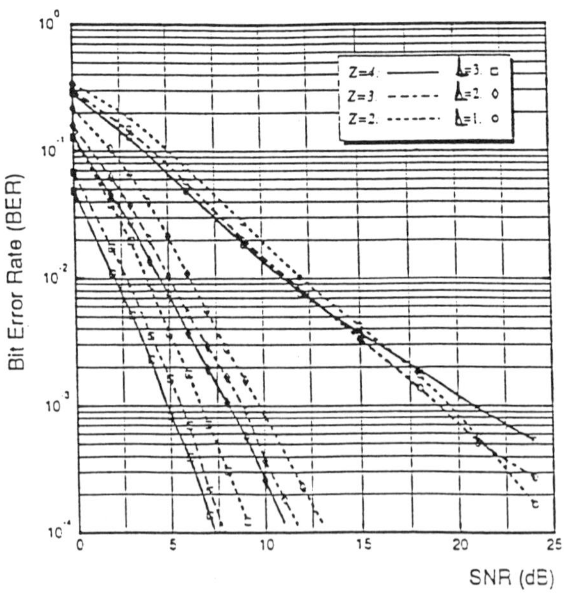

Figure 7: BER Performance curves of non-coherent receivers with and without Diversity for a 1/2 trellis coded $\frac{\pi}{4}$-shift QPSK in Ka-band average shadowing conditions.

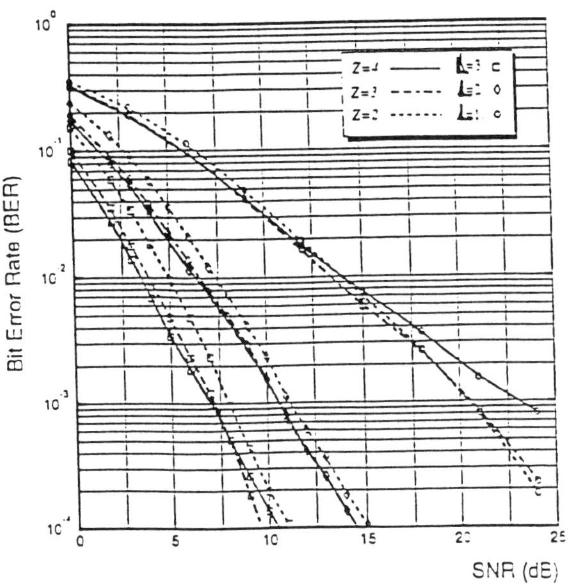

Figure 8: BER Performance curves of non-coherent receivers with and without Diversity for a 1/2 trellis coded $\frac{\pi}{4}$-shift QPSK in a Ka-band channel with heavy shadowing.

ential receivers and they are able to sustain reliable performance in a heavily impaired channel. At the same time, the non-coherent nature of the receiver structures provides relatively low cost/complexity and fast acquisition times which are the key factors in mobile and personal communication systems.

References

[1] IEEE Communications Magazine, Issue on Personal Communications Regulations, Vol. 29, Feb. 1991.

[2] D.J. Goodman, "Trends in cellular and cordless communications", IEEE Communications Magazine, Vol.29, pp. 31-40, June 1991.

[3] R. Olsen, D. Makrakis et. al., "The Canadian Olympus Propagation Experiment", Proceedings of GLOBECOM'90, pp. 6.5.1-6.5.6.

[4] Y. Akaiwa, et. al., "Highly Efficient Digital Mobile Communications with a Linear Modulation Method", *IEEE JSAC*, Vol. SAC-5, No. 5, pp. 890-895, June 1987.

[5] T.G. Jedrey, N.E Lay and W. Raferty, "An All Digital 8-DPSK TCM Modem for Land Mobile Satellite Communications", Proc. of IEEE International Conference on Acoustics Speech and Signal Processing, (ICASSP'88), pp. 1722-1725, May 1988.

[6] D. Makrakis, et. al., "Novel Receiver Structures for Systems Using Differential Detection", *IEEE Trans. on. Veh. Tech.*, Vol. VT-36, May 1987, pp. 71-77, also in Proceedings of ICDSC-7, May 1986, pp. 681-688.

[7] D. Makrakis, et. al., "A sequential decoder for the Differential Detection of Trellis Coded PSK Signals", Proceedings of ICC'88, pp. 1433-1438.

[8] D. Makrakis and P. Mathiopoulos, "Trellis Coded Non-Coherent QAM: A New Bandwidth and Power Efficient Scheme", Proceedings of VTC'89, May 1989, pp. 95-100.

[9] D. Makrakis and K. Feher, "Optimal Noncoherent Detection of PSK Signals", *Electronics Letters*, March 15, 1990, pp. 398-401.

[10] D. Makrakis , "*Generalized Non-Coherent Detection*", Ph.D. Dissertation Thesis, Dept. of Elect. Eng. University of Ottawa.

[11] "Improved
Non-Coherent Receivers. Parts I & II", CRC/NSERC/Communications Technical Reports No.1 & 2, May 1991.

[12] D. Makrakis et. al., "Performance Analysis of Asymptotically Optimal Noncoherent Detection of Trellis Coded Multi-Amplitude/Phase Modulation in Gaussian Noise and ISI Channels", IEEE Journal on Selected Areas in Communications, Vol. JSAC-13, No. 2, Feb. 1995.

[13] D. Makrakis, "Trellis Coded Modulations with Improved Non-Coherent Detection for Personal and Mobile Satellite Communications", International Journal of Satellite Communications (IJSC), Vol. 11, November 1993 , pp. 253-269.

[14] D. Makrakis et. al.,, "Optimal Decoding in Fading Channels: A Combined Envelope, Multiple Differential and Coherent Detection Approach"', IEEE Trans. on Comm., Vol. COM-42, Jan. 1994, pp. 63-75.

[15] C. Loo, "Measurements and models of a land mobile satellite channel and their applications to MSK signals", IEEE Transactions on Vehicular Technology, Vol. VT- 36, pp. 114-121, Aug. 1987.

[16] W. Zhuang, D. Makrakis, et al., "Modeling and Performance Analysis of EHF Land Mobile Satellite Channels - Signal Statistics ", Univ. of Ottawa Technical Report, 1992.

[17] W. Zhuang, J-Y. Chouinar and D. Makrakis, "Transmission Over Mobile Satellite Channels at L and Ka Bands", accepted in the Journal of Wireless Personal Communications.

[18] D. Makrakis et. al.,, "Improved Non-Coherent Receivers with Diversity for Personal and Mobile Satellite Communications", Technical Report, University of British Columbia, 1992.

[19] M.A. Weissberger, "*An Initial Summary of Models for Predicting the Attenuation of Radio Waves by Trees.*", U.S. Dept. of Defense, Report no. ESD-TR-81-101.

Interference Adaptive Multiple Access in a Multi Access Port Radio System

C. Roobol

S3/Radio Communications Systems, KTH, Stockholm, Sweden

Abstract

Access algorithms have been studied extensively for systems consisting of one base station, shared by a number of terminals. Interference from terminals in other cells is generally neglected. These studies therefore suggest conventional cellular systems, where different cells are allocated different frequencies. In this paper we will study a system consisting of a number of access ports. All terminals may access any of these ports using a common radio channel. It has been shown that such multi access port systems can obtain much higher throughput compared to conventional cellular systems. We propose two access algorithms. One of the algorithms is based on the principle of the Idle Sense Multiple Access (ISMA) algorithm. The algorithm takes interference from other terminals in consideration and is therefore called Interference Adaptive Multiple Access (IAMA). In the second algorithm first a selection of terminals that can transmit simultaneously is made, after which the selected terminals may transmit their data packets. We name this algorithm Multi Port Auction Aloha (MPAA). We compare the performance of the two algorithms to that of Slotted Aloha. We will show that both algorithms have a much higher throughput than slotted Aloha. The throughput of IAMA is clearly best.

1.Introduction

The ever existing problem when designing wireless local area networks is the fact that the radio spectrum is scarce. Therefore methods have to be found to use this spectrum efficiently. The field of access protocols has been studied extensively. Probably the most studied and well-known algorithm is slotted Aloha [4]. In this algorithm terminals can transmit at the beginning of a time slot, whenever they have a packet to transmit. When a packet does not arrive correctly at the receiver, due to a transmission from another terminal or some other cause, the terminal waits a random number of time slots until it retransmits the packet. In this protocol packets are transmitted independently. In general, messages consist of more than one packet. Then the slotted Aloha is not such a good algorithm. Preferable is that once a terminal is allowed to transmit one packet of a message, it may continue to transmit the rest or at least a part of the message. Algorithms as Carrier Sense Multiple Access (CSMA) [10] and Idle Sense Multiple Access (ISMA)[6] would be much more suitable for such a kind of traffic. In CSMA, terminals listen to the channel when they have a packet to transmit. If they hear that nearby terminals are transmitting they wait with their transmission. Only when they hear that the channel is empty, i.e. when no one in the vicinity is transmitting, they will start a transmission. A typical problem when using CSMA is the so-called hidden terminal problem. This problem occurs when two or more terminals transmit to the same access port but cannot sense each others transmission. This could be due to for instance walls shielding the two terminals.

A solution to this problem is the Busy Tone Multiple Access (BTMA) or a variation, the Idle Sense Multiple Access (ISMA). Both work according to the same principle. In ISMA, the access port transmits an idle signal when the channel is idle, i.e. when there is no terminal transmitting. In that case terminals are allowed to transmit their packets. Whenever the base station senses a signal, it stops transmitting the idle tone. Then the terminals know that they are prohibited to transmit.

Most of the studies dealing with aspects of the access protocols are done for systems consisting of one base station shared by a number of terminals. How to deal with the interference from terminals placed within the cell is often the main topic of these studies. Interference caused by terminals placed in other cells is generally neglected. Doing so, a conventional cellular architecture is assumed, where different cells are allocated different frequencies. An alternative to the cellular system is the multi access port system. The multi access port system consists of a number of access ports, which all terminals may access through a common radio channel. Such systems have been investigated in [1]-[3]. In [1], Linnartz and Kim show that such a system has a much better throughput than a conventional cellular system. Two terminals relatively close to each other may be able to transmit their packets simultaneously. If each terminal is near a different access port, the packets of both terminals may be received correctly. On the other hand, there is very often interference in the system, since there is very often someone transmitting.

In [3] the delay performance of slotted Aloha is investigated for a multi access port system. A mobile system in which terminals are assumed to move is compared to a stationary system in which the terminals are assumed to remain at the same position all the time. It is shown that the delay performance of a mobile system is much better than that of a stationary system. This is due to the fact that the terminals in the mobile system get into a good position every now and then. In the stationary system, a terminal placed in a bad position remains in that position all the time.

In this paper we will extend the investigation done in [3]. We will adopt the multi access port architecture, where all terminals use the same frequency and may transmit to any access port. For such a system we will propose access algorithms that can deal with messages consisting of a number of packets, rather than one packet. One of these algorithms is related to the ISMA protocol in the sense that terminals listen to the access ports before they start a transmission. Unfortunately, the multi access port architecture will bring some implications when using plain ISMA. We can illustrate this by means of the network in Figure 1.1. We see from this figure that terminal t2 is transmitting to access port 2, while access port 1 is idle. Access port 1 therefore transmits an idle signal, but access port 2 does not. The terminal t1 will receive the idle signal sent by access port 1 and will conclude that it is allowed to transmit. But, as mentioned, access port 2 is busy with terminal t2. If terminal t1 would transmit, its signal would interfere at access port 2, since only one frequency band is used.

To avoid such problems, we need to adapt the ISMA algorithm to the architecture. Access ports may need to transmit more information. Moreover, terminals need to take the status at other access ports into account. With the term 'status' we mean not only the idle tone, but also the levels of interference and signal to interference ratios. We call the algorithm Interference Adaptive Multiple Access protocol (IAMA).

We will in addition present another access protocol. In this protocol first a selection is made of terminals which can transmit simultaneously. After the selection period, the selected terminals may transmit a number of data packets. The selection process shows similarities with the auctioning in Resource Auction Multiple Access (RAMA) as it is presented in [7]. We call the second multiple access algorithm Multi Port Auction Aloha (MPAA). In the following we first describe the algorithms. Then we will describe the conditions under which we will investigate and compare the algorithms, followed by simulation results illustrating the difference between IAMA, MPAA and slotted Aloha. Finally we discuss the results in the conclusion and do some suggestions for further work.

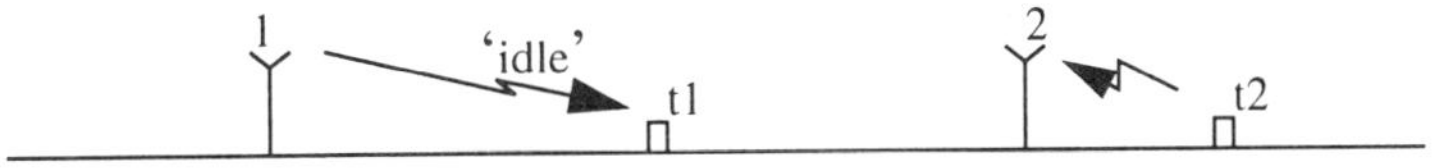

Figure 1.1 Scenario of an ISMA system, where access port 1 is idle, but access port 2 is not. The terminal t1 assumes it can transmit, but will very likely interfere with terminal t2.

2.Access algorithms

The performance of three access algorithms will be compared in this paper. The first algorithm is slotted Aloha. In slotted Aloha, the time axis is divided into time slots. When a message is generated at a terminal during a time slot, terminals start a transmission of the first packet at the beginning of the next time slot with a probability p. If the packet was not transmitted or if it was not received correctly at the access port because of a collision with a packet from another terminal, the terminal will attempt to transmit the packet in the following time slots also with probability p. When the first packet has been received correctly, the procedure for the following packets is similar to that of the first one. Thus, we assume independent transmissions of all the packets.

In the second and third algorithm consideration is taken to the fact that messages may consist of more than one packet. We will first describe the IAMA system in section 2.1, followed by the description of the MPAA system in section 2.2.

2.1 Interference Adaptive Multiple Access

We assume that the time axis is divided into time slots. Terminals may start a transmission of packets at the beginning of the time slots only. All terminals transmit on the same frequency. During every time slot, access ports transmit information about the status of the channel at the access port. We call this information status reports. A separate frequency channel is reserved for the transmission of these status reports. Access ports transmit the status reports by means of the Time Division Multiple Access protocol (TDMA)[5]. In this way, interference between different status reports from neighbouring access ports is reduced.

From a status report a terminal can derive whether a channel is 'idle' or 'busy' at the access port. When the channel is said to be idle at the access port, the access port is not receiving any packet. On the other hand, when the channel is said to be busy, the access port is receiving a packet from a terminal. In addition, the status reports also contain a quantification of the interference power at the access port if the status of the channel is 'idle'. Otherwise, if the status is 'busy', the value of the interference power and of the signal to interference ratio is included in the status report. In the following we may talk of idle and busy access ports, when the channel at the access port is idle respectively busy.

The above described situation is depicted in Figures 2.1 and 2.2. In Figure 2.1 a wireless system is depicted, in which terminals t2 and t4 transmit simultaneously to access ports 2 and 4, respectively. From Figure 2.2 we see that access ports transmit the status reports on a different frequency. Access ports need a small time period δ after the beginning of a time slot in order to estimate the received powers. After this period δ the ports transmit the status reports. As mentioned, the access ports share the frequency by means of TDMA. In this particular example access ports 1 and 4, 2 and 5 etc. may transmit their status reports simultaneously.

When a message is generated at a terminal, this terminal will evaluate whether it may start a transmission or not. This evaluation is done based on the information obtained from the status reports. In addition we assume that the terminals are able to estimate their own path loss to the respective access ports.

Two conditions have to be satisfied for a terminal to start a transmission. Let the received power at access port z of terminal i be P_{iz}. Let ξ be the set of busy access ports and Z_z be the interference power at access port z. A terminal i may transmit its first packet if the following conditions are satisfied.

1. We refer to γ_{t1} as threshold 1. There must be an idle access port z, at which the received interference power is such that

$$\frac{P_{iz}}{Z_z} > \gamma_{t1} . \tag{1}$$

2. The received power at all the busy access ports, $x \in \xi$, must be such that:

$$\frac{P_{vx}}{Z_x + P_{ix}} > \gamma_{t2} , \tag{2}$$

where the subscript v denotes the terminal of which access port x is receiving a packet, and γ_{t2} is threshold 2.

If the two conditions are satisfied, a terminal transmits its first packet with a transmission probability p. This last restriction avoids that packets of two terminals which are close to each other will collide all the time. Once the packet is received correctly, the terminal will continue to transmit the remaining packets, if any. After a terminal has transmitted N packets, it must first evaluate again whether it may continue to transmit. If the above described conditions are satisfied, the terminal is also again restricted to the transmission probability p. Therefore, a terminal with many packets will not automatically occupy the channel all the time, since after the transmission of N packets it has again to follow the same procedure as all other terminals.

In the ideal case the first condition avoids that a terminal would transmit while its power would not be sufficient. If this threshold is set high enough the probability that a terminal transmits with too low power diminishes. The second condition may assure that the received power of a packet from terminal i on top of the already existing interference power will not cause a destruction of another packet transmitted by terminal v.

Unfortunately, these conditions do not guarantee that there will not occur a collision resulting in a destruction of a packet. Two or more non transmitting terminals cannot know whether they can transmit simultaneously. A collision may occur when these terminals decide simultaneously to transmit the first packet.

Also the packets of an already transmitting terminal could be destructed due to interference. This may occur when two terminals in the vicinity of the already transmitting terminal have a message to transmit. The received powers of these two terminals separately may be sufficiently low. Hence, both terminals will conclude that they are allowed to start a transmission. However, the sum of the received powers of the two terminals may be too high causing the destruction of the packet coming from the already transmitting terminal.

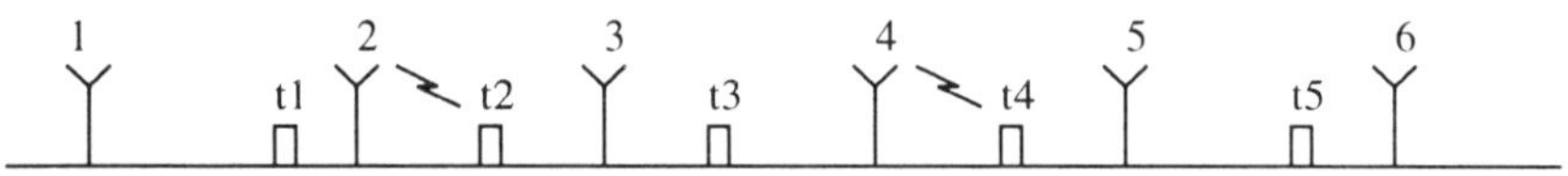

Figure 2.1 A corridor consisting of six access ports and 5 terminals. Terminals t2 and t4 are transmitting to access ports 2 and 4 respectively.

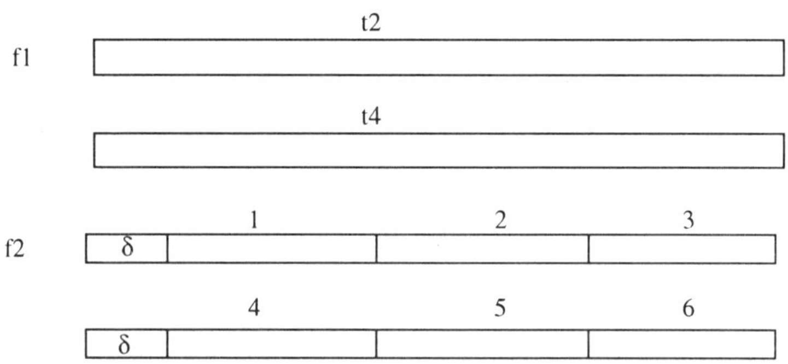

Figure 2.2 Representation of a time slot. At frequency f1 terminals t2 and t4 transmit simultaneously. After a small period the access ports transmit their status reports on the down link at frequency f2. The reuse distance is three, thus access ports 1 and 4, 2 and 5, 3 and 6 can transmit simultaneously.

2.2 Multi Port Auction Aloha

In the IAMA system, terminals base their decision on information about the interference and on their path loss. To obtain this information may be difficult, which may make the implementation of this algorithm complex. In this section we will describe another access algorithm, Multi Port Auction Aloha. We believe that this algorithm is easier to implement, since terminals do not need information about the interference power at the access ports.

Let us assume the time axis to be divided into frames. The duration of a frame corresponds to the duration of a number of packets, N, and a selection period $n\Delta$. Such a frame is shown in Figure 2.3. During the selection period, a number of terminals is selected from the total number of terminals that have packets to transmit. After the selection period has elapsed, all selected terminals may transmit simultaneously during the following N time slots.

The selection period consists of n periods of length Δ. By means of Figure 2.4 we will explain how the selection procedure works. Consider a set T, consisting of terminals that all have packets to transmit. At the beginning of a period Δ, these terminals will transmit their address with a probability p. Thus, only a subset of T will transmit their address. This subset is denoted R_1. The terminals of which the addresses are not received correctly due to collisions form the set $\bar{S}_1$. The terminals of which the addresses are received correctly due to the capture effect form the set S_1. If an access port receives an address of a terminal correctly, it will broadcast the address of that particular terminal. Access ports may do this according to a TDMA scheme. One period Δ corresponds to the time that a terminal transmits its address to the access ports followed by a TDMA frame during which the access ports may transmit an address on the down link. Figure 2.5 depicts a time period Δ. In the 'up link' box, terminals transmit their address. During the following periods, respectively access port 1 and 4, 2 and 5 etc. transmit addresses on the 'down link'. Thus after the first period Δ, terminals know whether they belong to the set S_1 or the set $\bar{S}_1$. During the next period Δ, the terminals that belong to S_1 will retransmit their address. The terminals that belong to $\bar{S}_1$ will transmit their address with a probability p_r. Now, the total number of terminals that transmit during the second period Δ form the set R_2. It is clear that the set R_2 is a subset of R_1. Consequently, the terminals that belong to S_1 will be guaranteed to transmit their address correctly again. In addition, there is a possibility that some terminals that belong to $\bar{S}_1$ will also transmit their address correctly. By repeating this procedure we hope to eventually increase the number of terminals that are able to transmit simultaneously.

An important parameter in this algorithm is the length of the selection period, $n\Delta$. The larger this period is, the more terminals may be selected. If this period is too large, the overhead due to the selection becomes significant. Other important parameters in the selection process are the probability of transmission p and the probabilities of retransmission p_r. In this study, no attempt is made to optimize these parameters, and the retransmission probability p_r is chosen equal to the probability p. Finally, the number of data packets per frame, N, also has a large impact on the performance of MPAA. If this number is chosen too large in comparison to the number of packets which the terminals need to transmit, there will always be unused slots, resulting in the degradation of the throughput performance.

Figure 2.3 Representation of a time frame, consisting of a selection period and N data packets.

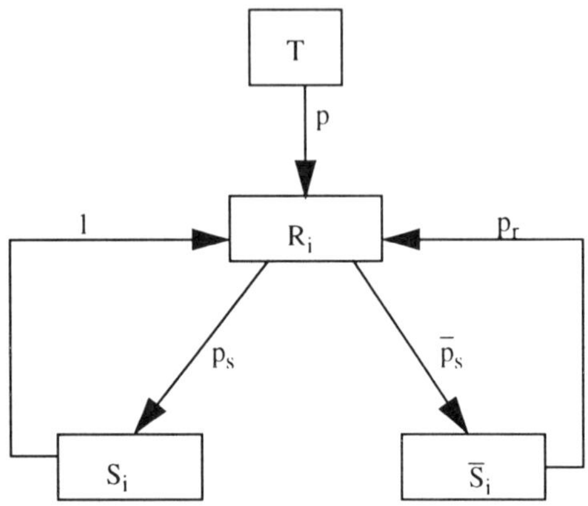

Figure 2.4 Diagram illustrating the selection procedure.

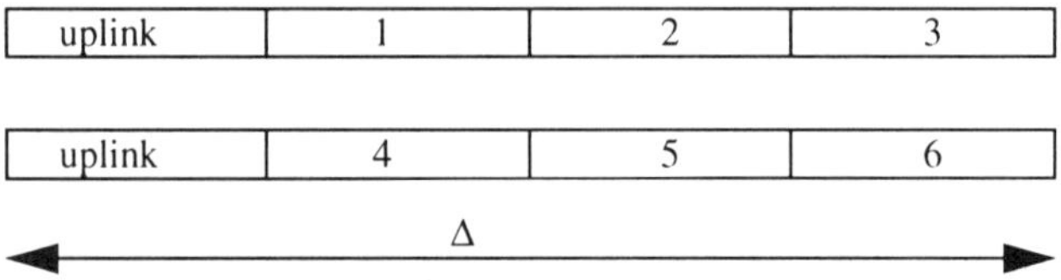

Figure 2.5 Representation of a time period Δ during which two terminals transmit their addresses on the up link and where access ports 1 and 4, 2 and 5 and 3 and 6 either acknowledge the reservation or are silent.

3.Models

We will compare IAMA and MPAA to slotted Aloha. To this end we adopt the following model. The studied system is a 1-dimensional system, which in practice could be a corridor. The 'corridor' contains access ports separated by a unit distance. In this corridor terminals are dispersed such that two consecutive terminals are separated by a distance which is exponentially distributed, with terminal density λ terminals per access port. We investigate a system in which the terminals are *stationary*. With *stationary* we mean that the terminals do not move at all. We consider the case where all terminals always have messages to transmit. Thus whenever a message of a terminal is received correctly, a new message is generated immediately and ready to be transmitted in the following time slot. A message consists of M packets. All terminals transmit over the same radio channel. Finally we assume that the access ports are slot synchronized.

3.1 Propagation and capture model

We assume that the amplitude attenuates with r^{α} where r is the distance and α is a constant. We denote the amplitude of a signal of terminal i, A_i. We assume Rayleigh fading, thus the received signal of terminal i at access port z, S_{iz} can be written as:

$$S_{iz} = \frac{A_i}{r_{iz}^{\alpha}} \cdot (\tilde{X}_{iz} + j\tilde{Y}_{iz}), \tag{3}$$

where $\tilde{X}_{iz}$ and $\tilde{Y}_{iz}$ are normal distributed random variables. The received power at access port z is:

$$P_{iz} = \frac{A_i^2}{r_{iz}^{2\alpha}} \cdot \left(\tilde{X}_{iz}^2 + \tilde{Y}_{iz}^2\right) \tag{4}$$

If K terminals transmit at the same time, the resulting interference signal at access port z is:

$$I_z = \sum_{i=1}^{K} \frac{A_i \tilde{X}_{iz}}{r_{iz}^{\alpha}} + j \sum_{i=1}^{K} \frac{A_i \tilde{Y}_{iz}}{r_{iz}^{\alpha}}. \tag{5}$$

The interference power caused by K interferers can now be written as:

$$Z_z = \left(\sum_{i=1}^{K} \frac{A_i \tilde{X}_{iz}}{r_{iz}^{\alpha}}\right)^2 + \left(\sum_{i=1}^{K} \frac{A_i \tilde{Y}_{iz}}{r_{iz}^{\alpha}}\right)^2 \tag{6}$$

The consequence of this model is that, given the distance r_{iz}, the received power of a terminal i at access port z is an exponentially distributed random variable. Also the interference power Z_z is an exponentially distributed random variable. We assume that the power declines with r^2. This implies that $\alpha = 1$.

In all the three systems terminals are assumed to transmit with the same constant power.

When a packet of a terminal i arrives at an access port z, the packet will be received correctly if the signal to interference ratio at the access port will exceed a certain threshold γ, the capture ratio. Thus, a packet of terminal i will be received correctly if:

$$\frac{P_{iz}}{Z_z} > \gamma \tag{7}$$

This capture model is extensively used in packet radio investigations[8][9].

3.2 Performance measures

The performance measures by means of which the two access algorithms are compared are the terminal delay and the throughput. The terminal delay is defined as the number of time slots that it takes for a terminal to get a packet transmitted successfully, counted from the first attempt.

Let W be the number of access ports. Ideally, all access ports should be busy serving all different terminals during a time slot. Thus the total number of available time slots is equal to the number of access ports W. Let B be the total number of access ports which are busy during one time slot, i.e. to which terminals are transmitting packets. Then we can define the throughput as:

$$U = \frac{B}{W} \tag{8}$$

4.Results

The performance of the three access systems have been studied by means of simulations. For the IAMA system the thresholds are all equal to the capture ratio, which is equal to 10, i.e. $\gamma = \gamma_{t1} = \gamma_{t2} = 10$. Furthermore, for the MPAA system the selection period is chosen equal to 2Δ. The number of packets in a time frame is chosen as $N = 10$. The length of the messages are of length $M = 10$. Thus in the MPAA system we will not have unused time slots. This will result in a somewhat optimistic throughput performance for the MPAA system. Also, in the MPAA system we have neglected the time lost due to the selection process. In the IAMA system we have not taken the extra cost due to the transmission of the status reports into account.

The delay of the three systems is plotted as a function of λ in Figure 4.1. From this figure we see that for all values of λ the delay performance of the IAMA system is clearly better than that of the S-Aloha and the MPAA systems. Since in the IAMA system terminals know more about the channel thanks to the status reports, they will be more considerate with respect to other terminals which are already transmitting. Hence there will not be as many collisions. In addition, the status reports enable a terminal to sense the channel in every time slot and therefore a terminal may start a transmission in every time slot. In the MPAA system a number of selected terminals may transmit during a whole frame. During this frame no other terminal may transmit. In the Aloha systems no information about other terminals is available and therefore terminals are not at all considerate with respect to each other. Consequently, not as many terminals will transmit simultaneously in the Aloha systems as in the IAMA system. Therefore we see a better delay performance for the IAMA system.

The fact that more terminals can transmit simultaneously in the IAMA system is illustrated by the Figure 4.2 In this figure, the throughput is plotted as a function of the transmission probability p. We see that the IAMA system has a much higher throughput than the MPAA and the S-Aloha system, for all values of the transmission probability p. As we can see, the throughput increases when the transmission probability p increases. However, as we know from [3], when the transmission probability p becomes too large, the delay will also become very large. In Figures 4.3 and 4.4 the throughput delay characteristics are shown. The assumption has been made that all terminals always have packets to transmit. Therefore, a low probability of transmission results in high delays. In addition, a low transmission probability results in a low throughput. Therefore, due to the full load condition, we see a high delay for low throughput. Increasing the transmission probability decreases the delay and increases the throughput. For a certain p a minimal delay is obtained. Increasing p even more will increase the delay again.

From the throughput delay characteristics for $\lambda = 1$, we can see that when S-Aloha has its minimal delay, the throughput of the IAMA system is approximately 1.7 times the throughput of S-Aloha and 1.4 times the throughput of MPAA. Due to the fact that in the MPAA system the terminals may attempt to obtain a reservation twice before the transmission of the frame starts, this algorithm works better than S-Aloha. However the improvement of the IAMA system over the S-Aloha system is most significant.

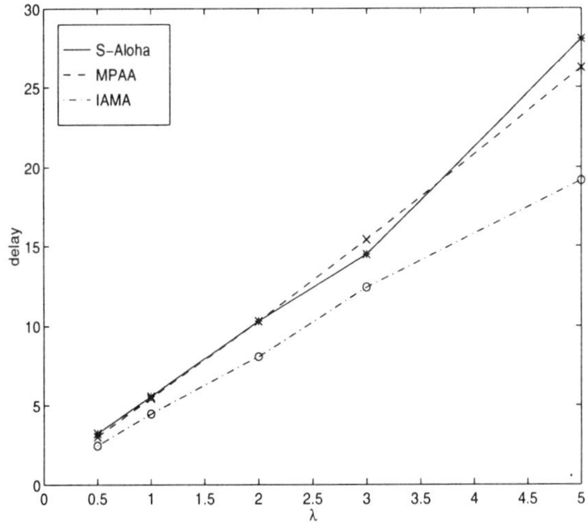

Figure 4.1 The packet delay as a function of the terminal density λ.

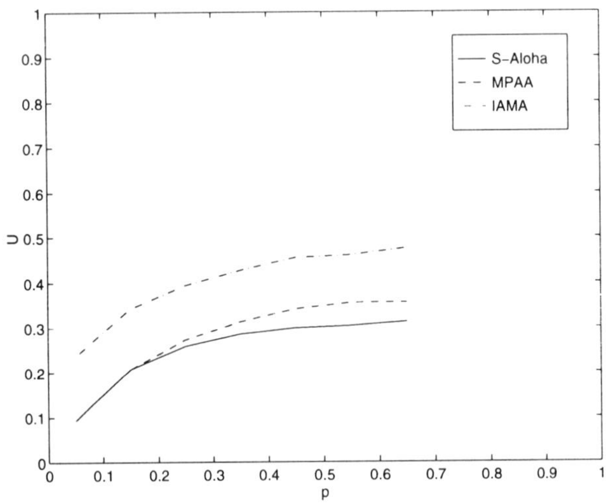

Figure 4.2 Throughput as a function of the transmission probability p, $\lambda = 2$.

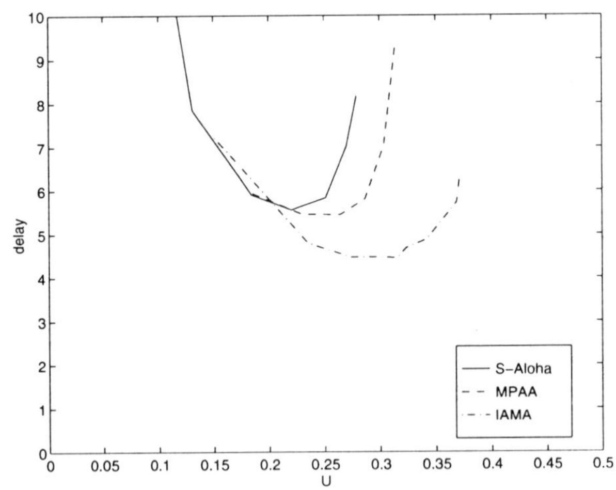

Figure 4.3 Throughput-delay curve, terminal density $\lambda = 1$.

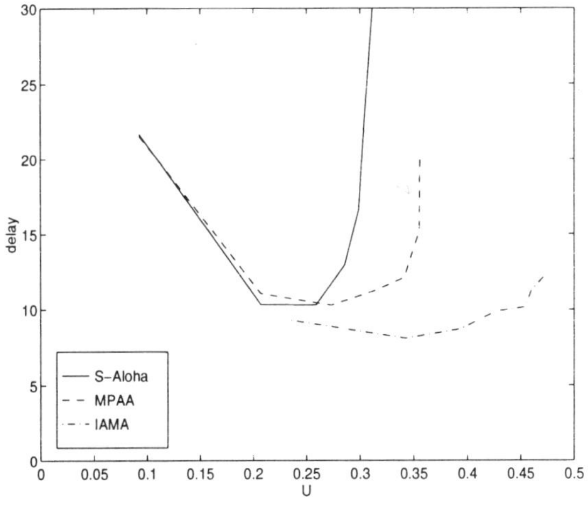

Figure 4.4 Throughput-delay curve, terminal density $\lambda = 2$.

5.Conclusions

In this paper we have studied access algorithms for a multi port environment. The main difficulty with the studied architecture is caused by the fact that all users are transmitting on the same frequency. Therefore access algorithms that depend on information sent by the access ports, as for instance ISMA, have to be adapted before they can be used in a multi access port environment. This has resulted in two new access algorithms named the Interference Adaptive Multiple Access protocol, based on the ISMA protocol and the Multi Port Auction Aloha. These algorithms are compared to slotted Aloha. Both the IAMA and the MPAA algorithms are designed such that they are suitable to transmit messages which consist of several packets. In the S-Aloha algorithm terminals transmit packets independently. Both algorithms have been shown to perform better than the conventional slotted Aloha, in particular in terms of the throughput. The delay performance of IAMA has been shown to be slightly better than the delay performances of the two Aloha systems. The throughput of IAMA is clearly better than that of both Aloha systems.

It should be mentioned that the complexity has not been taken into account. In the IAMA system terminals need to estimate the path loss to the access ports in order to be able to decide whether they may transmit or not. Also the access ports need to be able to estimate the signal to interferences and the interference powers. The

MPAA system seems to be more simple. In this system access ports only need to transmit the address on the down link of the selected terminals.

6.References

[1] Kim, H. J., Linnartz, J.P., "Virtual cellular network: a new wireless communications architecture with multiple access ports", IEEE Vehicular Technology Conference, Stockholm, 1994.

[2] Rosberg, Z., Sidi, M., "TDM policies in multistation packet radio networks", IEEE Transactions on Communications, vol. 37, no. 1, Jan. 1989.

[3] Roobol, C., "On the packet delay in wireless local area networks with access port diversity and power control", IEEE Personal Indoor Mobile Radio Conference, Toronto, 1995.

[4] Roberts, L.G., "Aloha packet radio system with or without slots and capture", Computer Communications Rev., vol. 5, Apr. 1975.

[5] Rom, R., Sidi, M., "Multiple Access protocols", Springer-Verlag, 1990.

[6] Linnartz, J.P., "Effects of fading and interference in narrowband land-mobile networks", Ph.D. thesis, Delft University of Technology, 1991.

[7] Amitay, N., "Distributed Switching and Control with Fast Resource Assignment /Handoff for Personal Communications Systems", IEEE Journal on Selected Areas in Communications, Vol. 11, No. 6, August 1993.

[8] Leung, C., Lau, V., "Capture models for mobile packet radio networks", IEEE Transactions on Communications, vol.40, no. 5, May 1992.

[9] Krishna, A., LaMaire, R.O., "A Comparison of Radio Capture Models and their Effect on Wireless LAN Protocols", 3rd International Conference on Universal Personal Communications, 1994.

[10] Tobagi, F., Kleinrock, L., "Packet Switching in Radio Channels: Part II- The Hidden Terminal Problem in Carrier Sense Multiple Access and the Busy-Tone Solution, IEEE Transactions on Communications, Vol. Com-23, No. 12, Dec. 1975.

TIME-FREQUENCY SLICING WITH DISTRIBUTED-QUEUEING REQUEST UPDATE MULTIPLE ACCESS (DQRUMA) FOR MULTI-RATE WIRELESS PACKET (ATM) NETWORKS

Zhao Liu, Mark J. Karol, Magda El Zarki*, Kai Y. Eng

Bell Labs, Lucent Technologies
101 Crawfords Corner Road, Holmdel, NJ 07733 USA

** Department of Electrical Engineering, University of Pennsylvania*
Philadelphia, PA 19104 USA

ABSTRACT

In a wireless packet (ATM) network that supports an integrated mix of multimedia traffic, a large variety of mobiles with different service rates will need to be accommodated efficiently in a unified bandwidth-on-demand fair-sharing manner. In this paper, we propose and study a Time-Frequency Slicing with Distributed-Queueing Request Update Multiple Access (DQRUMA) protocol for wireless packet (ATM) networks that support mixes of multi-rate traffics. The protocol incorporates Time-Frequency Slicing (an efficient multiplexing scheme for providing multi-rate packet transmissions) and DQRUMA (an efficient demand-assignment multiple access protocol for wireless access and scheduling) to form a unified bandwidth-on-demand platform for multi-rate wireless services. A slot-by-slot packet transmit scheduling policy named Bandwidth-on-Demand Fair-Sharing Round-Robin (BoD-FSRR) shares the system bandwidth amongst different rate classes of mobiles according to their demand on a fair-sharing basis. Simulation results show that the system provides close to ideal access performance for different rate classes of mobiles.

1 INTRODUCTION

In a wireless packet (ATM) network that supports an integrated mix of multimedia traffic, a large variety of mobiles with different service rates will need to be accommodated efficiently in a unified bandwidth-on-demand manner. In this paper, we propose and study the Time-Frequency Slicing with DQRUMA protocol as a spectrally efficient bandwidth-on-demand platform for multi-rate wireless packet (ATM) networks.

Time-Frequency Slicing [1] [2] is a spectrally efficient and cost effective technique that allows multiple users with different rate requirements to share a communications medium. Users are assigned different portions of the spectrum on a slot-by-slot basis, allowing them access to only the needed fractions of the shared resource. With a Time-Frequency Slicing system, time is slotted and users can modulate a signal in one or more of the available frequency bands over one or more time slots. Low-rate users can fill one frequency band over multiple time slots in a given time interval, while higher-rate users can fill one or more frequency bands or time slots in the same time interval. A *unit slice*, which consists of one frequency band allocation for one time slot, is the minimum amount of resource available to a user.

On the other hand, Distributed-Queueing Request Update Multiple Access (DQRUMA) protocol is an efficient demand-assignment channel access protocol for wireless *packet* access and scheduling [3]. Figure 1 shows a flow chart of the DQRUMA protocol at each mobile. Transmission requests are placed in short

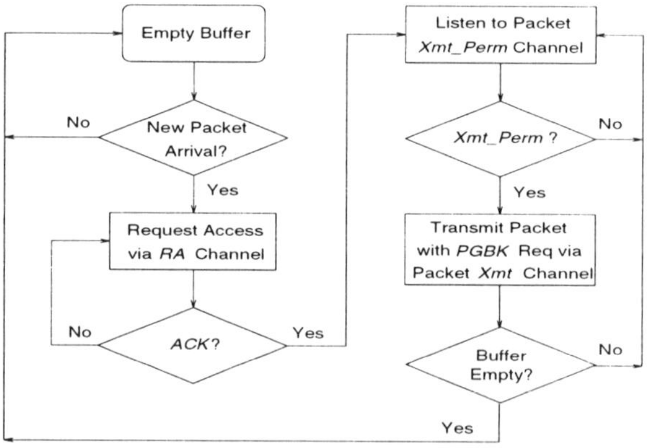

Figure 1 DQRUMA protocol.

mini-slots and can be piggybacked (*PGBK*) contention-free with packet transmissions to efficiently relieve contention on the Request-Access (*RA*) Channels under bursty and heavy traffic. Also, explicit slot-by-slot announcement of each packet Transmit-Permission (*Xmt_Perm*) allows the base station to implement an "optimal" scheduling scheme for integrated mixes of traffics with different rates and service requirements. DQRUMA works well with the Time-Frequency Slicing technique to provide a spectrally efficient bandwidth-on-demand platform for mobiles with different and even time-varying source rates in a wireless packet network.

In Section 2, we describe the system architecture of the Time-Frequency Slicing with DQRUMA protocol for multi-rate wireless packet networks. In Section 3, we address several important design issues of the system. In Section 4, we present some performance evaluation and simulation results. Section 5 concludes the paper.

2 TIME-FREQUENCY SLICING WITH DQRUMA

In the context of Time-Frequency Slicing, the word "rate" is used in two ways. First, a mobile's *traffic rate* (or *service rate*) is the average rate at which the mobile, as a source, generates packets. In a slotted system, it can be measured in terms of the average number of packet arrivals per time slot. Second, a mobile's *transmitter rate* corresponds to the maximum number of contiguous frequency bands the mobile can modulate in a time slot. In this study, we assume that a unit slice (i.e., one frequency band in one time slot) is designed to transmit one packet (ATM cell). Thus, a mobile's transmitter rate can be measured in terms of the maximum number of packets (ATM cells) it can transmit in a time slot. Note that a mobile's transmitter rate should be greater than its traffic rate.

Figure 2 shows a sample timing diagram of a Time-Frequency Slicing with DQRUMA protocol for wireless packet transmissions. It is assumed the uplink time slots have the same duration as downlink time slots. This diagram is an extension of the original DQRUMA timing diagram presented in [3]. We assume W is the total frequency bandwidth, which is divided into F contention-based access channels during each Request Access mini-slot and F unit slices during each Packet Transmission interval. A mobile will transmit over a single or multiple unit slice(s) according to the demand-assignment scheduler associ-

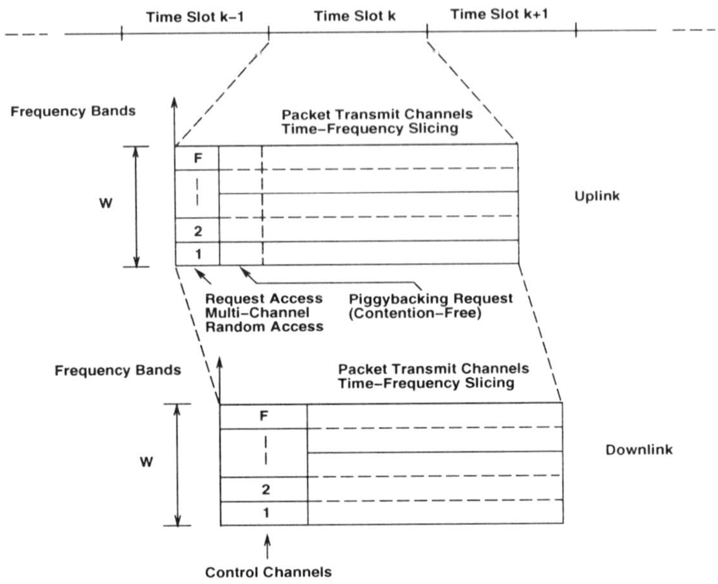

Figure 2 Time-Frequency Slicing with DQRUMA protocol for multi-rate wireless packet transmissions.

ated with DQRUMA. Also, associated with each uplink packet transmission is a Piggybacked-Request [3]. Finally, there are Control mini-slots on the downlink to announce: (1) the uplink packet Transmit-Permissions for the next time slot, (2) acknowledgments of successful transmission requests, and (3) *Listen-Notifications* for the next time slot. The base station uses Listen-Notifications to tell mobiles on which unit slices they will receive downlink packet transmissions. In Figure 2, the Control Channels are time-multiplexed (one per frequency) with the Packet Transmit Channels. In many scenarios, it is better to dedicate one or more frequency bands for the transmission of downlink control information.

In this section, we describe the various parts of the Time-Frequency Slicing with DQRUMA protocol for wireless packet networks. Detailed protocol design issues are addressed in Section 3.

2.1 Request Access

Each Request Access Packet from a mobile includes a field with the mobile's b-bit *Access Identification* (*Access ID*), which was assigned by the base station at call setup or call handoff, and a field of p bits that indicates the number of packets for which the mobile is requesting transmission. The size of the b-bit Access ID field is determined by the number of mobiles a base station can support. For example, a one byte Access ID ($b = 8$) will support up to 256 mobiles. For the p-bit request field, it can be chosen large enough to represent the number of packets that can be transmitted at full rate by the highest-rate mobile in the system (i.e., the largest number of unit slices a mobile can fill in a time slot). In other words, if m_{max} is the full rate of the highest-rate mobile in the system, a possible choice is $p = \lceil \log_2 m_{max} \rceil$. For example, if the highest-rate mobile in the system has a transmitter rate of $m_{max} = 32$, we can choose $p = 5$ bits.

For the Request Access mini-slot, we propose a multi-channel slotted ALOHA random access protocol. We choose this protocol mainly for its simplicity. Also, the contention-free Piggybacked-Requests associated with DQRUMA help relieve contention on the Request Access Channels [3], thereby reducing the access delay.

2.2 Piggybacking

With the DQRUMA protocol, the Piggybacking Request Field allows a mobile to make a contention-free Request Access each time it transmits a packet. As with the Request Access Packet, the size of this field can be taken as $p = \lceil \log_2 m_{max} \rceil$. For a mobile transmitting multiple packets in a given time slot (i.e., $\hat{m}$-rate packet transmission where $\hat{m} > 1$), the same p-bit piggybacking is included in each of the packets. This improves the system tolerance to packet errors and helps ensure the base station maintains up-to-date packet transmit request information.

2.3 Downlink Control Messages

The base station transmits control information to the mobiles using the downlink Control Channels. Short messages (perhaps variable-length) are sent to appropriate mobiles to inform them of one or more of the following: (1) a short Request-Access Acknowledgment of a successful transmission request; (2) permission to transmit using a specific uplink Time-Frequency Slice in the next time slot; and (3) notification to listen to a specific downlink Time-Frequency Slice in the next time slot. Each control message contains the intended mobile's *Access ID*. At call setup or call handoff, mobiles are told which Control Channel to listen to for their control messages. Mobile assignments are distributed across the set of Control Channels so that there is minimal delay in sending control information to the mobiles. If there are many mobiles, then multiple mobiles will receive control information on the same Control Channel. The base station can send a message to reassign mobiles to different Control Channels if a particular Control Channel becomes congested.

2.4 Uplink Packet Transmission

The uplink packet transmission uses Time-Frequency Slicing. When a mobile n receives a packet Transmit Permission, it configures its transmitter according to, for example, the lower frequency channel number F^l_{uplink} and upper frequency channel number F^u_{uplink} provided in the Transmit Permission: the packets are transmitted at center frequency (carrier) of $f_n = \frac{F^u_{uplink} + F^l_{uplink} - 1}{2} \times \frac{W}{F} + f_0$ and with the bandwidth $W_n = \frac{F^u_{uplink} - F^l_{uplink} + 1}{F} \times W$, where f_0 is the lower band limit of the Time-Frequency Slicing system, W is the overall bandwidth of the Time-Frequency Slicing system, and F is the number of unit slices per time slot.

3 DESIGN ISSUES

As in the original DQRUMA protocol [3], the design issues include selecting a random access protocol for the Request Access mini-slot and a packet transmission scheduling policy for the Packet Transmit Channels. In this section, we address these issues.

3.1 Request Access Protocol

As mentioned in Section 2.1, we propose a multi-channel Slotted ALOHA random access protocol for the Request Access mini-slot. Slotted ALOHA is simple and easy to implement, and the contention-free piggybacking associated with DQRUMA will help keep the Request Access Channels operating at a light load where Slotted ALOHA works efficiently.

During the Request Access mini-slot, a mobile n in the Request State (i.e., a mobile that just received a burst of packets to its buffer that was originally empty) will attempt request access and randomly choose one of the access frequency channels $(1, 2, \ldots, F)$ for its request. If more than one mobile choose a same frequency channel, then a collision occurs and all the mobiles involved in the collision will have to retransmit their requests in some latter access mini-slot. After failure in their Request Access, mobiles use a backoff algorithm in computing their retransmission probabilities (in Section 4, we assume a "harmonic" backoff algorithm with attempt probabilities 1, 1/2, 1/3, 1/4, ...). After receiving an *ACK* (on its assigned Control Channel) that its Request Access was correctly received, a mobile listens to its Control Channel until it hears its Transmit Permission (and the assigned Time-Frequency Slice).

3.2 Exploiting Idle Time-Frequency Slices

As with the original DQRUMA protocol [3], it is possible that during some time slots the number of Packet Transmit Requests in the Request Table is smaller than the number of unit slices in a time slot. As a result, some of the uplink *Xmt* unit slices will be (potentially) idle in the next time slot. To avoid wasting the valuable spectrum, the (otherwise) idle uplink *Xmt* unit slices are converted into multiple *RA* mini-slots (see Figure 3). The conversion to multiple *RA* mini-slots is announced over *all* downlink Control Channels, so that *all* mobiles will be notified. Also, as in the "normal" mode of operation (see Figure 2), Request-Access Acknowledgments are sent to mobiles over the appropriate downlink Control Channels.

As with the original DQRUMA protocol [3], even if there are a sufficient number of requests in the Request Table, it is still possible to dynamically convert some of the uplink Time-Frequency slices into multiple *RA* mini-slots whenever the base station senses there is a large backlog of mobiles in the Request State (e.g., by observing the number of collisions in the *RA* mini-slots). This adds one additional time slot delay for some of the uplink packet transmissions, but the benefit is that congestion on the Request-Access Channels is relieved and the base station gets a more accurate picture of which mobiles are waiting to transmit packets. This information may help the base station determine the order in which mobiles should transmit their packets. This is especially important for a system that supports mobiles with different traffic priorities.

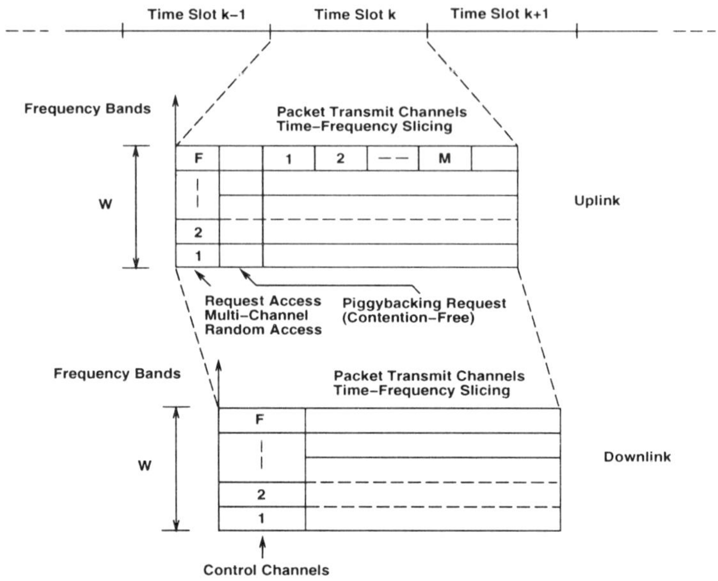

Figure 3 Idle packet transmission Time-Frequency slices are converted to multiple RA mini-slots.

3.3 Packet Transmission Scheduling Policy

In scheduling packet transmissions for the Time-Frequency Slicing with DQRUMA protocol, it is important to satisfy the Quality-of-Service (QoS) requirements of multi-rate mobiles, and also utilize the spectrum efficiently. In this section, we propose a packet transmission scheduling algorithm, called Bandwidth-on-Demand Fair-Sharing Round-Robin (BoD-FSRR), for Time-Frequency Slicing systems that support multi-rate packet transmissions. This policy schedules the packet transmissions by taking into account a mobile's transmitter capability and the traffic QoS requirements. The scheduler allows different traffic rate mobiles to share the system bandwidth in a bandwidth-on-demand and fair-sharing manner. Other service policies are also possible; we just present BoD-FSRR as one example.

The base station keeps a Request_Table that includes: (1) an index field n, (2) an Access ID field $ID[n]$ for the mobile's access ID (short ID), (3) a Request field $Req[n]$ for the mobile's packet transmit request in terms of the number of packets in the mobile's buffer, (4) a Rate Class field $m^s[n]$ for the mobile's declared service rate in terms of the average number of packets per time slot,

(5) a Transmitter Rate field $m[n]$ for the mobile's transmitter capability in terms of the maximum number of packets (or unit slices) it can transmit in a time slot, and (6) a Service Counter field $SC[n]$ for determining the order in which service should be given. Naturally, we assume $m^s[n] < m[n]$. In order to characterize a mobile's rate in terms of both the traffic rate and the transmitter capability, we define a mobile's effective rate as $\bar{m}[n] = \frac{m^s[n]}{m[n]}$.

Table 1 shows the pseudo-code of the BoD-FSRR algorithm. Each time slot, the scheduler decreases the Service Counter $SC[n]$ of a mobile n by the amount of its effective rate $\bar{m}[n]$, and checks the sign of the Service Counter. If $SC[n] > 0$, there is no need to give Transmit Permission to mobile n. Otherwise, there is a need to give Transmit Permission to mobile n if it has packets to transmit. The number of packets for which the Transmit Permission is given will depend on the current request, the transmitter rate of the mobile, and the remaining capacity of the system at the time. After announcing the Transmit Permission, the base station resets $SC[n]$ according to the actual transmission rate permitted relative to the mobile's transmitter rate.

In the table, F is the number of unit slices per time slot, N is the number of mobiles, and P_k and Q_k are pointers to the index field (with values between 1 and N, inclusive) in the Request Table. The code inside the first **repeat** loop is the Bandwidth-on-Demand Fair-Sharing part of the protocol. This part assigns packet Transmit Permissions, in a round-robin fashion, according to the Service Counter and the Request entry in the Request Table. If a mobile's Service Counter is negative and the Request entry is not empty, the protocol will give the Transmit Permission according to the mobile's Request, the mobile's transmitter capability, and the remaining channel capacity at the time. After all the mobiles with negative Service Counters are taken care of, if there is channel capacity left (i.e., $\hat{F} \geq 1$), it will go into the second **repeat** loop, which is the Best Effort part of the algorithm. In this part, the Request Table entry is looked into in a round-robin fashion. If a mobile's Request entry is not empty and its Service Counter is not equal to 1 (i.e., the mobile was not assigned bandwidth during the first **repeat** loop), then the protocol will give the Transmit Permission according to the mobile's Request, the mobile's transmitter capability, and the remaining channel capacity at the time. If there is still capacity left (i.e., $\hat{F} \geq 1$) after all the requests in the Request Table have been taken care of, the protocol then converts the remaining unit slices into mini-slots for Request Access (see Section 3.2). Notice that the Service Counter is updated according to the actual transmission rate for which the Transmit Permission is given. If a mobile is given a Transmit Permission at its full transmitter rate, the Service Counter will be updated to 1. Otherwise, the

```
BoD_FSRR (F, P_k, Q_k):
begin
    F̂←F, P̂_k←P_k, Q̂_k←Q_k;
end
repeat
    if SC[P_k]>0 then
        SC[P_k]←SC[P_k]−m̂[P_k];
    if SC[P_k]>0 then
        P_k←P_k mod_N +1;
    else
        if Req[P_k]>0 then
            begin
                m̂[P_k]←min{m[P_k], Req[P_k], F};
                Xmt_Perm(ID[P_k], m̂[P_k]);
                Req[P_k]←Req[P_k]−m̂[P_k], F←F−m[P_k];
                SC[P_k]← m̂[P_k]/m[P_k], P_k←P_k mod_N +1;
            end
until P_k=P̂_k or F=0;
if F>0 then
    repeat
        if Req[Q_k]>0 and SC[Q_k]≠ 1 then
            begin
                m̂[Q_k]←min{m[Q_k], Req[Q_k], F};
                Xmt_Perm(ID[Q_k], m̂[Q_k]);
                Req[Q_k]←Req[Q_k]−m[Q_k], F←F−m[Q_k];
                SC[P_k]←min{SC[P_k]+ m̂[Q_k]/m[Q_k], 1}, Q_k←Q_k mod_N +1;
            end
    until Q_k=Q̂_k or F̂=0;
if F̂>0 then
    Slot_Conv(F̂);
End pseudo-code.
```

Table 1 BoD-FSRR algorithm.

Service Counter will be updated according to the fraction of the transmission rate permitted relative to the transmitter's full rate. It should also be noted that actual Transmit Permission is given according to the mobile's demand.

4 PERFORMANCE EVALUATION

We use computer simulation to evaluate the performance of Time-Frequency Slicing with DQRUMA. In the simulation, we consider a system with 25 unit slices per time slot. There are 500 mobiles in the system, divided into two classes. The high-rate mobiles can transmit, at maximum, five unit slices during a packet transmission time slot, while the low-rate mobiles can transmit just one unit slice in a packet transmission time slot. In the simulation, we assume there are 250 1-rate mobiles and 250 5-rate mobiles. The burst arrival process at each mobile is assumed to be Bernoulli, with a deterministic number of packets per burst. We present our simulation results for two cases: one packet per burst and two packets per burst. The performance of Time-Frequency Slicing with DQRUMA becomes more "ideal" (as defined below) as the traffic gets "burstier," so we present simulation results for the two cases just mentioned in order to illustrate the *worst-case* performance of the protocol.

For comparison, we now define an Ideal Access Time-Frequency Slicing system. The Ideal Access Time-Frequency Slicing system is one in which the base station always has complete, perfect knowledge about the *Xmt_Reqs* of all mobiles in the system (with no overhead). In other words, the base station always knows which mobiles have packets to transmit (and how many) and the Request Table is always up-to-date. Note that an Ideal Access Time-Frequency Slicing system is one in which random access for requests is not needed, and therefore the "access delay" is zero. The only delay is the "service delay."

In the simulations, we assume that one unit slice can be converted into 25 *RA* mini-slots if there are not enough packet Transmit Requests to be scheduled (based on information in the Request Table). We also assume that the access and control overhead is 5%.

Figures 4 and 5 show the delay-throughput performance for the two classes of mobiles with one packet per burst and two packets per burst, respectively. The figures show the fair-sharing performance among the two different rate

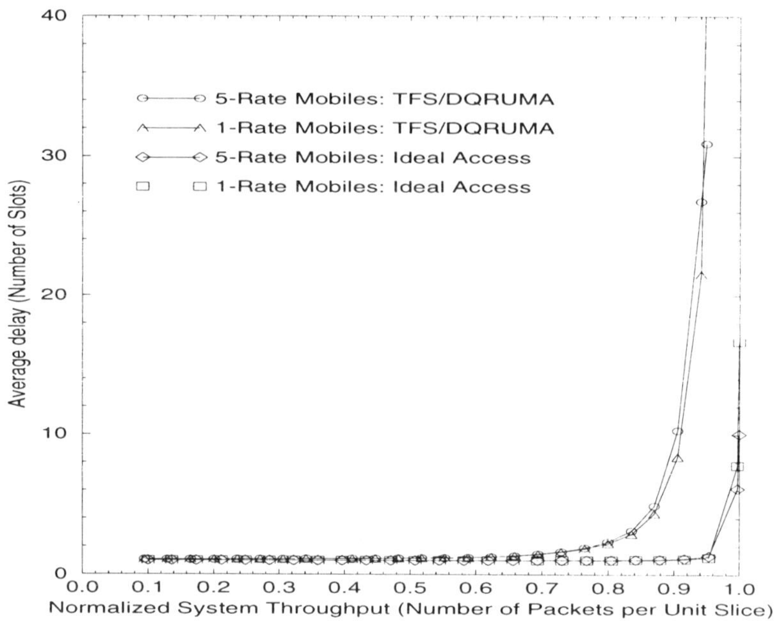

Figure 4 Average delay versus throughput for two rate classes of mobiles: 1-rate mobiles and 5-rate mobiles. The load of 250 5-rate mobiles is 5 times the load of 250 1-rate mobiles. Bernoulli burst arrival with 1 packet per burst for both classes.

classes of mobiles (*1*-rate mobiles and *5*-rate mobiles). Notice that the different rate classes of mobiles have relative similar average delay performance over the entire load range. Also, note that the Time-Frequency Slicing with DQRUMA protocol provides close to Ideal Access performance over the entire traffic load range. It can be observed that the traffic of one packet per burst is the worst case scenario for the Time-Frequency Slicing with DQRUMA protocol and when the traffics get burstier the performance gets closer to the Ideal Access performance. It is important to remember that all the results in Figs. 4 and 5 include 5% overhead for the Time-Frequency Slicing with DQRUMA protocol but no overhead for the Ideal Access system.

5 CONCLUSION

In this paper, we studied a Time-Frequency Slicing with Distributed-Queueing Request Update Multiple Access (DQRUMA) protocol for wireless packet (ATM)

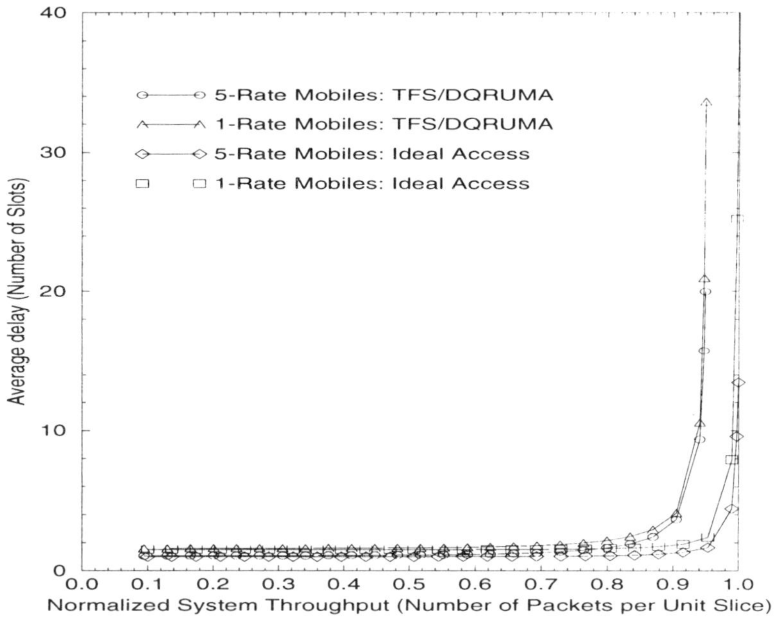

Figure 5 Average delay versus throughput for two rate classes of mobiles: 1-rate mobiles and 5-rate mobiles. The load of 250 5-rate mobiles is 5 times the load of 250 1-rate mobiles. Bernoulli burst arrival with 2 packets per burst for both classes.

networks. The protocol incorporates Time-Frequency Slicing (an efficient multiplexing scheme for providing multi-rate packet transmissions) and DQRUMA (an efficient demand-assignment multiple access protocol for packet access and scheduling) to form a unified bandwidth-on-demand fair-sharing platform for multi-rate wireless services. Thanks to the contention-free piggybacked requests of DQRUMA, the system delay performance is close to that of the ideal access system even after taking the overhead into consideration. A Bandwidth-on-Demand Fair-Sharing Round-Robin scheduler was proposed for the DQRUMA slot-by-slot packet transmit permission policy. The scheduler assigns system resources to different rate mobiles in a slot-by-slot bandwidth-on-demand fair-sharing manner. The simulation results show that for mobiles with mixes of different transmitter rates and traffic characteristics the system resources are allocated on a bandwidth-on-demand basis, and the different rate classes of mobiles experience similar delay performance that is close to ideal access performance.

REFERENCES

[1] C. Woodworth, M. Karol, Z. Haas, and R. Gitlin, "Spectrally Efficient Universal Time Slots Using Time-Frequency-Code Slicing," in *WCN Conference Record*, (The Hague, The Netherlands), pp. 1009–1013, ICCC, September 1994.

[2] M. Karol, Z. Haas, and C. Woodworth, "Performance Advantages of Time-Frequency-Sliced Systems," in *PIMRC'95 Conference Record*, (Toronto, Canada), pp. 1104–1111, IEEE, September 1995.

[3] M. Karol, Z. Liu, and K. Eng, "Distributed-Queueing Request Update Multiple Access (DQRUMA) for Wireless Packet (ATM) Networks," in *ICC'95 Conference Record*, (Seattle, USA), pp. 1224–1231, IEEE, June 1995.

Routing in an ATM-Based Mobile Network

Irene Katzela[1]
Columbia University
Center for Telecommunications Research
RM 801, 530 West 120th Str.
New York, NY 10027
Tel: (212) 854 -7378
Fax: (212) 316 - 9068
e-mail: irene@ctr.columbia.edu

Malathi Veeraraghavan
Room 4F-509
Bell Laboratories
101 Crawfords Corner Road
Holmdel, NJ 07733
Tel: (908) 949-8334
Fax: (908) 949-9118
e-mail:mv@research.att.com

Abstract

This paper presents a new routing algorithm for a wireless ATM LAN. The LAN combines positive aspects of connection-oriented and datagram communications schemes in its usage of the ATM VP/VC concept. In this scheme, a set of VPI trees, rooted at each node in the LAN, are preestablished. This has the effect of emulating "connectionless" networking which simplifies handoffs and other mobility management tasks. Given that the VPI sink trees are preestablished, there is a need for a procedure to compute and update the routes of these VPI trees with changing conditions in the LAN. Since traffic conditions are likely to change rapidly in mobile networks, we develop the Virtual Trees Routing Protocol (VTRP) to dynamically update the routing of VPI trees to handle congestion and/or topological changes in the network.

1 Introduction

ATM is emerging as a promising technology for multimedia applications in wired networks. To offer seamless interworking between wireless (mobile) hosts and wired ATM networks, many aspects of wireless ATM are currently under study [1][2][3][4][5][6][7]. This paper presents a new routing algorithm for a wireless ATM LAN called BAHAMA [4][5]. The LAN combines positive aspects of connection-oriented and datagram communications schemes in its usage of the ATM VP/VC concept. In this scheme, a set of VPI trees, rooted at each node in the LAN, are preestablished. This has the effect of emulating "connectionless" networking which simplifies handoffs, as compared to the handoff schemes presented in [2][8][9][10][11], and other mobility management tasks.

Fig. 1 shows a BAHAMA LAN. The network consists of two types of network elements: Portable Base Stations (PBSs) and mobile end points. Each PBS combines ATM switching functionality with wireless access interfaces (radio port functionality) providing microcell coverage. Mobile end points are assumed to be portable laptops with wireless ATM capability. The PBSs in Fig. 1 are represented by the numbered boxes and the mobile end points are represented by the X, Y, and Z circles. Also shown in Fig. 1 is the VP/VC usage in the BAHAMA LAN. For example,

1. This work was done during the author's summer internship at Bell Labs. in 1995.

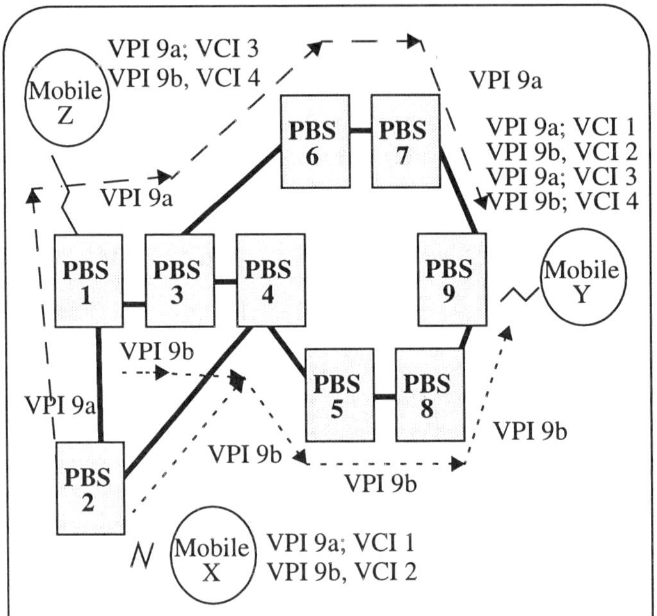

Fig. 1 BAHAMA LAN

VPI 9a and VPI 9b are assigned to PBS 9. In other words, all cells with VPI 9a and VPI 9b are routed to the destination PBS 9. Associated with each VPI is a unique route through the network (from each source), thus forming a tree with the destination as the root. To distinguish cells from different connections, each PBS controls the assignment of VCIs associated with VPIs that terminate at that PBS. For example, in Fig. 1, (VPI 9a; VCI 1) and (VPI 9b; VCI 2) denote X-to-Y connections, and (VPI 9a; VCI 3) and (VPI 9b; VCI 4) denote Z-to-Y connections.

VPI trees in BAHAMA are allowed to be partial, i.e., not all PBSs in the network are located on every VPI tree. Multiple VPIs are assigned to each PBS to ensure that every PBS is on at least two of the VPI trees (for reliability reasons) assigned to a given destination PBS.

Given that the VPI sink trees are preestablished, there is a need for a procedure to compute and update the routes of these VPI trees with changing conditions in the LAN. In other words, a routing protocol is needed for the BAHAMA LAN. Routing in ATM networks that use the traditional interpretation of VPI/VCIs is handled in many different ways.

The Least Loaded Routing (LLR) algorithms proposed in [12] and the PNNI routing protocol [13] for ATM networks are both not applicable to BAHAMA since the VPI trees are preestablished. During connection admission, there is no route determination; only a simple tree selection process is executed. This is unlike both the LLR approach and the PNNI routing protocol approach, which are based on route selection during connection setup. Since the BAHAMA LAN emulates connectionless networking, we studied the possibility of using the OSPF (Open Shortest Path First) routing algorithm [12], and RIP (Routing Information Protocol) [14], commonly used in IP networks, for BAHAMA. The former has the same drawback in that it is based on periodic topology state packet exchange. Route determination is made on a packet-by-packet basis (equivalent to a per-connection basis in ATM networks). This is counter to the approach of using provisioned VPI trees in BAHAMA. RIP, on the other hand, does determine routes a priori. However, the need for sequenced delivery, and multiple disjoint paths between every two nodes, does not permit the direct adaptation of RIP for the routing problem in BAHAMA.

In this paper, we propose the Virtual Trees Routing Protocol (VTRP) for dynamically updating the routing of the virtual trees in response to congestion and/or topological changes in the network. Specifically, VTRP is used to update data tables, stored in PBSs, that support the virtual tree structures of BAHAMA. As shown in Table 1, data table T_1, the *Transit VPIs data table,* in a PBS indicates the mapping of each VPI routed through the PBS to one of its outgoing ports. Data table T_2, the *Port/capacity data table*, stores the identifier of the PBS connected to each of its ports. Also, for each port, it stores the link capacity and the current usage value. Data table T_3, *Destination VPIs data table*, is required since VPI trees are allowed to be partial, in that only a subset of PBSs may be located on any VPI tree. At a given PBS P, this data table lists the identifiers of all the PBSs located on each of the VPI trees rooted at PBS P. Finally, given a destination PBS ID, a source PBS may be located on one or more VPI trees terminating at the destination PBS. The *PBS-to-VPIs mapping data table*, T_4, stores this information. The proposed VTRP algorithm updates data tables T_1, T_3, and T_4, while using data table T_2. The latter data table changes infrequently, and is hence maintained manually.

Table 1: PBS Data Tables

Data Table	Input	Output
T_1 :Transit VPIs	VPI	Port
T_2 : Port/capacity	Port	PBS ID and capacity (maximum and used)
T_3 : Destination VPIs	VPI	Identifiers of PBSs located on VPI tree
T_4 : PBS-to-VPIs mapping	PBS ID	VPIs

The VTRP algorithms are based on the assumption that the network is already configured with an initial set of VPI trees. We proposed off-line procedures to compute these initial trees in [15]. We provide a brief overview of these procedures here.

There are three criteria that the initial trees must meet. Primarily, we need to find a set of VPI trees rooted at each PBS such that there is *connectivity* between each pair of nodes. Secondly, for *reliability* reasons, we require that each PBS is on at least two of the VPI trees assigned to a given destination PBS. This implies that the two routes from a source PBS to the destination PBS must be disjoint. A third consideration is that the VPI trees should be routed in such a way that when these routes are traced during the admission procedure for connections with QoS requirements [4], the probability of success in finding the required resources along the route is maximized. This *load balancing* criterion requires spreading the VPI trees on as many network links as possible.

We identified *two options* to compute these trees given the above described criteria and constraints. The *first option* is to solve an optimization problem for an objective function, such as total blocked traffic, under the constraints stated above. However, this typically requires input data, such as predicted traffic patterns, which is not easily determined for new PCS applications.

The *second option* is to design heuristics to compute these initial VPI trees. We start by computing the all-pairs shortest-paths using the Floyd-Warshall algorithm [16], assuming equal weights on all links. This forms a set of primary trees, one rooted at each node. Three heuristics are proposed in [15] to determine the secondary trees rooted at each node. Based on the number of nodes, and the VPI space distribution, if more than four trees are required to be rooted at a destination node, then these heuristics can be extended.

2 Virtual Trees Routing Protocol (VTRP)

Procedures to automatically update the routing of the virtual trees are triggered by the occurrence of any one of the following events:

- A physical link in the network becomes congested;
- A physical link fails;
- A node (PBS[1]) fails or is removed from the system;
- A physical link is added to the system;
- A PBS is added to the system.

Due to space limitations, we only describe the algorithm for the congested link case in detail, while summarizing its operation for the remaining cases. Table 2 lists the notation used in the paper. A number of VTRP messages are exchanged between PBSs while handling each of these five cases. The mechanism by which these messages are transported is a detail that is ignored here under the assumption that most messages are sent on reserved "management" channels on VPIs rooted at the message destination node. Messages that are broadcast are sent on reserved broadcast management channels.

2.1 Congested Link

Many user-plane procedures have been defined to deal with congestion. These include cell tagging/dropping, rate control, etc. However, in a network in which routes are precomputed as in BAHAMA, changes in call arrival/departure patterns and/or user movements may lead to situations in which the precomputed routes can no longer support the incoming traffic. These situations can be observed by detect-

1. The terms *node* and *PBS* are used interchangeably in this paper.

Table 2:Notation

Symbol	Meaning
p_{im}	The mth port of the node i
$(p_{im}, p_{jm'})$	A physical link between port p_{im} on PBS i and port $p_{jm'}$ on PBS j
A_i	The set of nodes adjacent to node i
V_{jn}	The nth VPI tree that has PBS j as the root
$r_{ij,n}$	A route between nodes i and j on VPI tree V_{jn}
V_{pim}	Set of all VPIs routed through port p_{im} in the outgoing direction
$N(p_{im})$	Neighbor PBS on port p_{im}

ing consistent congestion patterns. Clearly, in such cases, a management-plane procedure to reroute the VPI trees is needed to prevent repeated congestion. In this paper, we do not focus on the procedures that detect consistent congestion patterns. Instead, our focus is on the algorithm for updating the routes of VPI trees, given that a consistent congestion pattern is detected.

Consider that a consistent congestion pattern is observed on link $(p_{im}, p_{jm'})$ in the direction from p_{im} to $p_{jm'}$. Node i detects the congestion and initiates a *four-phase* procedure. In the *first phase*, the node that detects congestion, PBS i, selects candidate VPIs for rerouting. *In the second phase*, PBS i initiates several route explorations through its neighbors for each candidate VPI. The "best" route is selected to reroute each such VPI. However, under some circumstances, one or more of these redirections may fail. In such cases, if node i determines that a sufficient number of reroutes have not been accomplished, it initiates *a third phase* to take further actions. These include node i requesting the "inactivation" of selected VPIs to prevent new connections from being admitted on these VPIs, or deleting itself from a set of VPI trees. The latter action will require a chain of redirections to be performed by downstream nodes. A *fourth phase* is needed to add a PBS to a VPI tree. This phase is required to keep alive the deletion option of phase III. In other words, since we require that a PBS be located on at least two VPIs to every destination for reliability reasons, if there is no counterpart addition procedure, a PBS will reach a point at which it can no longer exercise the deletion option.

Phase I: The goal of this phase is to determine a set of candidate VPIs for redirection. First, node i creates the set V_{pim} of all the VPIs that use the congested link. PBS i needs to select a subset of VPIs from the set V_{pim} for redirection. For example, in the network configuration of Fig. 2, we assume that link (p_{11}, p_{31}) experiences a consistent congestion pattern in the direction from p_{11} to p_{31}. Node 1 detects this condition and determines a suitable candidate set of VPIs for redirection.

Redirecting a VPI that carries live data on assigned VCIs while maintaining ATM cell sequence is a complex process. Hence, we avoid redirecting such VPIs

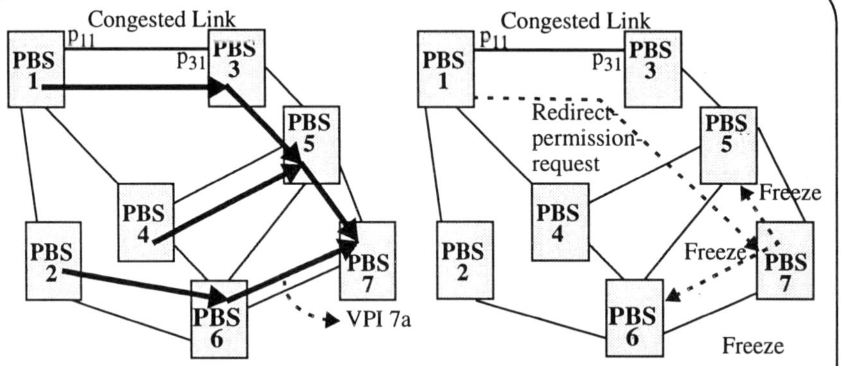

Fig. 2 Phase I: Selection of VPIs for redirection

and adopt an approach in which we only redirect *empty* VPIs. An *empty* VPI is one that does not have any assigned VCIs (existing connections). As mentioned above, the goal of this redirection procedure is not to minimize immediately-observed congestion (which is typically handled by user-plane procedures), but rather to minimize the probability of congestion by observing data over long periods of time. Thus, while an empty VPI currently does not carry traffic, its redirection will help by minimizing the number of VPIs on the congested link, and hence future congestion. If there are no empty VPIs in the set V_{pim}, then a procedure to gracefully "empty-out" VPIs before redirection is described later in Phase III. In some cases, the network may be required to start such procedures for non-empty VPIs even if there are one or more empty VPIs that are redirected.

We now address how a PBS determines which of its VPIs traversing the congested link are empty. In BAHAMA [4], two types of connection setup procedures are supported: setup of *best-effort* connections that do not require QoS guarantees, and the setup of connections that require QoS guarantees. The former procedure involves a simple end-to-end handshake between the two end (Local) PBSs with no involvement of transit PBSs. The latter type of setup further involves a node-by-node procedure to check for the availability of resources to meet the specified QoS parameters. Thus, a transit PBS has no knowledge of best-effort connections, but it does have a record of connections admitted with QoS guarantees. A transit PBS can use this information to immediately rule out "non-empty" VPIs. However, to ascertain that a VPI is empty, PBS i needs to communicate with the root of the VPI to ensure that no best-effort connections currently exist on this VPI. It does so by sending a *redirect-permission-request* message to the root l of each VPI V_{ln}, that may be potentially empty.

If root PBS l verifies that a VPI tree V_{ln} is empty, it broadcasts a *freeze* message notifying all PBSs of the redirection. As a parameter of this message, it includes the identifier of the node that is granted permission to perform the redirection. Upon receiving this broadcast message, all the nodes that have an entry for VPI V_{ln}

(including node i) will place this VPI in a "freeze" state. Based on the permissions granted, node i creates a set $V^e_{pim} \subseteq V_{pim}$ as the candidate set of VPIs for redirection. If the set $V^e_{pim} = \phi$, the null set, then node i skips phase II and proceeds directly to phase III.

In the example shown in Fig. 2, VPI 7a rooted at PBS 7 is shown to pass through link (p_{11}, p_{31}) in the direction from p_{11} to p_{31}. Assuming that VPI 7a is potentially empty (in that it has no connections admitted with QoS guarantees), PBS 1 sends a *redirect-permission-request* to root PBS 7 as shown in Fig. 2. Similar messages are sent to the root PBS of each such VPI that is routed on link (p_{11}, p_{31}) in the direction from p_{11} to p_{31}. Assuming that VPI 7a is empty, PBS 7 sends a broadcast *Freeze* message to all the PBSs as shown in Fig. 2.

Phase II: Having determined the candidate set of VPIs V^e_{pim} for redirection in Phase I, the goal of Phase II is to perform the redirections. For each VPI V_{ln} in V^e_{pim}, node i sends an *explore-route* message to all its adjacent nodes $k \in A_i$, $nk \neq l$ $k \neq j$, asking if the node is able to accommodate a redirection of the VPI V_{ln}. In the example, this message is sent to nodes 2 and 4 (see Fig. 3).

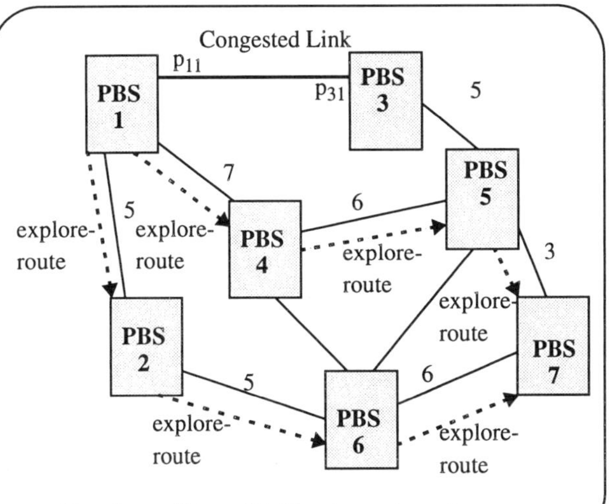

Fig. 3 Phase II: Route exploration

Upon reception of these requests, each node k checks its Transit VPIs data table T_1 to determine if it has an entry for the VPI in question.

- If it does not have an entry, it sends a *refusal* message to i.
- If it has an entry, it initiates a *Check-Resources-and-Cycles* process to determine if the path $r_{kl,n}$ has sufficient resources on the links along the path to permit this redirection, and to ensure that no cycles are caused in the tree structure of the VPI V_{ln}.

Check-Resources-and-Cycles Process

Node k reads its Transit VPIs data table T_1 (see Table 1) to determine the port identifier p_{km} corresponding to VPI V_{ln}. Then, it checks the entry for the port in the

Port/capacity data table T_2 to determine the spare capacity on the link attached to this port. It also checks for cycles by consulting the Port/capacity data table T_2 to determine whether the neighboring PBS on port p_{km}, denoted $N(p_{km})$, is PBS i.

- If there is no spare capacity on the link, the redirection along that path is not possible. Similarly, if node k determines $N(p_{km}) = i$, then redirection is not possible because a circuit or a loop is created, and the tree structure destroyed. In both these cases, node k stops the redirection, and sends a *redirection-reject* message directly to the root node l.
- On the other hand, if $(N(p_{km}) \neq i)$, and if there is spare capacity on the outgoing link that the VPI in question uses, the capacity field of the *explore-route* message is tagged with the minimum of its current value and the value of the spare capacity on this outgoing link. The *explore-route* message is then sent to the next node along the route $r_{kl,n}$.

In the example network, the spare capacities on relevant links are shown in Fig. 3. Given that both neighbors 2 and 4 are on the VPI tree being redirected (VPI 7a), they respond by performing the *Check-resources-and-Cycles* process. The *explore-route* message sent from node 1 to node 4 is shown in Fig. 4, which also explains the

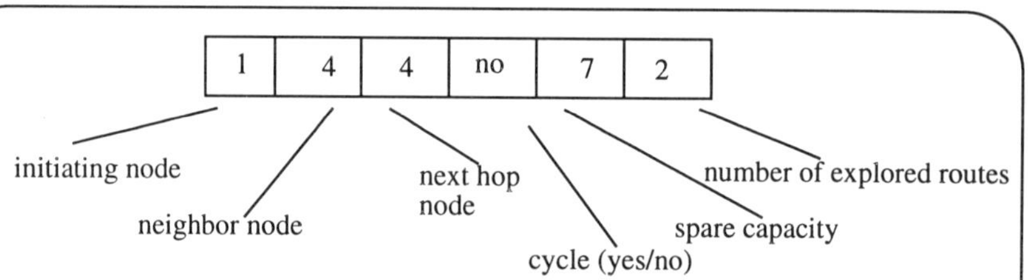

Fig. 4 Explore-route message

fields of the message. Other measures of resource availability, hop count, etc. may also be included as parameters in this message. Upon completing the *Check-Resources-and-Cycles* procedure, it sends the *explore-route* message to the next node (5). This message has the parameters 1, 4, 5, no, 6, 2 corresponding to the 6 fields of the message shown in Fig. 4. Note that the spare capacity is replaced by the smaller of the current value (7) and the spare capacity in the outgoing link to node 5 (6).

Each node along the route $r_{kl,n}$ that receives the *explore-route* message will perform the *Check-Resources-and-Cycles* procedure. If successful, this route exploration proceeds node-by-node until the root PBS l is reached. At any point along this exploration, if the *Check-Resources-and-Cycles* procedure fails, the corresponding PBS sends the *redirection-reject* message directly to the root PBS.

For each route $r_{kl,n}$ explored, the root PBS l will receive an *explore-route* message or a *redirection-reject* message. The node l selects one of these routes $r_{kl,n}$ through which redirection is possible. In the simple case shown above, where only spare capacity is part of the *explore-route* message, the root selects the path that has the most spare capacity. The root PBS l then sends a message to node i with the identifier of the neighbor PBS k through which the VPI V_{ln} must be redirected. If there is no neighbor that can accommodate the redirection, the node l sends a message to PBS i indicating that redirection of the VPI is not possible.

If node i receives a *redirection-success* message from PBS l specifying the identifier of the neighbor k through which the VPI V_{ln} must be redirected, node i changes the port identifier for V_{ln} in its Transit VPIs data table T_1 to indicate the port leading to its neighbor k. For example, as shown in Fig. 5, PBS 7 sends the

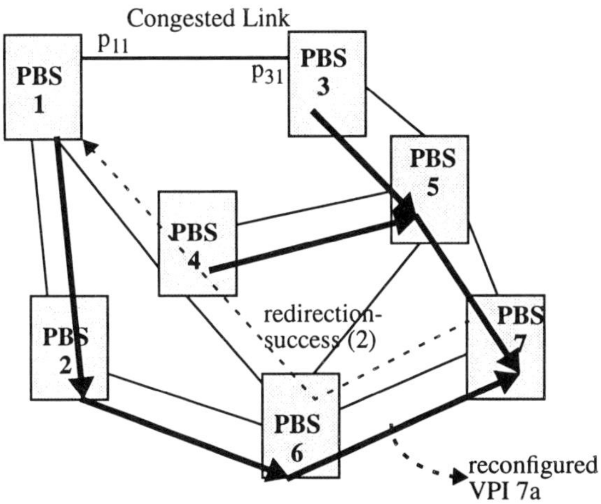

Fig. 5 Phase II: Redirected VPI 7a

redirection-success message to node 1 identifying 2 as the node through which VPI 7a must be redirected. It chooses PBS 2 instead of PBS 4, since the spare capacity on the path via node 2 is 5, while it is 3 for the path through neighbor 4. PBS 1 changes its Transit VPIs data table T_1 and the reconfigured VPI 7a is as shown in Fig. 5.

When node i completes this update in data table T_1, it sends a message to the root PBS l. Upon reception of the message, the root PBS checks if there are other queued requests for redirection on that VPI tree. If there are pending requests, root l gives permission for redirection to the next request in the queue. It then broadcasts a *freeze* message to all nodes informing them of the new redirection. If there are no more redirection requests in the root PBS l, then l sends a broadcast message to all PBSs that the redirection operations on the VPI tree have been completed. Upon reception of this message, each PBS that has an entry for VPI V_{ln}, will return the status of this VPI to "active".

On the other-hand, if node i receives a *redirection-failure* message, it makes no changes to its Transit VPIs data table T_1. It must then determine whether a sufficient number of other VPI redirections succeeded to alleviate the repeated congestion pattern observed. If not, it initiates Phase III of the algorithm.

Phase III: This phase is entered if one of the following two cases occurs:

- V^e_{pim} is empty, i.e., all VPIs in the set V_{pim} have active connections, or
- the number of successful redirections of the VPIs in the set V^e_{pim} is not sufficient to alleviate the repeated congestion pattern (a redirection may fail due to lack of spare capacity or the formation of cycles).

In the first case (V^e_{pim} is empty), node i arranges the VPIs in the set V_{pim} in the ascending order of N, where N is the number of active connections on each VPI. It then selects the first M VPIs from this set as candidates for redirection. For each VPI V_{ln} in this set, it sends a message to the root PBS l to inactivate the VPI V_{ln} which prevents it from admitting new connections on this VPI. This allows the VPI V_{ln} to gracefully "empty-out" as connections get released. In addition, the root l sends a broadcast message to all PBSs notifying them of the "inactivated" state of the VPI tree V_{ln}. When all existing connections (VCIs) on VPI V_{ln} terminate, the VPI is empty. Root l grants permission for redirection to the node that requested the original inactivation.

In the second case (many of the redirections fail), node i chooses to delete itself from a number of the VPIs in set V_{pim}. This option of deleting PBS i from a VPI tree helps reduce traffic on the VPI tree by eliminating data generated by mobiles located on this PBS. For reliability purposes, we require that each PBS be on at least two VPI trees that terminate at every other PBS in the network. If node i determines (by consulting its PBS-to-VPIs mapping data table T_4 shown in Table 1) that it is located on at least two other VPI trees terminating at the root PBS l, it can delete itself from the VPI tree V_{ln}. It starts the deletion process by sending a message to the root l asking for permission to delete itself from the tree.

PBS l temporarily freezes the VPI V_{ln}, thus preventing new connections from being admitted until the deletion is completed. In addition, it broadcasts a message to all the PBSs informing them that the VPI in question is in a "freeze" state due to a deletion operation being executed on it.

If granted permission to delete itself, PBS i must determine whether it is a leaf node on this tree or a transit node. In the latter case, downstream nodes will be sending data through PBS i on this tree, and hence it cannot simply delete itself without first having these nodes redirect the VPI tree to pass through some other node. PBS i determines whether it is a leaf node or not by sending a message to each of its adjacent nodes $k \in A_i, i \neq j$. For the example network, if we assume that both redirection attempts through nodes 2 or 4 failed, then node 1 may try this option of deleting itself from the VPI tree 7a. It does so by sending a *Check-if-leaf* message to its neighbor nodes, 8, 2, and 4, as shown in Fig. 6 (for purposes of illustrating this phase, we added a new node PBS 8 to the original example network shown in Fig. 2).

If the responses from the neighbors indicate that no node uses PBS i to route cells with the VPI V_{ln}, it means that i is a leaf node in the VPI tree. In this case, PBS i eliminates the entry that corresponds to VPI V_{ln} from its Transit VPIs data table T_1 and updates the PBS-to-VPIs-mapping data table T_4. It sends a message to

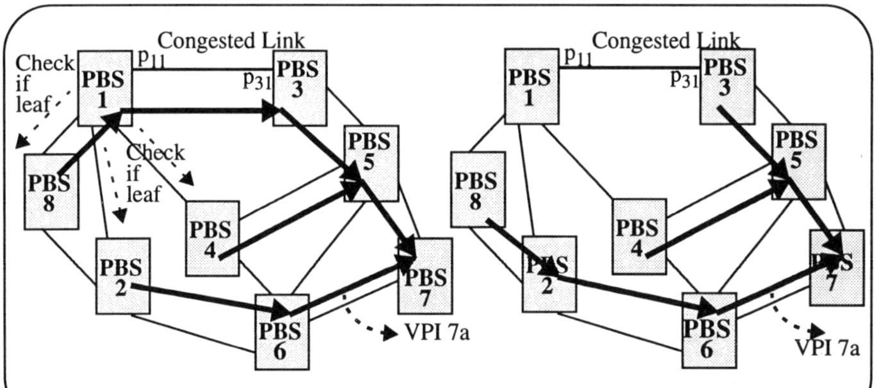

Fig. 6 Phase III: Node deletion

the root l notifying it of the completion of the deletion. PBS l eliminates i from the list of nodes that are included for V_{ln} in its Destination VPIs data table T_3 (Table 1), and re-enables the VPI V_{ln}.

On the other hand, if responses from neighbors indicate that PBS i is used to route cells with the VPI V_{ln} for some of the neighboring PBSs, then PBS i asks each such neighbor to redirect the VPI V_{ln} without using i. In the example, shown in Fig. 6, since only node 8 is downstream of PBS 1 for the VPI tree 7a, it will need to first redirect VPI 7a, before node 1 can delete itself from the tree. If all such neighbors successfully complete the redirection and notify node i, i eliminates the entry for VPI V_{ln} from its Transit VPIs data table T_1 and sends a message to the root l as before. In the example network, as shown in Fig. 6, after node 8 redirects VPI 7a (through node 2 instead of node 1), PBS 1 deletes itself from the VPI tree 7a.

If even one of the redirections of the downstream nodes fails, node i cannot delete itself, and must try other VPI trees for redirection/deletion. Note that the deletion option is only allowed if trees are allowed to be partial instead of global, which is one of our assumptions regarding the VPI trees as stated in Section 1.

Phase IV: As mentioned in Phase III, a PBS is able to delete itself from a VPI tree V_{ln} only if it has two other available VPIs rooted at l. If after a deletion procedure, PBS mphas only two active VPIs V_{lm} and V_{lp} to a destination node l, then it must initiate the addition procedure when a time period τ expires. Given that multiple (>2) VPI trees, are rooted at PBS it is likely that some of the neighbors of PBS i may be located on other VPIs, $V_{lq}, q \neq m, p$. PBS i obtains this information by sending a message to each of its neighbors. If a neighbor k responds indicating that it is located on a VPI $V_{lq}, q \neq m, p$, then node i simply joins this tree as a leaf node. It does so by making an entry in its Transit VPIs data table T_1 mapping V_{lq} to the port leading to neighbor k. Using this procedure node i can join a number of VPI trees leading to destination l.

If none of the neighbors are located on a VPI $V_{lq}, q \neq m, p$, then this phase terminates, and node i will attempt this addition procedure after a random amount of time.

2.2 Physical Link Failure

If a physical link $(p_{im}, p_{jm'})$ fails, and PBS i detects the failure, it first creates the set V^l_{pim} of VPI trees that use the failed link for each destination l. If there are more than two functional VPIs to a destination node l (*case1*), node i does not attempt to redirect the VPIs in the set V^l_{pim}. Instead, it merely performs a *selective freeze* of the VPIs at nodes downstream from itself. This prevents any of these nodes from sending cells with a VPI in the set V^l_{pim} to node i since these cells will be dropped. This temporary measure is taken based on the assumption that the failed link will be restored soon allowing for the reactivation of these VPIs. Otherwise, if node i has two or fewer functional VPIs to a node l after the link failure (*case 2*), then it redirects some of these VPIs in the set V^l_{pim} in order to have multiple routes to destination node l, for load-balancing and reliability reasons.

2.3 PBS Failure

A PBS failure is treated as being equivalent to the failure of all the physical links attached to this PBS. Thus, it is handled as a multiple physical link failure scenario. For each of the links attached to the failed PBS, the neighbor nodes initiate recovery procedures as described in Section 2.2.

2.4 Addition of a link

Assume that a new link $(p_{im}, p_{jm'})$ is added between PBSs i and j. The new link can be either a replacement for a failed link (*case1*), or a new link added between two ports for network growth (*case2*).

In *case 1*, PBS i finds the VPIs, $V_{lq} \in V^l_{pim}$, that were inactivated by the selective freeze procedure when the link failure occurred. For each VPI in the set PBS i performs a *selective-unfreeze* to the nodes downstream from itself in order to restore these VPIs back to their active states. In *case* 2, a new link is added between two ports that were previously not connected. In order to utilize the link, both nodes i and j redirect some of their VPIs on to the new link. A simple case of reconfiguration occurs if there are other links already in existence between nodes i and j. In this case, PBS i identifies a set of empty VPIs that traverse these parallel links, and switches some of them over to the new link. In case there are no such parallel links, node i needs to select a set of empty VPIs that pass through other links and redirect them on to the new link. The redirection procedure is as described in Section 2.1.

2.5 Addition of a PBS

This section describes the VTRP procedure to incorporate a newly-added node, PBS i, into the network. The addition of a PBS requires two phases. In *Phase I*, the VPI trees that node i should join in order to reach every other node in the network are determined. As per our reliability constraint, node i must join at least two VPI trees rooted at each destination node l. PBS i has a set of adjacent nodes A_i. It starts by asking each neighbor $k \in A_i$ for the list of VPIs that node k is located on. Each neighbor k sends a reply message back to node i with this list. If a VPI to a given destination node l is reachable through only one neighbor k', then node i simply attaches itself to this VPI tree through node k'. After receiving the replies from all the neighbors and joining as many VPI trees as possible, node i determines whether it meets the reliability and load-balancing criteria of being located on multiple (two or more) VPI trees to every destination. If it is, then this phase of the node addition procedure is complete. If not, then PBS i must select one or more of the VPI trees that are reachable through multiple neighbors by initiating a *resource-check* procedure to determine which of them has the most available resources.

In *Phase II*, we create the VPI sink trees rooted at node i, thus providing connectivity from all other nodes in the network to PBS i. Each node k adjacent to i has a number of sink trees with k as a root. Node i sends a message to each adjacent node k asking for the n least loaded sink trees rooted at k. A PBS can make this loading decision by simply determining how many VCIs are currently active on each of its VPIs. This gives a rough measure of the loading of each VPI tree.

Upon reception of the information from all adjacent nodes, PBS i selects a set of least loaded trees such that the union of all the nodes that are located on these trees is the set of all nodes in the network. It also ascertains that each node in the network has at least two paths to reach it. PBS i sends messages to the other nodes in the network to configure these nodes for node i reachability.

3 Summary

In this paper, we described a new routing algorithm for a wireless ATM LAN based on the VPI sink trees concept. This concept was developed to emulate "connectionless" networking in mobile ATM networks to simply handoffs and other mobility management tasks. Since it requires preestablished VPI trees rooted at each destination node, we identified the need for a routing protocol to manage the routes of these trees, while ensuring that there are at least two VPI trees from each source to each destination for reliability reasons, and that each destination node has multiple VPI trees for load-balancing reasons. To handle constant changes in the traffic and network conditions, caused by user mobility and link/nodes failures/restorals, we developed the Virtual Trees Routing Protocol(VTRP) to update the routes of the VPI trees in a dynamic and distributed fashion. VTRP procedures are triggered by congestion, link/node faults, and link/node additions.

REFERENCES

[1] D. Raychaudhuri and N. D. Wilson, "ATM-Based Transport Architecture for Multiservices Wireless Personal Communication Networks," *IEEE Journal on Selected Areas in Communications*, vol. 12, pp. 1401-1414, Oct. 1994.

[2] A. S. Acampora and M. Naghshineh, "An Architecture and Methodology for Mobile-Executed Handoff in Cellular ATM Networks," *IEEE Journal on Selected Areas in Communications*, vol. 12, pp. 1365-1375, Oct. 1994.

[3] J. Porter and A. Hopper, "An ATM based protocol for Wireless LANs," *Olivetti Technical Report*, April 19, 1994.

[4] K. Y. Eng, M. J. Karol, M. Veeraraghavan, E. Ayanoglu, C. B. Woodworth, P. Pancha, and R. A. Valenzuela, "A Wireless Broadband Ad-Hoc ATM Local-

Area Network," *ACM/Baltzer Journal on Wireless Networks*, vol. 1, no. 2, 1995.

[5] K. Y. Eng, M. J. Karol, M. Veeraraghavan, E. Ayanoglu, C. B. Woodworth, P. Pancha, and R. A. Valenzuela, "BAHAMA: A Broadband Ad-Hoc Wireless ATM Local-Area Network," *Proc. of ICC '95*, pp. 1216-1223, June 1995.

[6] P. Agrawal, A. Asthana, M. Cravatts, E. Hyden, P. Krzyzanowski, P. Mishra, B. Narendran, M. Srivastava and J. Trotter, "An Indoor Networked Wireless Computing System", *Proc. of ICC'95.*

[7] J. H. Condon, T. S. Duff, M. F. Jukl, C. R. Kalmanek, B. N. Locanthi, J. P. Savicki, J. H. Venutolo, "Rednet: A Wireless ATM Local Area Network using Infrared Links", *Proc of ACM Mobicom '95*, Berkeley, California, Nov. 95.

[8] T-H. Wu and L. F. Chang, "Architectures for PCS Mobility Management on ATM Transport Networks," *Proc. of ICUPC '95*, pp. 763-768, Nov. 1995.

[9] EIA/TIA IS-41 (Revision B): "Cellular Radio-telecommunications Intersystem Operations," Jan. 1992.

[10] GSM Recommendation 0.9.02, "Mobile Application Part (MAP) Specification."

[11] L. V. Hauwermelren, L. Vercauteren, A. Saidi. T. Van Landegem, "Requirements for Mobility Support in ATM," *Proc. of Globecom 94*, pp. 1691-1695, Nov. 1994.

[12] M. Steenstrup, Editor, "Routing in Communications Networks," Prentice Hall 1995.

[13] ATM Forum 94-0471R9, "Private-Network Node Interface Phase 1 Draft Specification."

[14] C. Hedrick, "Routing Information Protocol," *Internet RFC 1058*, Rutgers University, June 1988.

[15] I. Katzela and M. Veeraraghavan, "Virtual Trees Routing Protocol for a Mobile ATM LAN," *AT&T Internal Memorandum,* available upon request.

[16] D. Bertsekas and R. Gallagher, "Data Networks," Prentice-Hall, 1987.

OPTIMIZATION OF CAPTURE IN MULTIPLE ACCESS RADIO SYSTEMS WITH RAYLEIGH FADING AND RANDOM POWER LEVELS

Richard O. LaMaire*, Arvind Krishna*, and Michele Zorzi**

* *IBM T. J. Watson Research Center, P.O. Box 704, Yorktown Heights, NY 10598, USA*
fax: +1 914 784-6205; e-mail: {lamaire, krishna}@watson.ibm.com
** *DEI, Politecnico di Milano, Piazza Leonardo da Vinci 32, 20133 Milano, ITALY*
fax: +39 2 2399-3413; e-mail: zorzi@elet.polimi.it

ABSTRACT

To enhance the throughput of a slotted random access protocol in a radio communication system, we describe the use of a scheme in which multiple power levels are used at the transmitters. We first consider a situation in which n transmitters are simultaneously trying to send a packet to a central receiving station using a time-slotted access protocol, like slotted ALOHA. Each of these transmitters randomly chooses one of m discrete power levels during each attempt to send a packet. One of the simultaneously sent packets can often be successfully received due to the power capture effect. In this paper, we determine the optimal transmit probabilities for the power levels as well as the optimal values of the power levels themselves, when their range is constrained and in the presence of Rayleigh fading and power capture. After determining the precise optimal power levels and probabilities for maximizing the capture probabilities (i.e., for a given n), we propose a less complex, but nearly optimal, approximate approach based on using logarithmically equi-spaced levels. After demonstrating the closeness of our suboptimal results to the optimal results, we apply our approach to the problem of optimizing the throughput of the slotted ALOHA protocol for a case in which the input traffic is generated according to a Poisson process.

1 INTRODUCTION

The focus of this paper is the optimal choice of multiple power levels and of their associated transmit probabilities so as to maximize the performance of random access radio systems. Specifically, we consider a situation in which multiple transmitters send packets to a central receiving station using a slotted multiple access protocol. In a mobile radio system, the power of the signals received from the different transmitters will vary due to differences in the length of the transmission path and due to multipath fading effects. These physically-induced power variations have been observed to actually improve the throughput of random access protocols like slotted ALOHA [1, 2, 3]. This effect, known as power capture, occurs when differences in received power levels result in the successful reception of a packet in the presence of, typically weaker, contending transmissions. This effect has a significant impact on the throughput of random access protocols [1, 4, 5]. While slotted ALOHA is rarely used in its raw form, it is frequently used as a component of more complex medium access control protocols for wireless systems (e.g., [6, 7]).

In this paper, we consider the use of deliberate user-induced randomness in the transmitted power in an effort to improve the throughput of a radio system that uses a random access protocol. For the purposes of this paper we assume that the physically-induced power variations due to power roll-off with distance have been eliminated through power control.

We consider a situation in which n transmitters are communicating with a central receiving station using a slotted protocol. Each of these transmitters randomly chooses one of m discrete power levels during each attempt to send a packet. Two types of capture models will be considered in this paper: 1) one in which the transmitter with the largest received power captures the channel (referred to as the *perfect capture model* when introduced by Roberts [1]), and 2) one in which the transmitter captures the channel only if its signal-to-interference ratio at the central station is above some threshold, b.

While much work has been done on the use of the slotted ALOHA protocol in radio systems, the determination of the optimal probability distribution that the transmitters should use when choosing one of the m power levels has not been determined as a function of both the number of levels m and the number of transmitters n. In 1976, Metzner [5] assumed a Poisson traffic model (for n) and developed a recursive method for computing the optimal probability distribution of the m different power levels, in the presence of perfect capture. Shacham [8] analyzed the performance of a slotted ALOHA system that uses multiple discrete power levels for a perfect capture model and for a simplified version of the signal-to-interference ratio capture model. However, he did not consider the problem of how to choose the power levels and transmission probabilities. Lee [9] analyzed a slotted ALOHA system with random power levels for a signal-to-interference ratio type of capture model. Lee assumed linearly equi-spaced power levels and uniformly-distributed transmit probabilities.

The paper is organized as follows. In Section 2, we introduce and analyze the considered power capture models and state the optimization problems of interest. A formulation of the slotted ALOHA optimization problem is presented in Section 3. In Section 4, we present numerical examples to illustrate our analytic results, including a presentation of an accurate suboptimal methodology for choosing the power levels, before the paper concludes with Section 5.

2 SYSTEM MODELS AND ANALYSIS

In this section, we describe the two types of capture models that are treated in the paper. For both of these models we consider a system in which there are n transmitters, each using one of m power levels, indexed 1 through m from the smallest to the largest, to send a packet to a central receiving station. In this formulation, the n transmitters are all active and all attempt transmission within each slot. The possibility of choosing not to transmit, even if a packet is ready for transmission, is addressed at the end of this section.

2.1 Model 1: Perfect Capture Model

For the perfect capture model, we assume that if one of the n transmitters chooses a power level that is larger than those chosen by the remaining $n-1$ transmitters, then capture occurs and hence a packet is successfully received by the central receiving station. That is, capture occurs for transmitter i if:

$$S_i > S_j,\ \forall j \neq i, \quad \text{(Model 1)} \tag{1}$$

where S_i is the power received at the central receiving station from transmitter i. We denote by p_i the probability that a transmitter chooses power level i. The problem is to optimize the probabilities associated with the different power levels.

We introduce a more general notation than is necessary for this case, since it will be convenient to use the same notation later for the second capture model. The problem is to choose the vector of probabilities, $\underline{p} = (p_1, p_2, \ldots, p_m)$ so as to maximize

$$C_n = \sum_{\underline{k} \in \Omega_{n,m}} \delta_1(\underline{k}) \binom{n}{k_1, k_2, \ldots, k_m} \prod_{i=1}^{m} p_i^{k_i} \tag{2}$$

where $\Omega_{n,m} = \{(k_1, k_2, \ldots, k_m) : k_i \text{ nonnegative integers and } \sum_{i=1}^{m} k_i = n\}$, and where $\delta_1(\underline{k})$ is zero for all combinations of the exponents $\underline{k} = (k_1, k_2, \ldots, k_m)$ except for those that yield a successful capture. In (2), the multinomial coefficient $\binom{n}{k_1, k_2, \ldots, k_m}$ is the total number of combinations with n transmitters in which k_i transmitters choose power level i and $\prod_{i=1}^{m} p_i^{k_i}$ is the probability of each such combination. For the perfect capture model,

$$\delta_1(\underline{k}) = \begin{cases} 1, \text{ if } \exists \ell \text{ such that } k_\ell = 1 \text{ and } k_i = 0, \forall i > \ell, \\ 0, \text{ otherwise.} \end{cases} \tag{3}$$

For this perfect capture model, it is possible to find a recursive solution that yields the optimal $\underline{p}$ and the resulting capture probabilities, C_n (see [10]).

The size of the set $\Omega_{n,m}$ will limit the size of the problem that we will be able to study later in the numerical results section, and is given by

$$|\Omega_{n,m}| = \sum_{k_1=0}^{n} \sum_{k_2=0}^{n-k_1} \cdots \sum_{k_{m-1}=0}^{n-k_1-\ldots-k_{m-2}} 1 = \binom{n+m-1}{m-1}, \tag{4}$$

where the closed-form expression is obtained by noting that $|\Omega_{n,m}|$ is equal to the number of ways one can put n balls in m bins.

2.2 Model 2: Signal-to-Interference Ratio Model

In this subsection, we introduce a second type of capture model based on the ratio between the received signal power of a transmitter and the cumulative interference powers that are received from the other transmitters.

In order to describe this second capture model, we must first introduce some additional notation. Unlike in the perfect capture case, we must consider the actual values of the power levels, which we denote by v_i, $i = 1, 2, \ldots, m$, with $v_i < v_j$ for $i < j$. Here we assume a model in which the central receiving station successfully captures an incoming packet from transmitter i if the following is true:

$$S_i \geq b \sum_{j=1, j \neq i}^{n} S_j. \quad \text{(Model 2)} \tag{5}$$

The considered threshold capture model is simplified in that it does not account for the exact reception scheme; however, it is widely used in the literature (see [2] and the rich bibliography therein), and can be regarded as an approximation for noncoherent demodulation in the presence of coding. In narrowband systems, the capture ratio b typically ranges from 4 to 10 dB, but other values are possible including values that are less than 0 dB for code-division multiple-access systems. In this paper we assume that $b > 1$.

As a further constraint, we assume that there is a minimum power level $v_{\min}$ and a maximum power level $v_{\max}$. That is, the dynamic range of the power levels, $\gamma \triangleq v_{\max}/v_{\min}$, is constrained.

For the capture model of (5), it is only the relative values of the power levels that matter. Thus, throughout the remainder of the paper, we assume without loss of generality that $v_{\min} = 1$ and $v_{\max} = \gamma$. The problem we consider here is tc choose the values of the power levels, $\underline{v} = (v_1, v_2, \ldots, v_m)$, and their associated transmit probabilities, $\underline{p} = (p_1, p_2, \ldots, p_m)$, so as to maximize the capture probability for a given number of transmitters n and power levels m. If we do not assume that the power level dynamic range is limited, the problem degenerates to the problem that was solved for model 1, i.e., perfect capture [10]. In fact, if $\gamma \to \infty$, the power levels can be chosen such that $v_i > b(n-1)v_{i-1}$, so that (in the absence of Rayleigh fading) if one transmitter chooses power level v_i and all of the other transmitters choose even the next smaller power level, then the strongest transmitter will capture the channel.

We state the optimization problem for model 2 in the absence of Rayleigh fading for general n and m.

$$\max_{\underline{p},\underline{v}} C_n(\underline{p}, \underline{v}, b), \text{ subject to } \sum_{\ell=1}^{m} p_\ell = 1,\ p_i \geq 0,\ \forall i, \text{ and } 1 \leq v_1 < \ldots < v_m \leq \gamma, \tag{6}$$

$$C_n(\underline{p}, \underline{v}, b) = \sum_{\underline{k} \in \Omega_{n,m}} \delta_2(\underline{k}, \underline{v}, b) \binom{n}{k_1, k_2, \ldots, k_m} \prod_{i=1}^{m} p_i^{k_i}, \tag{7}$$

where

$$\delta_2(\underline{k}, \underline{v}, b) = \begin{cases} 1, & \text{if } \exists\, \ell \text{ such that } k_\ell = 1, k_i = 0\, \forall i > \ell, v_\ell \geq b \sum_{j=1}^{\ell-1} k_j v_j \\ 0, & \text{otherwise.} \end{cases} \tag{8}$$

In [10], a methodology for solving the above problem was given and several examples were computed.

2.3 Model 2: Signal-to-Interference Ratio Model with Rayleigh Fading

In the presence of a Rayleigh fading propagation model, if a user transmits with power S_T, the corresponding received power S_R at the central receiving station is given by $S_R = R^2 S_T$, where R is a Rayleigh distributed random variable with unit power, so that R^2 is exponentially distributed. Thus, given a set of transmitted powers $S_{T,i}$, $i = 1, \ldots, n$, the probability that transmitter 1 captures the channel, for model 2 with Rayleigh fading, is given by [3]

$$Pr\left[S_{R,1} \geq b \sum_{i=2}^{n} S_{R,i}\right] = Pr\left[R_1^2 \geq b \sum_{i=2}^{n} \frac{R_i^2 S_{T,i}}{S_{T,1}}\right] = \prod_{i=2}^{n} \frac{1}{1 + b\frac{S_{T,i}}{S_{T,1}}}. \tag{9}$$

Based on (9), for the case of Rayleigh fading and for general n and m, we seek to solve the optimization problem of (6) with

$$C_n(\underline{p}, \underline{v}, b) = \sum_{\underline{k} \in \Omega_{n,m}} \delta_{2R}(\underline{k}, \underline{v}, b) \binom{n}{k_1, k_2, \ldots, k_m} \prod_{i=1}^{m} p_i^{k_i}, \tag{10}$$

where

$$\delta_{2R}(\underline{k}, \underline{v}, b) = \sum_{i=1}^{m} k_i \left(\frac{1}{1+b}\right)^{k_i - 1} \prod_{\substack{j=1 \\ j \neq i}}^{m} \left(\frac{1}{1 + b\frac{v_j}{v_i}}\right)^{k_j}. \tag{11}$$

The solution of the above problem depends on the number of power levels considered. In particular, it is clear that the maximum capture probability is a non-decreasing function of the number of levels, m. However, for finite γ, we observed that there exists a value, M^*, such that for all $m \geq M^*$ the solution of the problem yields the same value of the capture probability. The above problem statement can therefore be generalized by considering the optimization over the number of levels as well, where it is understood that the *optimal* value of m is M^* as just defined. We state this optimization problem for model 2 in the presence of Rayleigh fading for general n and γ.

$$\max_{\underline{p},\underline{v}} C_n(\underline{p}, \underline{v}, b), \text{ subject to } m = M^*, \sum_{\ell=1}^{m} p_\ell = 1, \; p_i \geq 0, \; \forall i, \qquad \text{and } 1 \leq v_1 < \ldots < v_m \leq \gamma. \tag{12}$$

As a step towards understanding the behavior of the optimal solution, we consider the problem of determining when it is advantageous to use three as opposed to two power levels. As will later be shown in the numerical results, for the case of $n = 2$ transmitters, the power level probability mass function (pmf) is symmetric (when viewed on a logarithmic power level scale) about the geometric mean of the minimum and maximum power levels. For example, the power level masses at $v_{\min}$ and $v_{\max}$ are the same and for $m = 3$ power levels, the middle level is located at $\sqrt{\gamma}$, assuming $v_{\min} = 1$. Using these facts we developed an analytic method for determining the minimum power level dynamic range, γ, at which it is advantageous to use three rather than two power levels for the case of $n = 2$ transmitters. We denote this breakpoint value by γ_{23}. It can be shown that

$$\gamma_{23} = \{\gamma : 2\left(\frac{1}{1+b\sqrt{\gamma}} + \frac{1}{1+b/\sqrt{\gamma}}\right) - \frac{2}{1+b} - \left(\frac{1}{1+b\gamma} + \frac{1}{1+b/\gamma}\right) = 0, \gamma > 1\}, \tag{13}$$

where we observe that only one value of $\gamma > 1$ is found for the range of b of interest.

In Fig. 1, we show the ratio of v_2, the middle power level at the breakpoint value of the power level dynamic range γ_{23}, to the capture ratio, b. That is, we show $v_2/b = \sqrt{\gamma_{23}}/b$, which illustrates the relationship of the interlevel spacing, at the power level breakpoint, to the capture ratio. Note that $\lim_{b\to\infty} \sqrt{\gamma_{23}}/b = 1$ since for very large values of b the Rayleigh fading has little effect and we approach the case of model 2 without Rayleigh fading for which it was shown in [10] that the value of the dynamic range breakpoint after which three power levels should be used is simply b^2 for $n = 2$ transmitters, so $\sqrt{\gamma_{23}}/b = 1$. Fig. 1 demonstrates the large effect that Rayleigh fading has for small values of the capture ratio b.

2.4 Model 2 with Rayleigh Fading and Widely-Spaced Power Levels

In this subsection, we present a useful limiting result for model 2 with Rayleigh fading. If the values of the power levels are chosen such that they are widely-spaced (i.e., $\frac{v_{j+1}}{nv_j} \gg b, j = 1, \ldots, m-1$), then the interaction between distinct levels is negligible. In this case, the results are similar to model 1 (i.e., the case of widely-spaced levels for model 2) except that there is an improvement with Rayleigh fading since capture can occur even if two or more users transmit at the same power level whereas this is not true in the absence of Rayleigh fading. In the limit as $\frac{v_{j+1}}{bnv_j} \to \infty$, we use (10) and (11) to find

$$C_n = \sum_{i=1}^{m}\sum_{j=1}^{n} j\binom{n}{j} p_i^j \left(\frac{1}{1+b}\right)^{j-1} \left(\sum_{k=1}^{i-1} p_k\right)^{n-j}. \tag{14}$$

The above problem can be numerically optimized over the transmit probabilities. However, there will be a large number of terms to consider. We outline an alternative approach that is based on a recursion in the number of power levels m, and is similar to the approach described in [10]. As was done earlier, we denote by $C_{n,m}(p_1, p_2, \ldots, p_m)$, the probability of capture given n transmitters, m power levels, and a vector of power level probabilities, $p_1, p_2, \ldots, p_m$. Note that this is the same capture probability as in (14), where the dependencies have been made explicit in order to write the following recursion. We observe that the value of the capture probability that results from the use of the probabilities, $p_1, p_2, \ldots, p_m$, is related to the capture probability for $m-1$ power levels by the following:

$$C_{n,m}(p_1, p_2, \ldots, p_m) = \sum_{j=1}^{n}\binom{n}{j} p_m^j (1-p_m)^{n-j} j \left(\frac{1}{1+b}\right)^{j-1} +$$

$$(1-p_m)^n C_{n,m-1}\left(\frac{p_1}{\sum_{i=1}^{m-1} p_i}, \frac{p_2}{\sum_{i=1}^{m-1} p_i}, \ldots, \frac{p_{m-1}}{\sum_{i=1}^{m-1} p_i}\right), \; n \geq 1, \; m > 1. \tag{15}$$

This equation simply partitions the probability of capture for the m power level case into a part treating the case in which at least one transmitter chooses power level m and a case in which no transmitter chooses power level m. Hence, the result for the $m-1$ power level case can be used with the appropriate normalization of the probabilities. Since the value of $C_{n,m-1}$ is unaffected by the value of p_m, we can treat $C_{n,m-1}$ as a constant and solve for the optimal p_m by setting the first derivative equal to zero in (15). This yields that the value, p_m^*, to maximize $C_{n,m}$ is given by the root of the equation,

$$\frac{1}{1+b}\left(1 - p_m + \frac{p_m}{1+b}\right)^{n-2} (1 + b - bnp_m) - (1-p_m)^{n-1} C_{n,m-1} = 0. \tag{16}$$

It can be shown that the second derivative of (15) with respect to p_m is always negative for $n \geq 2, b \geq 1, 0 \leq p_m \leq 1$, thereby proving that the root of (16) maximizes (15).

These results suggest the following iterative algorithm for determining the optimal capture probability $C_{n,m}^*$. The starting conditions are given by $p_1^* = 1$ and $C_{n,1}^* = n\left(\frac{1}{1+b}\right)^{n-1}$. In general, assume that $C_{n,m-1}^*$ is known and that $(p_1^*, p_2^*, \ldots, p_{m-1}^*)$ are known. Then, use (16) to compute p_m^* followed by (15) to compute $C_{n,m}^*$. Then, modify p_i^*, $1 \leq i < m$, by scaling them all by $1 - p_m^*$. This approach is used in the numerical results shown later.

2.5 Other System Models

An alternative system model to consider is one in which each of the n transmitters can choose not to transmit with some probability. That is, a transmitter either chooses not to transmit with probability p_0, or it chooses to transmit at one of the m power levels, v_1, v_2, …, v_m. Thus, given that a transmission occurs, the transmit power is still constrained to be within the aforementioned range, $[v_{\min}, v_{\max}]$. We define p_i, $i > 0$, as the probability of choosing power level i *given that the transmitter chooses to transmit*. For this no-transmit case and model 2 with Rayleigh fading, the capture probability is given by

$$C_n^{\text{no-transmit}}(p_0, \underline{p}, \underline{v}, b) = \sum_{i=1}^{n} \binom{n}{i} p_0^{n-i}(1-p_0)^i C_i(\underline{p}, \underline{v}, b) \tag{17}$$

where C_i is given by (10), with $\underline{p} = (p_1, p_2, \ldots, p_m)$ and $\sum_{i=1}^{m} p_i = 1$. The behavior of this model is not fundamentally different from that of the previous model that did not include a zero power level. That is, transmitting at the lowest power level, $v_{\min}$, and 0 yield similar types of behavior for most cases.

2.6 A Continuous Power Level Distribution

Thus far in this paper we have only considered discrete power level distributions. Initially, we performed several investigations concerning the use of a continuous and a hybrid continuous/discrete distribution for the power levels. It is interesting to note that none of these attempts yielded a larger capture probability than a discrete distribution of sufficient order. We summarize our results and provide a concise analytic formulation in the Appendix.

In our first investigation, we formulated an optimization problem in which the power level distribution was assumed to be in the class of polynomial functions of some limited order (e.g., eight). Our optimization results yielded probability density functions that were attempting to match impulse functions at $v_{\min}$ and $v_{\max}$. In a second attempt, we assumed that the power level distribution was in the class of hybrid functions composed of a fixed number of discrete values and a continuous polynomial function. Whenever we allowed the number of discrete values to become as large as was required to achieve the optimal value in our previous purely discrete approach, the optimization program always yielded zero coefficients for the polynomial function. This evidence leads us to conjecture that the optimal power level distribution may be of a purely discrete nature.

3 ANALYSIS OF SLOTTED ALOHA

In the above, we considered the maximization of C_n, the conditional throughput given that there are n simultaneous transmissions. The optimal choice of the power levels, $\underline{v}$, and of the transmit probabilities, $\underline{p}$, depends on n, which may be unknown. In many system scenarios, n can be modeled as a random variable with some distribution function. For the case of a slotted ALOHA system, it is possible to compute the static throughput curves, $S(G)$, under the approximation that the global offered traffic (backlogged users plus fresh traffic) is Poisson of intensity G. This approximation, if used to compute the performance of slotted ALOHA, leads to a flawed analysis, since the protocol is known to be unstable for any positive arrival intensity, and therefore the static curves so computed are meaningless. However, it has been shown that, under certain conditions, there exist retransmission control algorithms which guarantee that all values of throughput on those curves can be achieved. More precisely, if S^* is the maximum of the curve $S(G)$, it is possible to show that there exists a control scheme that guarantees the system will be stable for all Poisson arrival rates less than S^* [11, 12, 3].

The static throughput curves, $S(G)$, can be readily computed from the capture probabilities, C_n, by averaging over the Poisson distribution for n, to obtain

$$S(G) = \sum_{n=0}^{\infty} \frac{G^n e^{-G}}{n!} C_n. \tag{18}$$

Note that this formally corresponds to having a large number of users in (17), each with accordingly small activity, so that the average number of transmitting terminals in a slot is equal to G. Therefore, at least in principle, the approach of Subsection 2.5 includes this case as well, for the appropriate values of the parameters. On the other hand, as pointed out in Section 4, this is not feasible since, even though the optimization problem is well defined, its solution is exceedingly difficult to compute for large n.

Therefore, because of the limits to the values of n that we can practically consider, we introduce the following approximation to (18) that restricts the values of n to those less than or equal to $n_{\max}$.

$$S_{n_{\max}}(G, C_\Delta, \underline{p}, \underline{v}, b) = \sum_{n=0}^{n_{\max}} \frac{G^n e^{-G}}{n!} C_n(\underline{p}, \underline{v}, b) + \left(1 - \sum_{n=0}^{n_{\max}} \frac{G^n e^{-G}}{n!}\right) C_\Delta, \quad (19)$$

where the value of C_Δ can be chosen as 0 to lowerbound $S(G)$ and as $C_{n_{\max}}$ to upperbound $S(G)$. That is, since $C_i < C_{n_{\max}}$, $i > n_{\max}$, we conclude that $S_{n_{\max}}(G, 0, \ldots) < S(G) < S_{n_{\max}}(G, C_{n_{\max}}, \ldots)$. These lower and upper bound forms of the truncated sum will be very close to one another and hence a good approximation of $S(G)$ for values of G that are much less than $n_{\max}$. Given this approximation, we state the optimization problem for model 2 in the presence of Rayleigh fading for a given γ and $n_{\max}$.

$$\max_{\underline{p}, \underline{v}} S_{n_{\max}}(G, C_\Delta, \underline{p}, \underline{v}, b), \text{ subject to } m = M^*, \sum_{\ell=1}^{m} p_\ell = 1, \; p_i \geq 0, \; \forall i, \text{ and}$$
$$1 \leq v_1 < \ldots < v_m \leq \gamma, \quad (20)$$

where M^* has the same meaning as in Subsection 2.3. A useful upper bound on the above optimal result can be found by using the optimal values of C_n obtained when each one is independently optimized. The resulting upper bound, $\overline{S_{n_{\max}}(G, C_\Delta)}$, is given by

$$\overline{S_{n_{\max}}(G, C_\Delta)} = \sum_{n=0}^{n_{\max}} \frac{G^n e^{-G}}{n!} C_n^* + \left(1 - \sum_{n=0}^{n_{\max}} \frac{G^n e^{-G}}{n!}\right) C_\Delta. \quad (21)$$

where C_n^* denotes the optimal capture probability value obtained from the solution of the optimization problem (6) with (10) and $C_\Delta = 0$ or $C_{n_{\max}}^*$ to upperbound each of the corresponding truncated sum forms of (20). Note that for $C_\Delta = C_{n_{\max}}^*$, (21) yields an upper bound on the non-truncated optimal $S(G)$ since C_n^* is a decreasing sequence.

4 NUMERICAL RESULTS

In this section, we describe the numerical results that we obtained for model 2, the signal-to-interference ratio model, for situations with and without Rayleigh fading. We demonstrate that for a Rayleigh fading propagation model, the use of logarithmically equi-spaced power levels is nearly optimal. We will use a capture ratio of $b = 10$ dB for most of our examples, but will also include a final example with $b = 3$ dB to illustrate that the same qualitative behavior holds.

4.1 Optimization Techniques

In order to solve the previously stated optimization problems over the power levels and transmit probabilities, we made use of the LANCELOT (Large And Nonlinear Constrained Extended Lagrangian Optimization Techniques) package [13]. The capture probability function of (10) falls into the category of functions that are *group partially separable* [13, p. 109]. The LANCELOT program is designed to take advantage of this fact enabling us to efficiently compute a solution to the problem of (10) given a choice of power level values. We developed an automated interface to LANCELOT to determine the optimal probabilities given the power levels. A somewhat more difficult problem is the simultaneous determination of both the power levels and the transmit probabilities. We only treat small cases of this problem (i.e., $n \leq 4$ and $m \leq 4$).

4.2 Capture Probability Examples

The results in the next two figures treat the case of model 2 with widely-spaced power levels as was analyzed in Subsection 2.4 for the Rayleigh fading case and in [10] for the case without Rayleigh fading. In Fig. 2, we show the optimal capture probability that is achieved as the number of power levels is increased, for different numbers of transmitters. These results were computed using the methodology outlined in Subsection 2.4 for the Rayleigh fading case.

A direct approach as an optimization problem (e.g., as is treated in Subsection 2.3) would not allow us to treat the large case sizes that are considered here (note that $|\Omega_{20,20}| = 6.9 \times 10^{10}$). In Fig. 2, note that most of the gain is achieved with only a relatively small number of power levels (e.g., ten). For comparison, we show in Fig. 3, the optimal capture probabilities for model 1 without Rayleigh fading (i.e., effectively model 2 with widely-spaced power levels) and for model 2 with Rayleigh fading and widely-spaced power levels. The results for the case without Rayleigh fading were computed using an iterative method that is derived in [10]. Note that in the case of widely-spaced power levels, Rayleigh fading always helps since it allows capture to occur, with some probability, when two or more stations transmit at the same power level whereas the model 1 case does not.

In the next several figures, we consider the form of the optimal power distribution for the Rayleigh fading case of model 2 in which we limit ourselves to a discrete power distribution. In Figs. 4 and 5, we show the optimal power probability mass functions (pmfs) for various values of the power level dynamic range, γ. For the case of $n = 2$ transmitters shown in Fig. 4, the power pmf is always symmetric about the geometric mean of the minimum and maximum power levels when viewed on a logarithmic scale. Note how the number of power levels increases as the power level dynamic range is increased. Further, note that for a given case, the power levels are approximately equi-spaced on the logarithmic scale. Due to the complexity of this optimization problem in which we are computing both the optimal power levels and their masses, we limited ourselves to four power levels. In can be seen from Fig. 4 that for a power level dynamic range of 10^8, the power pmf approaches a discrete uniform distribution as is expected since the model 1 case, which is model 2 with widely-spaced power levels, yields a discrete uniform distribution for $n = 2$ transmitters. If we had not limited the number of power levels, the optimal choice for the case with a power level dynamic range of 10^8 would have been to use more than four levels. In Fig. 5, we show the power pmf for a case of $n = 4$ transmitters. For this case and all other cases for which $n > 2$, the pmf is not symmetric. When $n > 2$, it is optimal to skew the pmf towards the lowest power level as shown in Fig. 5. Note again that the number of power levels increases as the power level dynamic range increases, however, we have again limited our problem to a consideration of a maximum of four power levels. We note that the optimal power levels are again approximately equi-spaced on the logarithmic scale although the approximation is less accurate for the case of four transmitters than for that of two transmitters. This observation concerning the placement of the optimal power levels suggests an approximation that we have found to be quite accurate. Namely, we assume logarithmically equi-spaced power levels and then compute the power level probabilities as opposed to solving for both the precise optimal power levels and their associated probabilities.

We formally state this log equi-spaced power level approximation since we will use it throughout the rest of the paper.

$$\max_{m,\underline{p}} C_n(\underline{p}, \underline{v}^{[m]}, b), \text{ subject to } \sum_{\ell=1}^{m} p_\ell = 1,\ p_i \geq 0,\ \forall i, \text{ and}$$
$$\underline{v}^{[m]} = (1, \gamma^{1/(m-1)}, \ldots, \gamma^{(m-2)/(m-1)}, \gamma),\ 1 \leq m \leq \lceil \log(\gamma)/\log(b) \rceil + 1. \tag{22}$$

where C_n is as defined in Subsection 2.3. In the above, we sought to use as small a number of power levels as possible to limit the problem complexity, but at the same time we tried to use a sufficient number so that the resulting suboptimal solution is practically indistinguishable from the actual optimum. In our previous results (e.g., Fig. 1), we have evidence that we should not consider a number of power levels that is greater than the number of multiples of b that can fit into γ. Concerning the above approximation, we note that the maximizing index m is usually chosen to be at or near the limit, $\lceil \log(\gamma)/\log(b) \rceil + 1$, that we have placed on it.

In Fig. 6, we show the optimal capture probabilities at the symbol points (i.e., the circles and squares) for an optimal determination (using LANCELOT) of both the power levels and probabilities where we have again limited our problem size to four power levels. In this same figure, we have also shown the suboptimal capture probabilities obtained by using the log equi-spaced power level approximation given by (22). As can be seen from Fig. 6 this yields an excellent match with the precise optimal results for $\gamma < 10^6$. For power level dynamic ranges above approximately 10^6, the effect of our limiting the optimal number of power levels to four for the symbol points causes them to be less than the values of our log equi-spaced approximation in which the number of power levels has been allowed to exceed four as dictated by (22). In Fig. 6, we have included two horizontal lines that indicate the optimal capture probability for the case of model 2 with Rayleigh fading and widely-spaced power levels, as analyzed in Subsection 2.4, where we have assumed the limit of four power levels. As can be seen from the figure, the value of the symbols approach the limiting values indicated by these horizontal lines, whereas the approximation, which is an accurate representation of the exact optimal and which uses more than four power levels, is able to achieve larger capture probabilities.

To understand the effect of Rayleigh fading on model 2, we include in Fig. 7 a comparison of the Rayleigh and non-Rayleigh cases for $n = 2$ and 4 transmitters. For both of these values of n, the conclusions are similar - Rayleigh fading increases the capture probability for small power level dynamic ranges, relative to the capture ratio b, but decreases the capture probability for dynamic ranges that are large relative to b. This can be explained by noting that Rayleigh fading increases the effective power level dynamic range of the problem since transmitting at $v_{\min}$ means that the received power, after fading, will often be smaller than $v_{\min}$. Further, when two transmitters choose the same power level, there is the possibility of capture occurring whereas this is not true for the non-Rayleigh case. The magnitude of this second effect is illustrated by Fig. 3 in which the widely-spaced power level assumption removes any level to level interaction. These positive effects of Rayleigh fading mean that, particularly for small values of γ, Rayleigh fading can yield larger capture probabilities than the non-Rayleigh case in which capture is often prohibited by the limited dynamic range, e.g. for $\gamma < 10 = b$ in Fig. 7. The reason Rayleigh fading yields smaller capture probabilities for large values of the power level dynamic range is because the spreading effect of Rayleigh fading actually reduces the number of power levels that are used in the optimal power level pmf relative to the non-Rayleigh case. For large values of power level dynamic range, γ, it is this effect and the associated lowering of the capture probabilities that more than compensates for the aforementioned positive effects of Rayleigh fading.

4.3 Slotted ALOHA Examples

In this subsection, we present numerical examples of throughput maximization in a slotted ALOHA system as was considered in Section 3. In the first example, shown in Fig. 8, we use a capture ratio of $b = 10$ dB. In this figure, we show a graph of the approximate optimal slotted ALOHA throughput $S(G)$ that was computed using the log equi-spaced approximation of (22) applied to the optimization problem of (20). In this figure, this approximate optimal throughput has been shown for both the lower and upper bound forms of the truncated sum (i.e., with reference to (19), both $S_{n_{\max}}(G, 0, \ldots)$ and $S_{n_{\max}}(G, C_{n_{\max}}, \ldots)$). Note that there is a noticeable difference between these bounds at G values of three and larger since we only use $n_{\max} = 8$ transmitters to limit the complexity of our problem. In Fig. 8, we also show the upper bound, $\overline{S_{n_{\max}}}(G, C_\Delta)$, given by (21) where both the lower and upper bound forms of the truncated sum have also been shown for this case. Note how large the peaks of the static throughput curves become as the power level dynamic range γ is increased. The curve for $\gamma = 1$ corresponds to standard slotted ALOHA (all transmitters choose the same power) where, in the presence of Rayleigh fading, we have $C_n = n\left(\frac{1}{1+b}\right)^{n-1}$ and $S(G) = G\exp\left(-\frac{bG}{1+b}\right)$.

In Fig. 9, we show the same type of results, except for a capture ratio of $b = 3$ dB. Comparing Figs. 8 and 9, we note first that the $\gamma = 1$ curve has a higher peak for the $b = 3$ dB case than the $b = 10$ dB case, since the capture probabilities are larger for the smaller capture ratio and the maximum of the throughput curves is equal to $(1 + 1/b)e^{-1}$. For the $b = 10$ dB case, the Rayleigh fading alone raises the throughput by 10% above that of slotted ALOHA without fading. Thus, the addition of transmitter power level variations is able to yield a significant increase in throughput for this case. For the $b = 3$ dB, the gain due to adding transmitter power level variations on top of Rayleigh fading (which by itself yields a 50% gain), is relatively less than for the $b = 10$ dB case. Comparing these figures, note that a capture ratio of $b = 3$ dB yields a slightly larger peak throughput. This is most evident for the case of $\gamma = b^2$.

As a final point, we emphasize that given a fixed power level dynamic range γ, a radio system with the smaller capture ratio $b = 3$ dB will often be able to achieve a much higher throughput than a system with $b = 10$ dB because of its ability to fit many more power levels. For example, if $\gamma = 32$, the b^6 curve (with maximum throughput larger than 0.7, as shown in Fig. 9) can be achieved for $b = 3$ dB, but the $b = 10$ dB case will not even be able to achieve the 10^2 curve, i.e., will yield a maximum throughput smaller than 0.55 (see Fig. 8).

5 CONCLUSIONS AND FUTURE WORK

In this paper, we have considered and analyzed the problem of optimally choosing power levels and the probability distribution with which they are used in a slotted random access system in which Rayleigh fading is present. The significant gain in throughput that occurs due to such an approach was demonstrated through numerical examples. These numerical examples suggested a useful approximation for the nearly optimal placement of the power levels using a logarithmically equi-spaced rule. The results of this work can be used to formulate new types of access control protocols for radio systems based on using multiple random power levels to avoid contention.

An ongoing goal of this research is the investigation of optimal choices of power levels and transmit probabilities for situations in which distance attenuation is included in the propagation model, in addition to the Rayleigh fading effects that have already been considered. The completion of this work would allow the development of optimal access control protocols for use in the land mobile radio environment.

Appendix: Summary of Continuous/Discrete Rayleigh Case

In this appendix, we concisely present the formulation of the optimization problem that was described in Subsection 2.6 for the hybrid continuous/discrete probability density function (pdf). We assume a power pdf, $f(x)$, defined on $[1, \gamma]$ and of the form

$$f(x) = \sum_{i=1}^{N_c} c_i x^{i-1} + \sum_{i=1}^{N_d} c_{N_c+i}\delta(x - v_i), \tag{23}$$

where $\delta-$functions are used to denote the discrete levels, N_c is the number of coefficients for the continuous part of $f(x)$, N_d is the number of discrete power levels, and we assume $1 \leq v_i \leq \gamma, i = 1, \ldots, N_d$. Since $f(x)$ is a pdf, it must satisfy

$$\int_1^\gamma dx\, f(x) = \sum_{i=1}^{N_c} \frac{\gamma^i - 1}{i} c_i + \sum_{i=1}^{N_d} c_{N_c+i} = 1. \tag{24}$$

Further, we have the constraint that $f(x) \geq 0$, which restricts the choice of coefficients of the polynomial part of $f(x)$.

The capture probability, given n transmitters, is given by

$$C_n = n \int_1^\gamma dx\, f(x) \left(\int_1^\gamma \frac{f(y)dy}{1 + b\frac{y}{x}} \right)^{n-1}, \tag{25}$$

for a continuous power pdf. The optimization problem is to find the values of the coefficients, $\underline{c} = (c_1, c_2, \ldots, c_{N_c}, \ldots c_N)$, with $N = N_c + N_d$, so as maximize $C_n(\underline{c})$. We can express C_n as a function of the coefficients as follows.

$$C_n(\underline{c}) = n \sum_{\underline{k} \in \Omega_{n,N}} \psi(\underline{k}) \prod_{i=1}^{N} c_i^{k_i}, \tag{26}$$

where

$$\psi(\underline{k}) = \sum_{i=1}^{N} \theta(\underline{k} - \underline{e_i}, i), \tag{27}$$

and where $\underline{e_i}$ is a vector of all zeros except for the i-th component, which is equal to one. Also,

$$\begin{aligned}
\theta(\underline{k}, i) &= k! \int_1^\gamma dx\ x^i \prod_{j=1}^{N} \frac{A_j^{k_j}(x/b)}{k_j!} \qquad \text{for } 1 \leq i \leq N_c \\
\theta(\underline{k}, i) &= k! \int_1^\gamma dx\ \delta(x - v_{i-N_c}) \prod_{j=1}^{N} \frac{A_j^{k_j}(x/b)}{k_j!} \\
&= k! \prod_{j=1}^{N} \frac{A_j^{k_j}(v_{i-N_c}/b)}{k_j!} \qquad \text{for } N_c < i \leq N \\
\theta(\underline{k}, i) &= 0 \qquad \text{if } k_j < 0 \text{ for some } j
\end{aligned} \tag{28}$$

where $k = \sum_{i=1}^{N} k_i$, and

$$A_i(\alpha) = \int_1^{\gamma} \frac{y^{i-1}dy}{1+y/\alpha}$$
$$= \sum_{j=1}^{i-1} \frac{(-1)^{j+1}(\gamma^{i-j}-1)\alpha^j}{i-j} + (-1)^{i-1}\alpha^i \log\left(\frac{\gamma+\alpha}{1+\alpha}\right) \quad \text{for } 1 \le i \le N_c$$
$$A_i(\alpha) = \int_1^{\gamma} \frac{\delta(y - v_{i-N_c})dy}{1+y/\alpha} = \frac{1}{1+v_{i-N_c}/\alpha} \qquad \text{for } N_c < i \le N. \tag{29}$$

Acknowledgments — The authors gratefully acknowledge Andrew Conn for his help in using the LANCELOT package.

REFERENCES

[1] L. G. Roberts, "Aloha packet system with and without slots and capture broadcast channels," *Computer Communications Review*, vol. 5, pp. 28–42, April 1975.

[2] J. P. Linnartz, *Narrowband Land-Mobile Radio Networks*. Norwood, MA: Artech House, 1993.

[3] M. Zorzi and R. R. Rao, "Capture and retransmission control in mobile radio," *IEEE J. Select. Areas Commun.*, vol. 12, pp. 1289–1298, Oct. 1994.

[4] A. Krishna and R. O. LaMaire, "A comparison of radio capture models and their effect on wireless LAN protocols," in *Proc. IEEE 3rd International Conf. on Universal Personal Communications*, (San Diego, CA), pp. 666–672, Sept. 1994.

[5] J. J. Metzner, "On improving utilization in ALOHA networks," *IEEE Trans. Commun.*, pp. 447–448, April 1976.

[6] R. O. LaMaire, A. Krishna, and H. Ahmadi, "Analysis of a wireless MAC protocol with client-server traffic and capture," *IEEE J. Select. Areas Commun.*, vol. 12, pp. 1299–1313, Oct. 1994.

[7] F. Borgonovo, L. Fratta, and M. Zorzi, "Performance analysis of capture-division packetized access (CDPA) for cellular systems with partial reuse and power control," in *Proc. IEEE GLOBECOM '95*, pp. 1141–1146, Nov. 1995. (see also *IEEE JSAC*, May 1996).

[8] N. Shacham, "Throughput-delay performance of packet-switching multiple-access channel with power capture," *Performance Evaluation*, vol. 4, pp. 153–170, August 1984.

[9] C. C. Lee, "Random signal levels for channel access in packet broadcast networks," *IEEE J. Select. Areas Commun.*, vol. 5, pp. 1026–1034, July 1987.

[10] R. O. LaMaire and A. Krishna, "Maximizing throughput in a random access radio system by optimal power level choice," in *Proc. IEEE International Conf. on Commun. (ICC '96)*, (Dallas, Texas), June 1996, to appear.

[11] B. Hajek and T. van Loon, "Decentralized dynamic control of a multiaccess broadcast channel," *IEEE Trans. Automatic Control*, vol. AC-27, pp. 559–569, June 1982.

[12] K. S. Kwak and R. R. Rao, "Controlled ALOHA with geometric capture probabilities," *IEEE Trans. Automatic Control*, vol. AC-35, pp. 1270–1274, Nov. 1990.

[13] A. R. Conn, N. I. M. Gould, and P. L. Toint, *LANCELOT: A Fortran Package for Large-Scale Nonlinear Optimization (Release A)*. New York, NY: Springer Verlag, 1992.

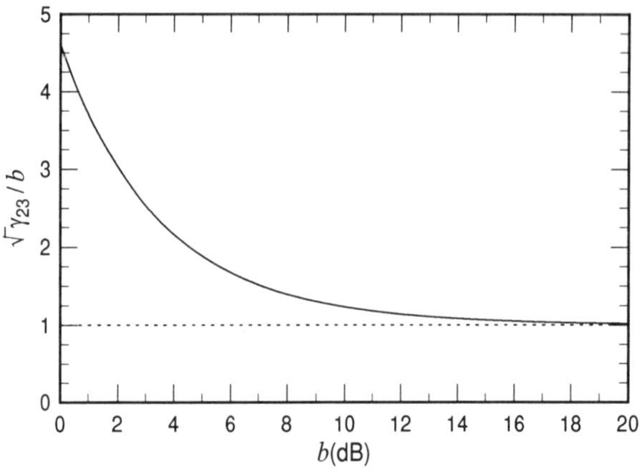

Figure 1 Model 2 with Rayleigh fading - Illustration of two-to-three level dynamic range break-point, $\sqrt{\gamma_{23}}/b$, vs. the capture ratio, b; $n = 2$ transmitters.

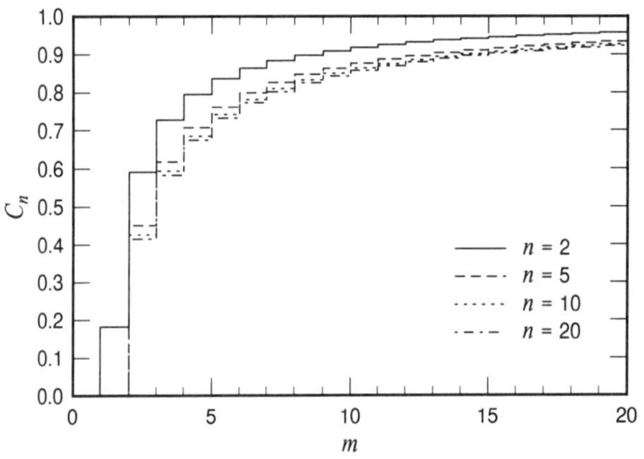

Figure 2 Model 2 with Rayleigh fading and widely-spaced power levels - Optimal capture probability, C_n, vs. number of power levels, m; $b = 10$ dB, and $n = 2$, 5, 10, and 20 transmitters.

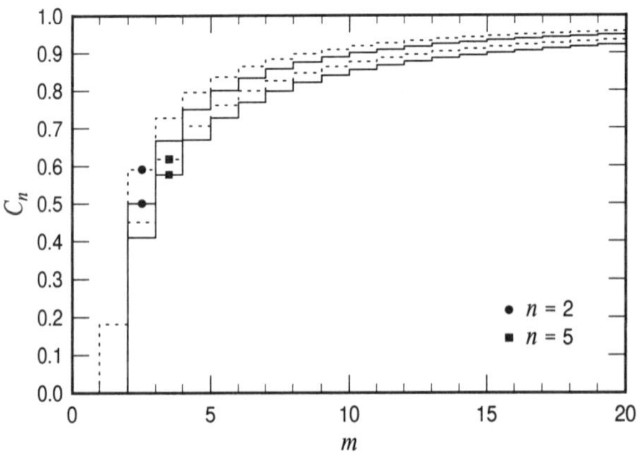

Figure 3 Widely-spaced power levels models - Optimal capture probability, C_n, vs. number of power levels, m, without (solid) and with (dotted) Rayleigh fading; $b = 10$ dB, and $n = 2$ and 5 transmitters.

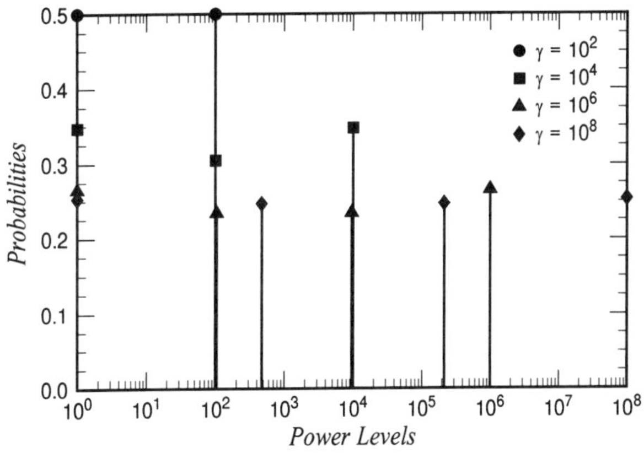

Figure 4 Model 2 with Rayleigh fading - Optimal power mass function for different power level dynamic ranges where at most $m = 4$ power levels have been used; $b = 10$ dB and $n = 2$ transmitters.

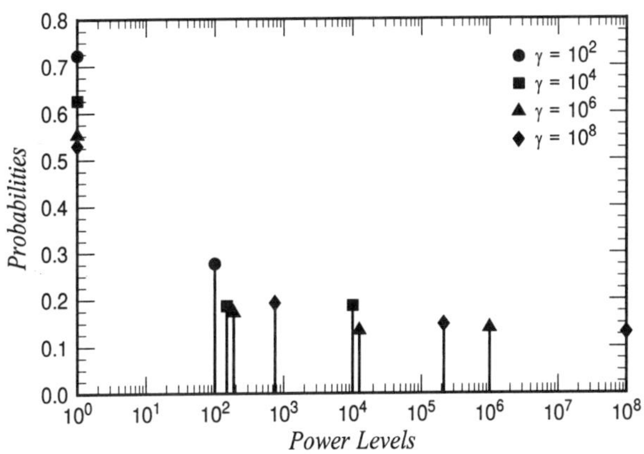

Figure 5 Model 2 with Rayleigh fading - Optimal power mass function for different power level dynamic ranges where at most $m = 4$ power levels have been used; $b = 10$ dB and $n = 4$ transmitters.

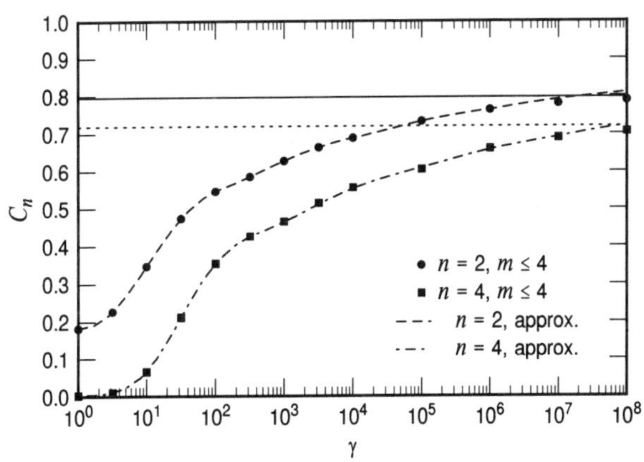

Figure 6 Model 2 with Rayleigh fading - Optimal capture probability, C_n, vs. power level dynamic range for the exact case, but with the number of power levels limited to $m = 4$, and for the log equi-spaced approximation; $b = 10$ dB, and $n = 2$ and 4 transmitters.

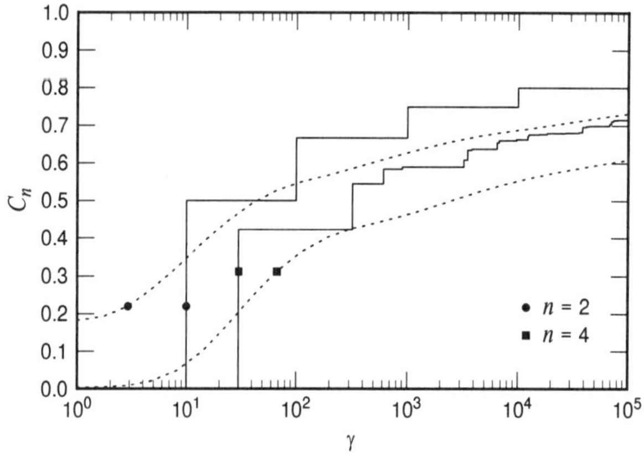

Figure 7 Model 2 - Optimal capture probability, C_n, vs. power level dynamic range without (solid) and with (dotted) Rayleigh fading (using log equi-spaced approximation); b = 10 dB, and n = 2 and 4 transmitters.

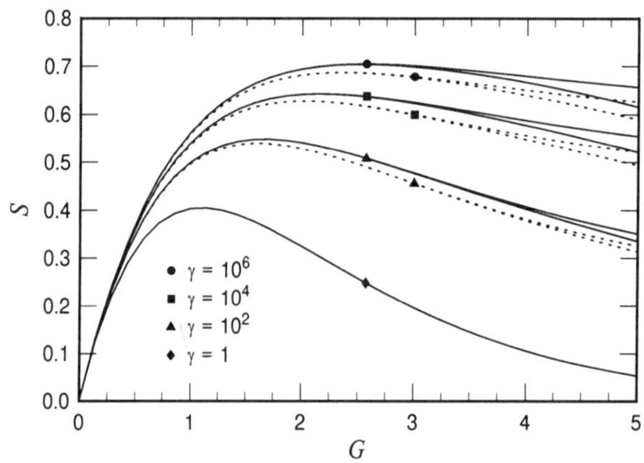

Figure 8 Model 2 with Rayleigh fading - Slotted ALOHA throughput, $S(G)$, vs. offered load, G for the upper bound (solid) and optimal (dotted) cases using the log equi-spaced approximation for both cases; b = 10 dB, and n_{max} = 8 transmitters; all cases show upper and lower truncation bounds.

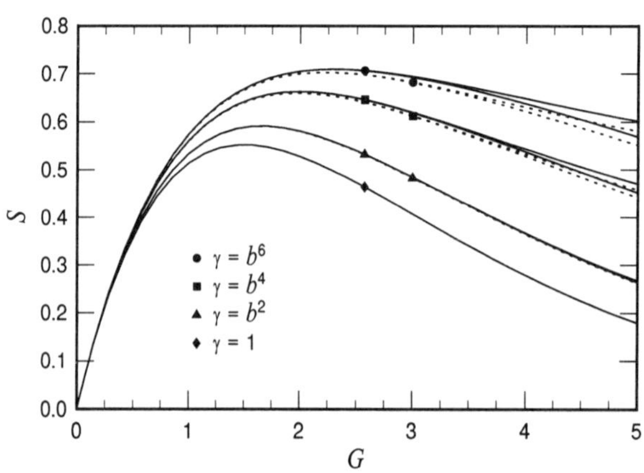

Figure 9 Model 2 with Rayleigh fading - Slotted ALOHA throughput, $S(G)$, vs. offered load, G for the upper bound (solid) and optimal (dotted) cases using the log equi-spaced approximation for both cases; $b = 3$ dB, and n_{max} = 8 transmitters; all cases show upper and lower truncation bounds.

A NEW MEDIUM ACCESS CONTROL PROTOCOL FOR INTEGRATED TRAFFIC PERSONAL COMMUNICATION NETWORKS *

Dimitrios Makrakis †
Dept. of Electrical Eng.
Lakehead University
955 Oliver Road
Thunder Bay, Ont.
P7B 5E1
CANADA

Ranjeet S. Mander
Luis Orozco-Barbosa
Dept. of Electrical Eng.
University of Ottawa
161 Louis Pasteur
Ottawa, Ont., K1N 6N5
CANADA

Panayota Papantoni-Kazakos
Dept. of Electrical Eng.
317 Houser Hall, Box 870286
University of Alabama
Tuscaloosa, AL 35487-0286
U.S.A.

Abstract

The future success of the personal communications service depends on its ability to accommodate efficiently integrated traffic and service a variety of applications and communication sources with different quality of service requirements. For doing so, efficient algorithms are needed to support integrated services in the personal and mobile communications environment. In the present work, we propose a Spread Slotted Random Access protocol with a multi-priority service mechanism incorporated in its structure. The priority mechanism is implemented and controlled explicitly by the users and without any involvement from the network. This makes the allocation of priority transparent to the network infrastructure and adds to the simplicity and practical value of the proposed scheme. We examine the proposed protocol under reservation and non-reservation policies and evaluate it for integrated and data traffic. Our results show that the protocol is able to achieve high levels of utilization and is capable to accommodate relatively large volumes of integrated traffic.

---

*This research was supported by a Start-up Research Grant from Lakehead University, by the Natural Sciences and Engineering Research Council of Canada (NSERC) under Research Grant (Individual) OGP0184108 and the Telecommunications Research Institute of Ontario (TRIO).

†Also Adjunct Professor with the Dept. of Electrical Eng. of the Technical University of Nova Scotia (TUNS), Halifax, Nova Scotia.

1 Introduction

Future generations of Personal Communication Networks (PCNs) should support multimedia applications, thus they should be able to service integrated traffic generated by different types of sources such as voice, video, data. It is essential that the integrated traffic should be accommodated in a power and spectrally efficient way. It is also important that the network is able to satisfy the Quality of Service (QoS) Requirements of the various types of sources and applications.

Code Division Multiple Access (CDMA) is quite popular in the personal communications field [1], due to its robustness to several serious impairments of the wireless personal communications channel and system (multipath fading, interference, frequency selectivity, synchronization errors), its ability to take advantage of the voice activity factor with relatively simple hardware, and the higher peak power efficiency it demonstrates as compared to Time Division Multiple Access (TDMA). In addition to that, CDMA networks can share the same bandwidth with other narrowband signals, which allows for a more efficient use of the available radio frequencies spectrum. Presently, several hybrid networks considering the simultaneous operation of narrowband and CDMA systems over the same frequency bandwidth have been proposed. (See for example [2].)

In [3], [4] we examined a relatively simple version of CDMA network called Spread Slotted Aloha (SSA). Our analysis showed that the SSA network is able to perform well, while its overall structure remains simple. However, similarly to other proposed CDMA networks, SSA has not taken into consideration the possibility of servicing more than one type of traffic. Also, SSA uses the Slotted Aloha (SA) protocol to handle the competition between packets. SA, though simple, it becomes unstable when dealing with large user populations and highly dynamic traffic loads. (A scenario that is expected to be a seen in future PCNs.) In the present work, we propose a number of modifications to SSA that resolve these two weaknesses. First, we introduce a multi-priority mechanism into the structure of the protocol, in order to be able to accommodate nonhomogeneous traffic. We also replace SA with a stable Collision Resolution Algorithm (CRA) to handle the collisions occurring between competing packets. It is worth mentioning that the priority mechanism is controlled exclusively from the users themselves, thus, there is no need for central control and intervention of the network. This reduces considerably the complexity of the Base Station (BS) and of the network in general. Several other works have appeared in the literature, dealing with the application of stable collision resolution algorithms to CDMA. (See for example [5]). However, these works do not consider integrated traffic, neither they have included into the protocol the ability to service users and applications requiring different priorities. In addition to that, the CRA adopted in this work is different from algorithms examined in the past. This algorithm has several advantages from other CRA proposed in the past, in terms of practical value and implementation efficiency.

In the present work we describe an improved version of SSA and the multi-priority Spread Slotted Random Access (SSRA) protocol in a general fashion and examine their performance when servicing voice and data traffic. For the remainder, this paper is organized as follows. In section 2 we discuss briefly the system. Section 3 describes the priority mechanism that is incorporated into the protocol. Sections 4 and 5 describe the SSA and the proposed SSRA protocol respectively. Section 6 deals with performance issues. Section 7 concludes this work.

2 Description of the System

Mobile and portable terminals located within a cell communicate their information to a BS. After receiving the information, the BS is responsible for directing it (through the wired and wireless network) to its destination. We define as active user, a user with a packet available for transmission.

We consider a slotted system with the size of the slot being equal to the size of the packet. The feedback is binary (Collision (**C**) versus Non-Collision (**NC**)) and the Feedback Channel (**FC**) is considered error-free. Let us assume that the network is allocated a total of N_c spreading codewords. In the present work we assume that a packet transmission attempt fails (thus the base station returns a **C** response through **FC**) if either or both of the following two conditions occurs: i) there is at least one more user that transmits during the slot using the same codeword; ii) a total of more than N_{IL} users attempt transmission during the slot. The second condition applies an interference limitation to the network. Users generate interference to each other [6]. The interference level increases as the number of users becomes larger. We assume that if there are more than N_{IL} users transmitting simultaneously, the interference increases to a level that produces a Bit Error Rate (BER) which is unacceptably high for the service. N_{IL} depends on the spreading gain, the operating Signal to Noise Ratio (SNR), the requirements of the network and the service regarding BER performance, the type of spreading codewords and the synchronization properties of the transceivers.

In several applications, (for example voice), when a user is active, it produces packets at regular time intervals. This knowledge has been used in the past to implement some hybrid TDMA protocols which combine random access and reservation characteristics (see for example [7], [8]). The main principle of these protocols is the following. When a user switches from the inactive to the active state, it starts producing packets. The first generated packet starts competing and continues doing so, until it is withdrawn for reasons associated with the nature of the application or the operation of the system (i.e. due to the aging of the packet or because of exceeding the maximum allowable number of attempts). Reception of the first successful packet initiates a process which marks certain slots as reserved (**R**) and makes them available to the user. The reservation policy eliminates the need of competition for a large percentage of packets, thus it reduces the number of slots that end up wasted due to collisions. When a user becomes inactive, it loses its reservation rights. In the present work we examine both, reservation and non-reservation versions of the SSA and SSRA protocols. Both, the codeword and the slot are reserved for the active user. Below we explain how priority is allocated to the users of different classes and how the priority mechanism is incorporated into the protocol.

3 Description of the Priority Mechanism

Let us assume that the network wants to accommodate M_g different classes of users. (The segregation of the total user population to classes is based on the type of application supported by the user and/or the nature of the user.) We define a number of N_c groups, $\mathcal{G}_1, \mathcal{G}_2, \ldots, \mathcal{G}_{N_c}$. Each group corresponds to a different level of priority. The groups are ordered in an upwards increasing priority level (i.e. $\mathcal{G}_1$ has the lowest and $\mathcal{G}_{N_c}$ the highest priority). The system does not have to utilize all the groups (i.e. M_g could be less than N_c.) Let us assume that a user which belongs to the $\mathcal{G}_m$ group ($1 \leq m \leq N_c$), becomes active during slot $k-1$. The user accesses the network as follows. The user chooses randomly m out of the N_c available codewords. The user follows the evolution of the m competition processes (that take place between users spreading their packets with one of the m chosen codewords), by monitoring the information coming through the feedback channel. This allows the user to identify when is the earliest time that (according to the rules of the protocol) it can join one of the m competition processes. (The rules that qualify a user to join a competition process will be described at a latter point.) In several cases, the user might be qualifying to join the competition process of more than one codewords. In this case, the user chooses randomly and equally probably one of these codewords and joins its competition process. It should be clear at this point that users with higher priority have more choices since they are authorized to access more codewords and join the one that allows them to start competing at the earliest time possible. Notice that this priority mechanism is totally controlled by the users themselves, without any intervention from the network. It is fair to claim that the priority assignment process is totally transparent to the functionality of the network. In [11] a similar approach was used in a two-priority scheme. The differences between [11] and the scheme described in the present section are the following. The present scheme can implement multiple (more than two) priorities and can accommodate more than two classes of users. [11] deals with narrowband channels, whereas the present work deals with wideband channels, able to service more than one users at the same time. This has implications on the algorithmic operation and the performance of the protocol. In addition to that, [11] examined the performance of the two-priority scheme strictly under limit Poisson traffic. In the present work, we examine the performance using more realistic traffic models for the wireless network.

4 Description of the Spread Slotted Aloha with Priority

The operation of the SSA with user priority is as follows. Let us assume that a user which belongs to the mth priority class is active during the $(k-1)$ slot (k is an integer). Let us also assume that $\mathcal{L}$ out of the N_c codewords are under reservation during the k slot. The user decides to transmit the packet in the k slot with probability $P(m, l, \Delta\tau_g)$, where m indicates the class of the user, l the number of times this packet was involved in failed transmission attempts in the past and $\Delta\tau_g$ the number of slots elapsing between the moment the packet was generated and the present slot. As mentioned earlier, in several applications, a packet becomes useless for the application, if it arrives latter than a certain time after its generation. In this case, if $\Delta\tau_g$ grows over a certain limit (its value depending on the application) the packet is withdrawn ($P(m, l, \Delta\tau_g) = 0$). Similarly, the protocol strategy could be such that if a packet exceeds a number of failed attempts, it is withdrawn. (A large number of consecutive collisions indicates that the network is overloaded.) If the user decides in favor of transmitting during the k slot, it chooses randomly (and equally probably) $n = min\{m, N_c - \mathcal{L}\}$ out of the $(N_c - \mathcal{L})$ available codewords and afterwards it selects and uses one of them for spreading its signal. The user receives through the **FC** information about the luck of its transmission. If a Collision (C) is received, the user repeats at slot $k+1$ the process described earlier.

5 Description of the Spread Slotted Random Access Protocol with Multi-Priority

The behaviour and performance of the SSRA protocol is determined by the policies followed by the users on: i) how they select the codeword(s) they plan to use for spreading their signal; ii) how a user decides to join a competition process; iii) what is the algorithmic process to be followed by a competing user. The procedure followed by the user for the choice of spreading codeword(s) was discusses in section 3. In the present section we shall address the second and third items.

In the present work, the collision resolution protocols chosen to replace SA is the 2-C algorithm and modifications of it. Modifications have been applied to incorporate additional priority mechanisms in the structure of the 2-C protocol. The 2-C algorithm is stable and provides relatively high throughput. Under limit Poisson traffic, it can accommodate traffic load of 0.429 packets/slot guaranteeing finite average packet delay. It can be deployed with a limited sensing policy. 2-C has been chosen over a number of other stable protocols for the following reasons: i) its simplicity of implementation; ii) it is robust to errors in the feedback channel [10]; iii) it allows the users to be able to identify the ends of competition processes by simply observing the feedback channel (more about this will follow later). While several other protocols allow users to enter the competition process right away (after they become active), 2-C prevents them from doing so. This forces a more orderly evolution of the competition process. Below we describe briefly the 2-C algorithm.

5.1 Description of the 2-C Protocol

The collision resolution algorithm of the 2-C protocol can be described by using the two-cell concept, presented in Fig. 1. We assign to the system two cells. The first is called Transmission Cell (**TC**) and the second Waiting Cell (**WC**). Users participating in a competition process reside in one of these cells. The content of the cells is updated from one slot to the next and the updating is performed according to the result coming trough the feedback channel (i.e. **C** versus **NC**). Users that during a slot belong to the **TC** are allowed to transmit in this slot. Users residing in the**WC** have to refrain from transmission.

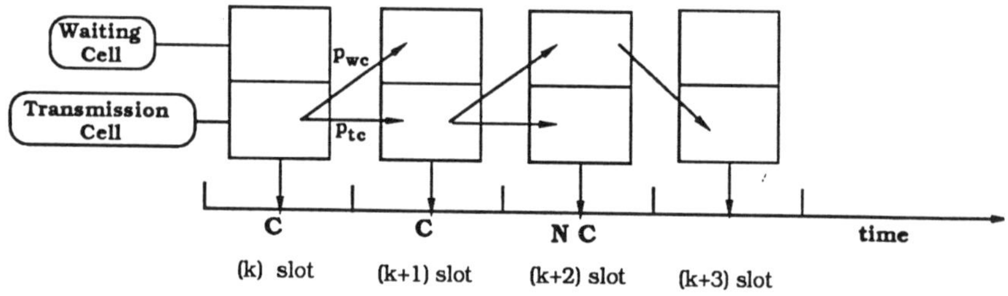

Figure 1: Description of the 2-C Algorithm, using the two-cell concept.

As mentioned earlier, one of the characteristics of the 2-C protocol is that it does not allow new users to enter an ongoing competition. They have to remain idle until all the packets engaged presently in competition are accommodated. (The mechanism that allows active but idle users to identify that a competition process is in progress or that it was completed will be explained at a latter point). When the on going competition process terminates, (let us assume that this occurs at slot $(k-1)$) each active but idle user that is able to identify its termination, decides if it qualifies to join the next to be formed competition process. The qualification is determined by a windowing or a probability of transmission policy. In the windowing case, the user examines the time that elapsed between the moment the packet was generated and the end of the competition process. If the difference is less than a certain value $d_w(m)$ (window size for the mth priority class users) the user is allowed to transmit. If not, the user refrains from transmission, however it updates the value marking the time of the packet generation by an amount equal to the window size (i.e. if the packet was marked as being generated at τ, it updates the time of its generation to $\tau + d_w(m)$). If the probability of transmission process is used, every active but idle user that belongs to the mth class, decides with probability $P_d(m)$ to join the new competition process.

If the decision is positive, the user attempts transmission in the coming slot (i.e. k^{th}). All users deciding to join the competition process, starting at k slot, position themselves in the **TC**. If a collision occurs in the kth slot, the **FC** carries a **C** status to the users. Then, every user that participated in the collision, chooses with probability $p_{tc} = 0.5$ to remain in the **TC** and with probability $p_{wc} = 0.5$ to move in the **WC**. The users residing into the **WC** refrain from transmission during the $(k+1)^{th}$ slot. Those remaining in the **TC** attempt transmission. If a collision occurs, the process is repeated again, and (statistically speaking), more users move in the **WC** for the $(k+2)^{th}$ slot. If instead, a **NC** is received through the **FC**, (i.e. during the $(k+1)^{th}$ slot at most one user was residing in the **TC**) all users residing in the **WC** move to the **TC** and transmit at $(k+2)^{th}$ slot. The reader can verify that when following this policy, the only way to receive two consecutive **NC** (i.e. **NC-NC**) is when both, **TC** and **WC** are empty, in which case, all the competing users have been accommodated. The **NC-NC** pattern is the marker which permits the idle users to identify the termination of an on-going competition process.

The algorithmic steps of the 2-C protocol can be implemented easily by the individual users as follows. Each user is equipped with a binary counter, its value being either 1 or 2. We represent the value of the counter of a specific user during the kth slot as $\mathcal{Z}_k$ ($\mathcal{Z}_k \in \{1,2\}$). We represent the feedback status received by the user during the kth slot as $\mathcal{F}_k$ ($\mathcal{F}_k \in \{\mathrm{NC}, \mathrm{C}\}$). A user attempts transmission during the kth slot if $\mathcal{Z}_k = 1$ and withholds transmission if $\mathcal{Z}_k = 2$. A packet is successfully transmitted if $\mathcal{Z}_k = 1$ and $\mathcal{F}_k = \mathrm{NC}$. When a user becomes active, it sets its counter to $\mathcal{Z}_k = 2$ and starts monitoring the feedback channel. An idle user does not change the content of its counter until it identifies two consecutive NC. When this happens, the user checks if it meets the qualifying criteria for joining the new competition process. If the decision is positive, the user switches the value of its counter to 1 and transmits at the first slot that follows the two consecutive slots marked with NC. Let us assume that this is the nth slot. From that point on, the user updates its counter as follows.

For $l \geq n$ and until the packet has been successfully transmitted:

1. if $\mathcal{F}_l = \mathrm{NC}$ and $\mathcal{Z}_{l-1} = 2$, then $\mathcal{Z}_l = 1$.
2. if $\mathcal{F}_l = \mathrm{C}$ and $\mathcal{Z}_{l-1} = 2$, then $\mathcal{Z}_l = 2$.
3. if $\mathcal{F}_l = \mathrm{C}$ and $\mathcal{Z}_{l-1} = 1$, then

$$\mathcal{Z}_l = \begin{cases} 1 & \text{with probability } p_{tc} = 0.5 \\ 2 & \text{with probability } p_{wc} = 0.5 \end{cases} \tag{1}$$

A packet that expired (i.e. becomes obsolete due to the violation of a delay bound) is replaced by the next packet in line. This packet continues the algorithmic process from the point that the expired packet left it.

5.2 The 2-C protocol with Imbeded Priority

Let us assume that a user that belongs to the mth priority class is active but idle. Let us also assume that the user sees a **NC-NC** pattern, marking the $k-1$ and k slots. Instead of attempting transmission, the user waits for an additional $(M_g - m)$ slots (i.e. the user waits to see a pattern consisting of $\{M_g - m + 2\}$ **NC** before it attempts a transmission). If a collision occurs before the user identifies $(M_g - m + 2)$ consecutive **NC**, it starts counting from the beginning (right after the slot marked as **C**). This waiting ensures that no users of higher priority want to use the network resources. We call this policy Class Blocking (**CB**). A second level of priority can be included as follows. If the user joins a competition process and right after the first attempt it ends up with a **C**, it refrains from re-transmitting for $(M_g - m + 1)$ slots. If the **FC** marks these slots as **NC**, the user joins (again) and executes the CRA which will send eventually the packet through the network. If a **C** appears in one of these slots, the user refrains from further attempts, returns back into the idle mode and starts the process from the beginning. We call this strategy Double Class Blocking (**DCB**) policy. **CB** and **DCB** affect the priority by giving to users with higher priority the opportunity to join and form competition process earlier as well as compete mainly within themselves.

As mentioned earlier, several TDMA based protocols [7], [8] use a hybrid random access and reservation policy to accommodate users which generate packets periodically when in the active state. We have included the reservation option in SSA and we do the same for the proposed SSRA protocol as well.

6 Performance Evaluation Results

We evaluate the SSA and SSRA protocols when transporting voice and data traffic. We assume that voice terminals are equipped with "*fast*" Voice Activity Detectors (VADs). "*Fast*" VADs are able to distinguish between mini-talkspurts and mini-silence periods that reside within a talkspurt. The traffic generated by a voice user equipped with a fast VAD can be described by a 3-state Markovian model [12]. The average duration of the silence and talkspurt periods have been set equal to 1.35 sec. and 1 sec. respectively. The average duration of mini-silences and mini-talkspurts, occurring within a talkspurt are set equal to 50 msec. and 0.275 sec. respectively. Digital voice is relatively insensitive to channel errors and packet losses. On the other hand, voice is sensitive to delays. Packets should satisfy a certain delay bound in interpacket arrival distance. If a packet violates this limit, it becomes obsolete and the transmitter should discard it in favor of a more recently produced packet. In the present work, we assume that the maximum tolerable delay for voice packets is 32 ms, and the maximum acceptable packet dropping rate is 1%.

Data packets should be delivered to the destination error free, however, data traffic is insensitive to delays. We assume that the aggregate data traffic follows a Poisson distribution. This assumption is based on our conviction that in the future, we will see a large number of data terminals supporting non-real time applications (i.e. paging, e-mail etc.), each one producing a small traffic activity. Poisson traffic represents quite accurately the aggregate traffic produced by a large number of low activity users.

The simulated SSA and SSRA configurations use $N_c = 4$ codewords and $N_{IL} = 10$. We remind the reader that N_{IL} specifies the maximum number of simultaneous transmissions which can occur without resulting in loss of packets due to unacceptable interference level. Higher values of N_{IL} do not provide any substantial improvement. Voice users are allocated to $\mathcal{G}_4$ and data users to $\mathcal{G}_1$. The duration of every slot is being set equal to 1.6 msec, thus a voice packet becomes obsolete 20 slots after the slot it was generated (i.e. for voice, $\Delta\tau_g = 20$).

As mentioned earlier, SSA has been studied with and without reservation policy. In section 4 we described a version of SSA that is able to provide multiple priority (thus, it can accommodate users with different requirements), by applying a codeword selection policy. For the special case of voice and data traffic we examine in the present section, this policy applies as follows. Voice users, when in active state and before committing themselves to a specific codeword, examine which codewords have slots that are non-reserved and they randomly choose one of these unreserved slots. They repeat this procedure every time a voice packet has to go through competition. On the contrary, data users commit themselves to a particular codeword, and keep using this codeword until either, the packet has being successfully transmitted or it is withdrawn. We will be referring to this scheme as SSA/Conf-1. In the case of SSA without reservation, we consider two different schemes. In the first, a voice user chooses at random a codeword every time it has a packet available for transmission. A data user commits itself to a codeword at its first try and keeps it until either the packet goes through or is rejected. This scheme will be referred as the SSA/Conf-2. In the second scheme, both voice and data users choose a codeword at the beginning of their active period, and keep using it until the end (i.e. the talkspurt finishes or the data packet goes through or is rejected). This scheme will be called the SSA/Conf-3.

We investigated the performance of the various SSA and SSRA configurations for various numbers of voice users and various values of data load. Due to the space limitations in the present work, we will present results only for data load $\lambda = 0.2$ packets/slot. For SSA, the following values of $P(1, m, \Delta\tau_g)$ (data) and $P(4, m, \Delta\tau_g)$ (voice) are used:

$$P(1, m, \Delta\tau_g) = \begin{cases} 1 & \text{for } m = 0 \\ 0.02 & \text{for } 1 \le m \le 9 \\ 0 & \text{for } m > 9 \end{cases} \tag{2}$$

$$P(4, m, \Delta\tau_g) = \begin{cases} 1 & \text{for } m = 0 \text{ and } \Delta\tau_g \leq 20 \\ p_{ret} & \text{for } m \neq 0 \text{ and } \Delta\tau_g \leq 20 \\ 0 & \text{for } \Delta\tau_g > 20 \end{cases} \quad (3)$$

where p_{ret} is the re-transmission probability of collided voice packets. When reservation policy is used $p_{ret} = 0.15$. In the absence of reservation policy $p_{ret} = 0.5$. These values have been chosen after thorough investigation of the influence of $P(1, m, \Delta\tau_g)$, $P(4, m, \Delta\tau_g)$ on the performance of the various SSA configurations.

SSRA has been examined with and without reservation policies as well. In both cases, we examine configurations with and without **CB** policy. Below we will provide results for the following four configurations. SSRA with reservation and **CB** policy (SSRA/Conf-1). SSRA with reservation but without **CB** policy (SSRA/Conf-2). SSRA without reservation but with **CB** policy (SSRA/Conf-3). SSRA without reservation and without **CB** policy (SSRA/Conf-4). In the reservation schemes, the data users use a window size $d_w(1)$ equal to 5 slots. In the non-reservation schemes, $d_w(1) = 2.5$ slots. In all the examined cases of the SSRA protocol, no windowing is applied to the voice users, however, voice packets are dropped if they have not been transmitted successfully within a time frame of twenty slots from the time they were generated. When **CB** policy is not applied, both, voice and data users are allowed to join the new competition process right after they have identified the termination of the previous one. Users detect the termination of a competition process by identifying two consecutive **NC** through the **FC** (see section 5.1). When **CB** policy is applied, voice users are allowed to transmit after two **NC**, while data users have to wait for three **NC** before attempting transmission.

The results concerning the performance of the various SSA and SSRA configurations are displayed in Figs. 2 to 6. The results have been acquired through computer simulations, which were implemented using the OPNET software package. Fig. 2 shows the Packet Loss Rate (PLR) of voice packets versus the number of voice users. Fig. 3 presents the average delay experienced by data packets versus the number of voice users. Fig. 4 shows the blocking probability of data packets. Fig. 5 presents the utilization of the network. From the four curves, the following points are evident.

SSRA is evidently superior to SSA in all accounts. For a 1% dropping rate of voice packets, SSA with reservation (SSA/Conf-1) accommodates 34 voice users. SSRA/Conf-1 accommodates 61 voice users (a 79.4% improvement over SSA/Conf-1). SSRA/Conf-2 accommodates 54 voice users (a 58.8% improvement over SSA/Conf-1). Notice that the **CB** policy offers a considerable advantage. SSRA/Conf-1 accommodates 7 more voices users as compared to SSRA/Conf-2, i.e. 61 as compared to 54 (a 12% improvement). Our results have shown that the benefits offered by **CB** increase as the data load λ becomes higher. For the non-reservation schemes the numbers stand as follows. SSA/Conf-2 can accommodate up to 11 voice users without having the packet dropping rate exceed the 1% bound. SSA/Conf-3 accommodates only 2 users. The non-reservation SSRA schemes achieve the following levels. SSRA/Conf-3 can accommodate up to 24 voice users (offering an additional 1.182 capacity to the network, when used instead of SSA/Conf-2). SSRA/Conf-4 accommodates up to 18 users (offering an additional 0.636 capacity to the network, when used instead of SSA/Conf-2). Compared to SSA, SSRA achieves higher level of channel utilization as well. This becomes evident from the curves of Fig.5.

It is evident from Figs. 2, 3 that SSRA accommodates the data packets with less delay and achieves lower blocking probability as compared to SSA. Our results have shown that the advantage of SSRA over SSA increases as the data load becomes higher. Comparison of the curves corresponding to SSRA configurations with **CB** policy (SSRA/Conf-1 and SSRA/Conf-3) with those of SSRA configurations that do not use **CB** policy (SSRA/Conf-2 and SSRA/Conf-4) shows that **CB** policy penalizes the data traffic. This is expected, since **CB** policy forces data to withhold transmission in favor of voice packets.

Reservation improves considerably the performance of both, the SSA and SSRA schemes. For voice packet dropping rate of 1%, SSRA/Conf-1 accommodates 61 voice users versus 28 voice users accommodated by SSRA/Conf-3 (an improvement of 154%). This improvement is due to the reservation policy followed by SSRA/Conf-1. Reservation policy is absent from SSRA/Conf-3. Advantages exist for SSA as well. SSA/Conf-1 accommodates 34 users versus 11 users accommodated by SSA/Conf-3 (an increase of 238%). The same conclusions can be reached by observing the channel utilization levels achieved by the various configurations (see Fig. 5). SSRA shows a channel utilization close to 71% and SSA close to 51%. In the absence of reservation, the utilization of both schemes drops. Without reservation, SSRA reaches a maximum utilization close to 38% and SSA close to 28%.

Fig. 6 gives an idea of how the value of N_{IL} affects the dropping rate of voice packets and blocking rate of data packets. The curves displayed in Fig. 11 correspond to SSRA with reservation but without **CB** (SSRA/Conf-2). The number of voice users in the channel has been set equal to 70. We see that practically, all the improvement in terms of lowering the loss rate of voice and data packets is achieved for $N_{IL} = 8$. Further increase of N_{IL} offers only minimal additional improvements, that do not justify the additional complexity and/or waist of bandwidth that would result from an increase in the value of N_{IL}.

Summarizing this section, we can claim that our results demonstrated the superiority of the SSRA protocol as compared to SSA. In addition to that, considerable gains in capacity and channel utilization are possible by adopting a reservation policy.

7 Conclusions

We have proposed a multi-priority Spread Slotted Random Access Protocol suitable for servicing integrated traffic. The priority mechanism is implemented and controlled solidly by the users, without requiring the intervention of the network. This adds to the simplicity and practical value of the proposed scheme. The protocol was evaluated for voice and data traffic and was compared to several variations of the Spread Slotted Aloha. Configurations with and without reservation were examined for both protocols. The results verified the superiority of the proposed protocol: high channel utilization and ability to service effectively high volumes of integrated traffic. The simplicity and high performance level of the proposed scheme makes it valuable for supporting integrated traffic wireless applications.

References

[1] D. L. Schilling et. al., "Broadband CDMA for Personal Communication Systems", IEEE Communications Magazine, Vol.29, No. 11, pp. 86-93, Nov. 1991.

[2] Chih-Lin I, L. J. Greenstein and R. D. Gitlin, "A Microcell/Macrocell Cellular Architecture for Low- and High- Mobility Wireless Users", IEEE Journal on Selected Areas in Communications, Vol. SAC- 11, No. 6, pp. 885-891, Aug. 1993.

[3] D. Makrakis and K.M.S. Murthy, "Spread Slotted Aloha Techniques for Mobile and Personal Satellite Communications", IEEE Journal on Selected Areas in Communications, Vol. SAC- 10, No. 6, pp. 985-1002, Aug. 1992.

[4] D. Makrakis and K.M.S. Murthy, "Performance Analysis of Spread Slotted Aloha Networks in the Presence of Timing and Frequency Uncertainties", Proceedings of the 2nd International Conference on Universal Personal Communications, Ottawa, Ont., Oct. 1993.

[5] Ming-Chih Hu and Jin0Fu Chang, "Collision Resolution Algorithms for CDMA Systems", IEEE Journal on Selected Areas in Communications, Vol. SAC- 8, No. 4, pp. 542-554, May 1990.

[6] L. S. Lehnert and M. B. Pursley, "Error Probability for Binary Direct-Sequence Spread Spectrum Communications with Random Signature Sequences, IEEE Transactions on Communications, Vol. COM- 35, No. 1, pp. 87-98, Jan. 1987.

[7] D. J. Goodman, R. A. Valenzuela, K. T. Gaylaird and B. Ramamurthi, "Packet Reservation Multiple Access for Local Wireless Communications", IEEE Transactions on Communications, Vol. COM- 37, No. 8, pp. 885-890, Aug. 1989.

[8] N. M. Mitrou, T. D. Orinos and E. N. Protonotarios, "A Reservation Multiple Access Protocol for Microcellular Mobile Communication Systems", IEEE Transactions on Vehicular Technology, Vol. VT- 39. No. 8, pp. 340-351, Aug. 1990.

[9] M. Paterakis and P. Papantoni-Kazakos, "A Limited Sensing Random-Access Algorithm with Binary Success-Failure Feedback", IEEE Trans. on Comm., Vol. COM-37, No. 5, pp. 526-530, May 1989.

[10] M. Paterakis and P. Papantoni-Kazakos, "A Simple Window Random Access Algorithm with Advantageous Properties", IEEE Trans. on Inf. Theory, Vol. IT-35, No. 5, pp. 1124-1130, Sept. 1989.

[11] P. Papantoni-Kazakos et. al., "Transmission Algorithms for Multi-Channel Packet Radio System with Priority Users", submitted for publication.

[12] D. J. Goodman, S. X. Wei, "Efficiency of Packet Reservation Multiple Access", IEEE Transactions on Vehicular Technology, Vol. VT- 40, No. 1, pp. 170-176, Feb. 1991.

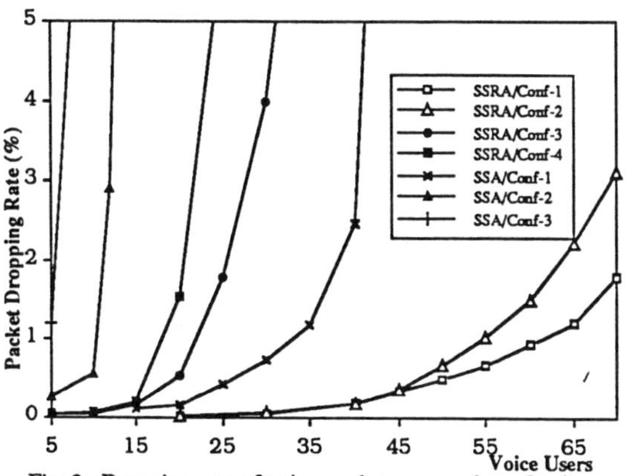

Fig. 2: Dropping rate of voice packets vs number of voice users.

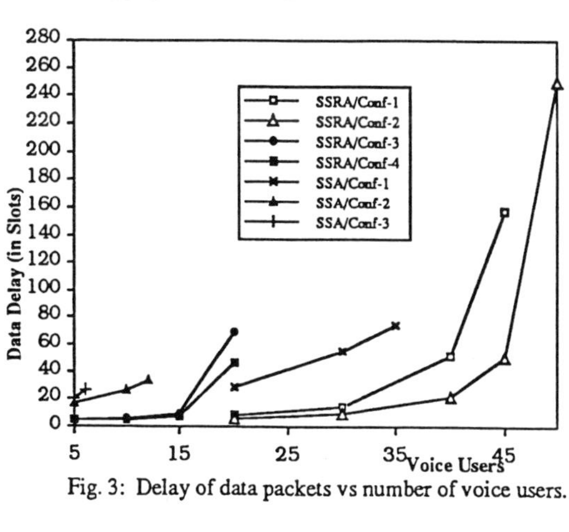

Fig. 3: Delay of data packets vs number of voice users.

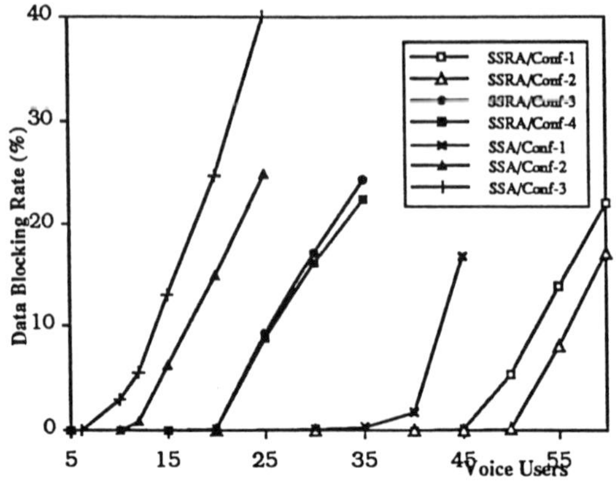

Fig. 4: Blocking rate of data packets vs number of voice users.

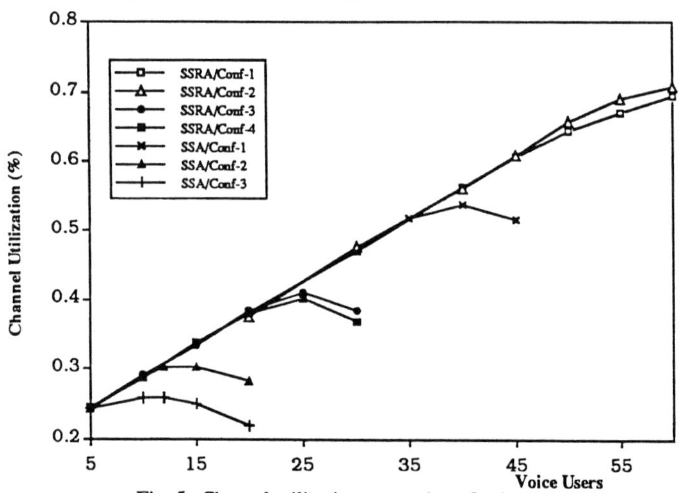

Fig. 5: Channel utilization vs number of voice users.

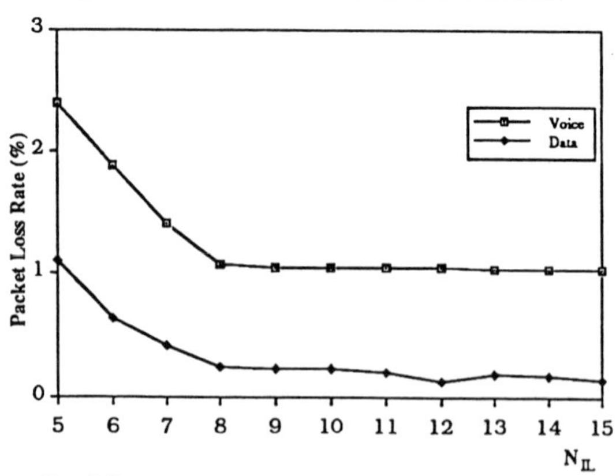

Fig. 6: Dropping rate of voice packets and blocking rate of data packets vs the maximum acceptable number of simultaneous transmissions. Number of voice users: 70.

A CALL-LEVEL ACCESS CONTROL STRATEGY FOR INTEGRATED SERVICES WIRELESS PACKET NETWORKS

Franco Davoli, Piergiulio Maryni, Carlo Nobile

Department of Communications, Computer and System Sciences

DIST-University of Genova
Via Opera Pia 13,
16145 Genova, Italy
email:franco@dist.unige.it

ABSTRACT

A simple call admission control system for multiservice wireless networks is presented. By means of it, connection requests coming from different "classes" of traffic (i.e., voice, data) are regulated in order to meet a set of Quality of Service (QoS) constraints. The admission mechanisn works on two levels: a static and a dynamic one. The static level looks at "packet-level" QoS, such as cell loss and delay, while the dynamic level takes care of call dynamics and allows to drive the load of the system with respect to the various traffic classes . Whenever a connection request for the n-th traffic class appears, both levels of control intervene and the call is accepted if both of them do not reject it. The decision at the static level is deterministic, and it is based on the concept of Schedulable Region; the decision at the dynamic level is probabilistic and it is based on the concept of Input Rate Region, which is explained in the paper. Results that show the performance of this control mechanism are presented.

1 INTRODUCTION

Third generation wireless systems represent one of the key topics in Telecommunications. If one wants to exploit the possibilities offered by the technology, the whole scenario can become quite complex, since it heavily involves the use of different media such as video, voice and data that have to be coordinated in real time together with the user mobility. In this respect the sharing of the channel can be done in several ways [1, 2], along with access control mechanisms that allow to dynamically allocate the bandwidth among services with possibly very different Quality of Service (QoS) constraints [3]. In this paper, we investigate a connection admission control scheme that operates at two different levels, namely, a static and a dynamic one. As regards the static level, we rely on the concept of "Schedulable Region" [4], which is the region in the connection space where QoS requirements are satisfied. As regards the dynamic level, the maximum load, in terms of arrival rates that the system can bear, is computed by considering all the arrival rates configurations (for each traffic class) for which the probability of being on the boundary of the "Schedulable Region" is below a given threshold. We call this second load information "Input Rate Region": when an arrival rate configuration lies inside the Input Rate Region, it is defined to be "acceptable". Whenever the real configuration of arrival rates lies outside the Input Rate Region, the call admission control system computes the values of a set of filters which bring back the rates to an acceptable level. Since many choices are possible, the "best" one is chosen by maximazing a revenue function that takes into account both the probability of rejecting an incoming call and the revenue of the various traffic classes. The Input Rate Region, togheter with a suitable revenue function, ensures that a "low revenue" call is blocked only when a "high revenue" one is likely to arrive shortly. The paper is organized as follows. In Section 2 we give an overview of the system model. In Section 3 we describe the call admission control mechanism, which relies on the knowledge of the Input Rate Region. The Input Rate Region is also explained in Section 3 along with the steps for computing it. In Section 4 we present simulation results, which highlight the effect of the control mechanism compared to the uncontrolled situation.

2 THE SYSTEM MODEL

We consider a cellular system, where the base station and the mobile stations within a cell share a common channel. A finite number of types of mobile stations is envisaged. We might think, for example, of stations which generate voice as stations of the same type. Mobile stations of the same type generate the same kind of traffic, and require the same QoS constraints.

Each mobile station generates packets either singularly or in bursts, which are supposed to be transmitted on the channel following a multiple access control mechanism.

The number of mobile stations inside the cell is not constant. New mobile stations appear with a given "arrival" rate and disappear with a given "departure" rate. Whenever a station appears, it asks the base station for the ability to transmit. This ability can be either granted or refused. New mobile stations that appear within the cell represent either stations which have been "turned on" or stations which come from other cells. Notice that, for the sake of simplicity, the base station does not make any difference between the two cases, and therefore there is no real concept, in this model, of handoff. Indeed, this problem could be solved by introducing a handoff flow as third traffic class. Arrival and departure rates of mobile stations are not constant too. Indeed, these rates are supposed to change in an unpredictable way. In the following we will refer to a connection request as a call request.

Stations are not given a portion of the channel for the whole duration of their life, but only for the duration of a burst. Multiplexing takes therefore place at burst level. In order for this to work it is essential for stations to transmit their bursts according to a given multiple access protocol: in our case, we utilized, both in the simulations and in the computation of the Schedulable Region, the Reservation Random Access-Independent Stations Algorithm (RRA-ISA) [5]. It consists of an algorithm that controls the sharing of the channel in a random access protocol over a Time Division Multiplexing (TDM) structure, in which each slot of the frame accomodates one packet. The algorithm maximizes the one-step throughput [6], by computing a partition of the stations over each frame, and by assigning each slot a given set of the partition. The partition is computed by keeping into account the burst transmission probability of each mobile station.

Our goal, however, is to perform call admission control independently of the particular multiple access protocol being utilized. For this reason, we adopt the concepts of Schedulable Region [4] and Input Rate Region. The Schedulable Region is the set of call configurations for which a number of QoS contraints are guaranteed. It represents the maximum load of a system in terms of number of connections of the various traffic types. It can be easily computed for simple systems (e.g., those working with peak rate allocation, like TDMA) and it is usually estimated with simulations for more complex ones. Its computation (or estimation) has to be performed only once, for a given set of burst-level traffic characteristics (e.g., burstiness, average rate), performace requirements (e.g., maximum packet delay, loss probability, delay jitter), and system parameters (e.g., channel capacity, multiple access protocol).

Figure 1 shows an example of Schedulable Region, of a system in which the RRA-ISA scheme controls the access of two types of mobile stations: voice, or

Class 1, stations, which generate bursts of packets and require continuation of service within the burst, and data, or Class 2, stations which generate single packets.

Notice that within the Schedulable Region a given configuration of connections is supposed to last forever, i.e., there is no concept of call dynamics. The Schedulable Region is a static concept. The set of Quality of Service requirements the Schedulable Region refers to is related to the "packet level", that is packet delay, packet loss probability and so forth.

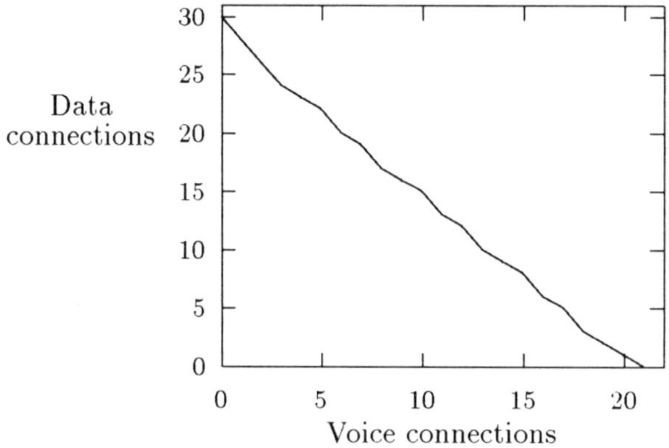

Figure 1 The Schedulable Region

3 THE CALL ADMISSION CONTROL MECHANISM

The Schedulable Region of a system gives an insight of the maximum load achievable in terms of number of connections for each traffic class. If a given configuration of calls is inside the Schedulable Region, then we know that packet level QoS's are guaranteed. The *complete sharing* admission policy [7] only relies on the Schedulable Region in order to decide if a new incoming conection has to be accepted or not. Such an admission control, even if simple and straightforward, has several problems. It may be unfair, in the sense that some classes might monopolize the system resources; it may lead to poor resource utilization and poor long-run average revenue. Furthermore, the system is burstly reactive: it starts rejecting connections only at the very last moment, when the network is "full", regardless of the load which is being offered. In essence, all these drawbacks can be summed up in the impossibility to drive the system load: the controller, if there is enough space, accepts connections independently of the traffic class they belong to.

Hence, a more refined admission control mechanism would be desirable (see, for instance [8] for a general discussion on this point). A possible solution comes from controlling not only the number of connections, but also the connection request dynamics. In the following, we will describe a possible, very simple, parametric strategy along with a method for computing a set of suitable parameters.

3.1 The Connection Admission Control Strategy

A control on connection request dynamics can be easily achieved with a set of input rate "filters", as depicted in Figure 2.

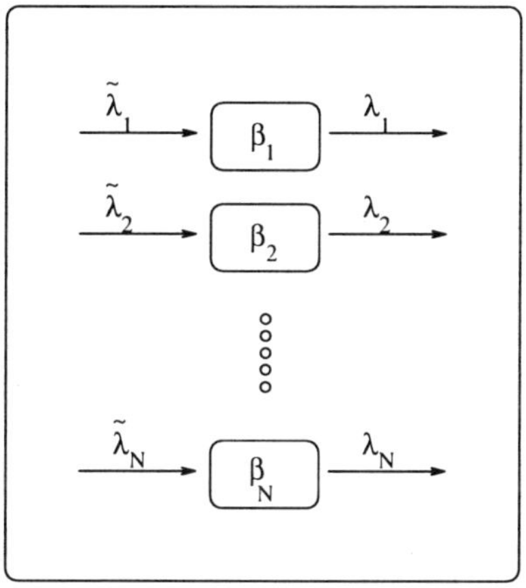

Figure 2 The Second Step of the Call Admission Control Mechanism

More specifically, whenever an incoming connection request from traffic Class i comes up, the connection control mechanism performs two steps:

1. check the Schedulable Region to see if the new connection can be accepted; if so

2. accept the connection with probability β_i and reject it with probability $(1 - \beta_i)$.

The first step ensures that by accepting the new connection, the QoS's of the whole system are still guaranteed. If a connection passes the first step the we call it "eligible". By means of the second step, if Class i has an incoming call rate $\tilde{\lambda}_i$ (calls/s), then the effective rate "seen" by the system for that Class is $\lambda_i = \beta_i \tilde{\lambda}_i$. By means of this mechanism, input rate can be appropriately driven.

Notice that, this filtering operation does not replace the complete sharing admission policy, based on the Schedulable Region, but it comes with it.

The problem now is how to choose the "target" input rates λ_i, (and therefore the parameters β_i) given that customers are generating an offered load $\tilde{\lambda}_i$. To do this we introduce a higher level concept of the system load, able to give us an insight on call dynamics: the Input Rate Region (IRR).

3.2 The Input Rate Region

While the Schedulable Region describes the maximum system load in terms of number of connections, the Input Rate Region (IRR) describes it in terms of call arrival rates. In other words, given a set of service rates, the IRR is the set of call arrival rate configurations for which a number of contraints are guaranteed. In [9] the IRR, which is called Admissible Load Region, is computed by means of an algorithm, based upon the knowledge of the Schedulable Region, which considers as QoS constraints a set of call blocking probabilities that have not to be exceeded. In our case, we set a constraint on the probability, for the state of the system, of being on the boundary of the Schedulable Region.

Let us define $\underline{n} = (n_1, n_2)$ as the state of the system, where n_1 and n_2 represent the number of connected stations of Class 1 and 2, respectively. Let us further define the state probability

$$\pi(i, j) = P\{n_1 = i, n_2 = j\}; \tag{1.1}$$

In a system with Schedulable Region Υ, following the method proposed in [7], we have:

$$\pi(i,j) = \frac{(\rho_1^i/i!)(\rho_2^j/j!)}{\sum_{(k,l)\in\Upsilon}(\rho_1^k/k!)(\rho_2^l/l!)} \tag{1.2}$$

where $\rho_1 = \lambda_1/\mu_1$ and $\rho_2 = \lambda_2/\mu_1$; $1/\mu_1$, $1/\mu_2$ represents the average durations of connections of Class 1 and 2, respectively. Obviously, if $(i,j) \notin \Upsilon$, then $\pi(i,j) = 0$.

Figure 3 shows the probability distribution given the Schedulable Region Υ depicted in figure 1 and $\lambda_1 = 0.04$, $\lambda_2 = 0.03$.

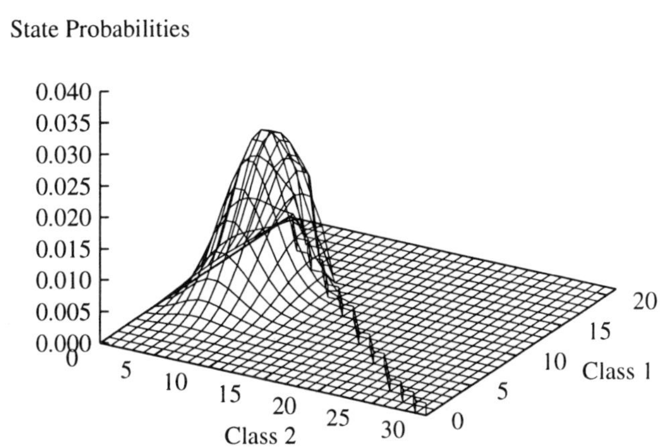

Figure 3 The computed state probabilities over the Schedulable Region.

In our case, we define the IRR the region in the conenction arrival rates such that the sum of all the probabilities of the states on the boundary of Υ do no exceed a given threshold ϵ.

In other words, a point (λ_1, λ_2) belongs to the IRR if

$$\sum_{\underline{n} \in \Upsilon_B} \pi(\underline{x}) < \epsilon \ . \tag{1.3}$$

where Υ_B represents the boundary of Υ.

Figure 4 has been obtained with $\epsilon = 0.1$, $\mu_1 = 0.006$, $\mu_2 = 0.003$.

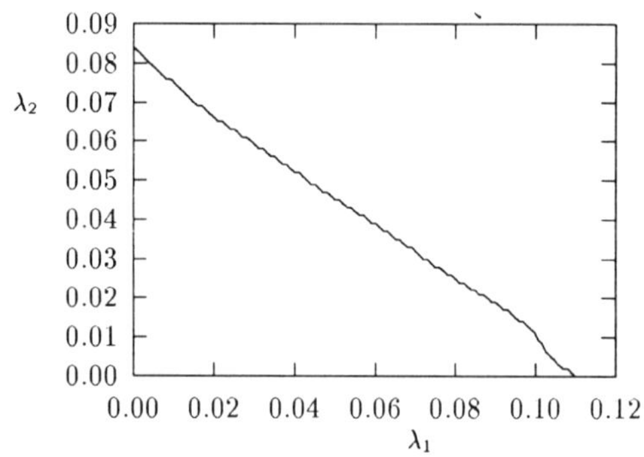

Figure 4 The Input Rate Region

Loosely speaking, the parameter ϵ defines the degree of saturation that we are willing to accept. Larger values of ϵ indeed give rise to a larger IRR where, however, the probability distribution of the states can be "stuck" against the boundary of the Schedulable Region. In our case, as will be clear in the following, this situation is undesirable because it would make our controllers start their operation only when the system is already heavily loaded. This fact, in turn, would decrease the ability to drive the system load towards a desired distribution. On the other hand, too small values of ϵ might lead to a poor utilization.

3.3 A parametric optimization procedure

We turn now to describe a possible way to choose the values of the filtering parameters β_i, given a IRR as defined above.

Let us denote with N the number of traffic classes and with $\tilde{\underline{\Lambda}}$ the set of input rates, $\tilde{\underline{\Lambda}} = \text{col}[\tilde{\lambda}_1, \tilde{\lambda}_2, \ldots, \tilde{\lambda}_N]$. If such a configuration $\tilde{\underline{\Lambda}}$ lies inside the IRR, then each eligible incoming call request is filtered in, that is $\beta_i = 1$ for each $i = 1, \ldots, N$. If a given configuration exceeds the IRR, then the filtering probabilites $\beta_1, \ldots, \beta_n$ are used to bring the system back to an acceptable level.

More specifically, suppose that, at a given time, the load $\underline{\tilde{\Lambda}}$ is outside the IRR, and suppose that we want to bring it back to the point $\underline{\Lambda}$ inside the IRR. Then we choose the filter $\underline{\beta} = \text{col}\,[\beta_1, \ldots, \beta_n]$ is such a way that $\beta_i \tilde{\lambda}_i = \lambda_i$ for each $i = 1, \ldots, N$, as depicted in Figure 5.

Given that a point is outside a region, there are many ways to bring it inside. We will only consider the set of points that represent the boundary of the region, bringing back to one of them the working point of the system. This choice is natural, since we have to bring back the working point to an acceptable situation, but still mantaining the highest possible load. Hence, given a IRR Ψ, let us denote with Ψ_B its boundary.

Moreover, since a filtering operation can only decrease rates, only a subset of the IRR boundary will be eligible to become the working point of the system. More specifically, if at a given time the load of the system, in terms of input rates, is $\underline{\tilde{\Lambda}}$ and this configuration lies outside the IRR, then a point $\underline{\Lambda} \in \Psi_B$ will be eligible to become the target working point if and only if $\lambda_i \leq \tilde{\lambda}_i$ for each $i = 1, \ldots, N$.

The task is therefore to choose which one, among the eligible ones, is going to become the effective working point of the system. This can be achieved by means of the maximization of a simple revenue function that takes into account both the external rates (i.e., $\tilde{\lambda}_i, \quad i = 1, \ldots, N$) and the input rates of the target working point (i.e., $\lambda_i, \quad i = 1, \ldots, N$). Once the best working point has been found, it is straightforward to compute the values of the filtering probabilities $\underline{\beta}$, by taking into account that $\beta_i \tilde{\lambda}_i = \lambda_i$. Given an external rate $\underline{\tilde{\Lambda}}$ and a target rate $\underline{\Lambda}$, we define the revenue function as follows:

$$J(\underline{\Lambda}|\underline{\tilde{\Lambda}}) = \left(\sum_{i=1}^{i=N} K_i \lambda_i \right) + W \frac{1}{1 + \sqrt{\frac{\sum_{i=1}^{i=N} (\tilde{\lambda}_i - \lambda_i)^2}{\sum_{i=1}^{i=N} \tilde{\lambda}_i^2}}} , \tag{1.4}$$

where $K_i, i = 1 \ldots N$ and W are (positive) weighting coefficients.

The part of the revenue function proportional to $\underline{\Lambda}$ (i.e., $\sum_{i=1}^{i=N} K_i \lambda_i$) allows to drive the load of the system by means of the weighting coefficients K_i. For class

i, the coefficient K_i can be thought as the amount of money we can get for a single call during each time period of one second. The second term represents the weitghed distance between $\tilde{\underline{\Lambda}}$ and $\underline{\Lambda}$, and drives the system towards the value of $\underline{\Lambda}$ which is closest to $\tilde{\underline{\Lambda}}$. The aim of this second term is to give the possibility to mantain the input load not too different from the offered one.

For a given Input Rate Region Ψ, the maximization problem becomes therefore:

$$\max_{\underline{\Lambda}\in\Psi_B} J(\underline{\Lambda}|\tilde{\underline{\Lambda}}) \quad , \tag{1.5}$$

with the constraint

$$\beta_i \leq 1 \quad i = 1, \ldots, N$$

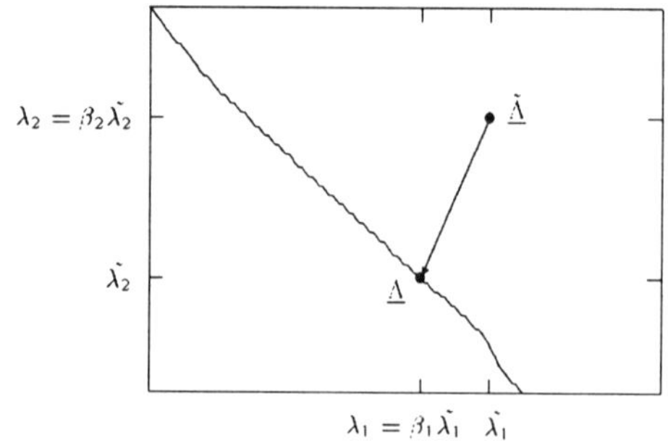

Figure 5 The filtering operation based on the IIR.

4 NUMERICAL RESULTS

In this section we present results from simulation experiments. More specifically, we compare the same system with and without the filtering control explained above. Within the controlled scenario we considered two different cases obtained by giving different weights to voice and data. In both the controlled examples we set $W = \frac{2}{3}$; in the first controlled case $K_1 = 1$ and $K_2 = 100$, while in the second one $K_1 = 100$, $K_2 = 1$. All simulations have a length of 20,000,000 steps; each step corresponds to a frame which, in turn, lasts 11 ms, and is composed of 11 slots. With these figures, the simulation covers a period of about 60 hours.

The values of μ_1 and μ_2 are respectively 0.006 and 0.003 conn./sec, which correspond to a situation in which, on the average, a voice call lasts for about 3 minutes and a data one lasts for about 5 minutes.

The values of λ_1 and λ_2 change randomly during the simulation. Both λ_1 and λ_2 are randomly chosen from uniform distributions: both λ_1 and λ_2 range between 0.03 and 0.07. During a simulation, the values of λ_1 and λ_2 change 1000 times. We may note at this point that our control system is based on the certainty equivalence principle. In the simulations, whose aim is to evaluate the effectiveness of the control with respect to the desired goals, we suppose to know the exact values of λ_1 and λ_2. In practice, the offered loads should be estimated, and the accuracy of the estimation would obviously influence the overall behavior (see [10] on this point).

Figure 6 shows the estimated probability of call refusal, obtained by dividing the number of blocked calls by the total number of arrived calls. The percentage of refused calls is quite high due to the overload conditions we imposed to the system. We point out again that in the uncontrolled case, calls are only refused if they drive the state outside the Schedulable Region (which is plotted in Figure 1), while in the controlled case, calls are blocked, both with the uncontrolled criteria and by the filters, β_1 and β_2, described above.

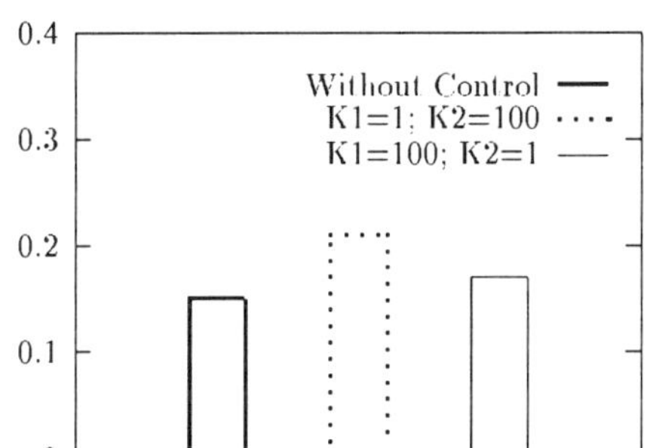

Figure 6 Percentage of refused calls in the three cases.

In Figure 7 the estimated blocking probabilities of voice and data calls are shown, obtained by dividing the number of blocked voice (data) calls by the total number of arrived voice (data) calls. The effect of the filtering operation is evident: when voice calls have a lower weight (i.e., the case $K_1 = 1$ and $K_2 = 100$), then more than 30 % of voice calls are blocked while about 10 % of data calls are blocked; on the other hand, when voice calls have a higher weight (i.e., the case $K_1 = 100$ and $K_2 = 1$), almost 10 % of voice calls and more than 20 % of data calls are blocked.

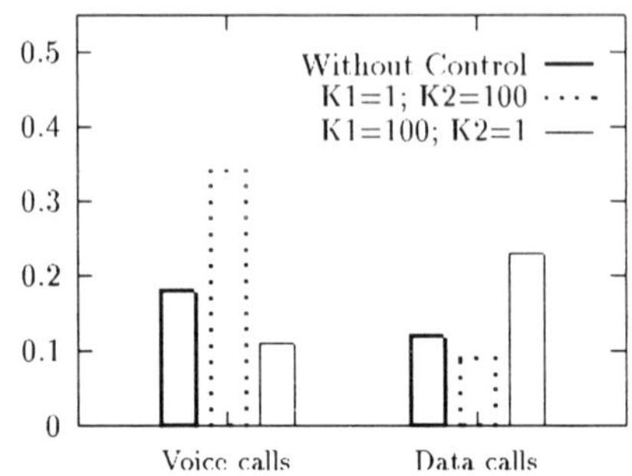

Figure 7 Voice and data refused calls in the three cases.

Figure 8 shows the values of the throughput of the system. This has been computed by counting the slot with a successful transmission and dividing it by the total number of slots. Figure 8 suggests that the filters do not decrease the total throughput of the system, which, by considering the multiple access nature of the wireless system can be considered quite high.

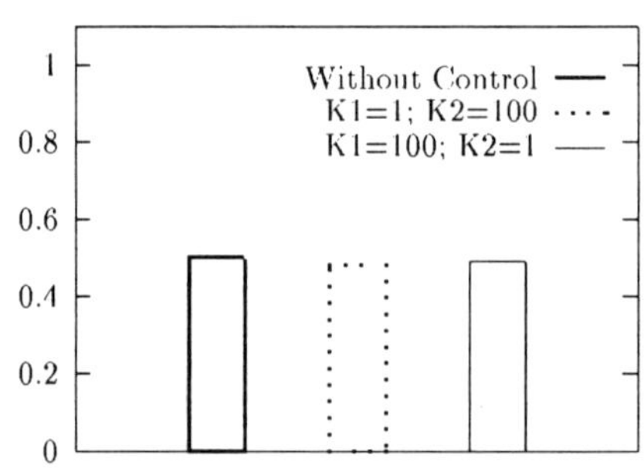

Figure 8 Throughput of the system in the three cases.

Figure 9 shows the estimated state probabilities of the system respectively controlled with $K_1 = 100$ and $K_2 = 1$, and not controlled. These results are, in essence, normalized histograms of the states in which the system run during the simulations. State probabilities in the two cases are quite stucked against the Schedulable Region because of the high load conditions, but the effect of the control, in driving the system load, is evident. The small wall in the plot represents the Schedulable Region boundary. Notice the by the effect of the static level, no connections that would drive the system load outside the Schedulable Region are accepted.

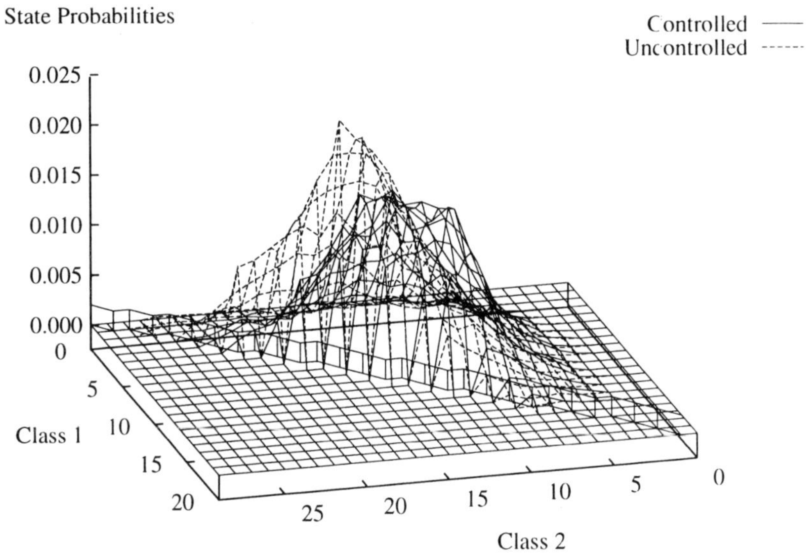

Figure 9 Estimated state probabilities.

5 CONCLUSIONS

A control framework able to drive the load of a wireless multiple access system has been presented. The control scheme is composed of two parts. The first one refuses an incoming call if this would drive the system outside its Schedulable Region. The second acts on estimated input rates by pre-filtering incoming traffic and driving it towards different load situations. Filtering is achieved by a probabilistic call acceptance mechanism which is computed on per-class basis by minimizing a cost function. The action of the second controller becomes stronger as the offered input load increases. Results have been presented which show a good behavior of the above framework, in terms of controllability of the load of the system without affecting the total throughput.

REFERENCES

[1] D. J. Goodman. Trends in cellular and cordless communications. *IEEE Communications Magazine*, 29(6):31–40, June 1991.

[2] D. J. Goodman. Cellular packet communications. *IEEE Communications Magazine*, 38(8):1272–1280, August 1991.

[3] J.E. Wieselthier and A. Ephremides. Fixed- and movable-boundary channel-access schemes for integrated voice/data wireless networks. *IEEE Trans. Communication*, 43(1):64–74, January 1995.

[4] J.M. Hyman, A.A. Lazar, and G. Pacifici. Real-time scheduling with quality of service constraints. *IEEE Journal on Selected Areas in Communications*, 9(7):1052–1063, September 1991.

[5] R. Bolla, F. Davoli, A. Iscra, C. Nobile, and S. Zappatore. Analysis and performance comparison of the RRA-ISA multiple access protocol for packet voice and data cellular systems. In *Proceeding of PIMRC'95*, pages 378–382, Toronto, September 1995.

[6] M. Aicardi, G. Casalino, and F. Davoli. Independent Stations Algorithm for the maximization of one-step throughput in a multiaccess channel. *IEEE Transactions on Communications*, COM-35(8):795–800, August 1987.

[7] K.W. Ross. *Multiservice Loss Models for Broadband Telecommunication Networks*. Springer, 1995.

[8] S. Jordan and P.P. Varaiya. Control of multiple service, multiple resource communication networks. *IEEE Trans. Communication*, 42:2979–2988, November 1994.

[9] J.M. Hyman, A.A. Lazar, and G.Pacifici. A separation principle between scheduling and admission control for broadband switching. *IEEE Journal on Selected Areas in Communications*, 11(4):605–616, May 1993.

[10] R. Warfield, S. Chan, A. Konheim, and A. Guillaume. Real-Time traffic estimation in ATM networks. In *Proceeding of ITC 14*, pages 907–916, 1994.

The Use of Information Theory in Wireless Networks

Robert G. Gallager

MIT, Cambridge, MA, US

Abstract - The essence of wireless networks is the combination of multiaccess communication with fading multipath channels. In more traditional communication over point to point wire or fiber channels, the use of information theoretic concepts is widespread. There is a spectrum of research and development, from the most abstract information theoretic to the most detailed system design; system needs are recognized by theoreticians, and theoretical ideas are encorporated by system designers. In wireless networking, on the other hand, there is a considerable gap between actual systems and information theoretic work.

The purpose of this talk is to explore theoretical issues that seem almost ripe for system exploration, and to explore system issues that are ripe for theoretical development. In fact, we shall suggest that the multiaccess coding theorem says something useful about wireless networks, and that it provides insight into ways of dealing with both interference and multipath fading.

Watch the received chips!

James L. Massey

Signal and Information Processing Lab., Swiss Federal Inst. of Tech. (ETH), Zurich, Switzerland

Abstract - It will be stressed that on additive-white-Gaussian-noise channels, whether for one sender or for many senders, the transmitted symbol waveforms (e.g., CDMA signature sequences) are only of interest in how they affect the statistics of the received chips. This point of view will be pursued relentlessly and will be shown to lead to a much simplified and more insightful proof of the Massey-Rupf result that, in the equal-user-powers case, a signature (multi)set achieves symmetric capacity if and only if it is a Welch-bound-equality sequence (multi)set. A new "trick" will then be introduced that allows this result to be generalized to arbitrary power distributions for the users. The effect of multipath distortion of the signals will also be considered.

Wireless Packet Access for the last Mile

Norman Abramson

ALOHA Networks, San Francisco, CA

Abstract - Wireless networks transmitting packet data from large numbers of users are available throughout the world. Today these networks are almost always based upon some version of an ALOHA protocol. Some of these networks are terrestrial (such as Mobitex, ARDIS, ReFlex, Multipoint mpNET and ARIA System III) while others are satellite based (ATT/Tridom, Nextar and almost all two way VSAT networks). Still other systems employ ALOHA channels in the request channel of a demand assigned multiple access (DAMA) system (Inmarsat, IS-95).

In all of these cases the data rate of the ALOHA channel is limited to at most a few tens of kilobits per second. In this talk we explain this practical limitation on conventional ALOHA. We then introduce a second generation version of conventional ALOHA, Spread ALOHA. Spread ALOHA combines the proven simplicity and operational flexibility of conventional ALOHA with the high bandwidth of a spread spectrum channel, required for the applications of today.

We describe a Spread ALOHA network now under construction which will operate in a true random access packet switched mode at a data rate of 3 Mbs.

Array Processing with Multiuser Detection

Raymond L. Pickholtz

Deparment of Electrical Engineering and Computer Science
The George Washington University, Washington, DC

Abstract - Array processing of signals in a multiuser radio environment affords the possiblity of radical interference reduction by spatial discrimination; and threby increase capacity significantly.

In a CDMA system, there are many choices for exploiting the benefits of arrays. These include, but are not limited to, the following: Preprocessing with the array and postprocesing with a conventional receiver, using a retrodirective beamformer followed by a multiuser detector and a matched filter array followed by a multiuser detector. In this talk we will derive the structure of the latter, evaluate its BER performance and compare the results with the other methods.

Using adaptive arrays also introduces an entirely new dimension to the problems of call setup and handoffs.